Also in the Variorum Collected Studies Series:

MAURICE CROSLAND
Scientific Institutions and Practice in France and Britain, c.1700–c.1870

DONALD CARDWELL
The Development of Science and Technology in Nineteenth-Century Britain
The Importance of Manchester

RICHARD YEO
Science in the Public Sphere
Natural Knowledge in British Culture 1800–1860

D.E. ALLEN
Naturalists and Society
The Culture of Natural History in Britain, 1700–1900

ROY M. MACLEOD
The 'Creed of Science' in Victorian England

DAVID M. KNIGHT
Science in the Romantic Era

JACK MORRELL
Science, Culture and Politics in Britain, 1750–1870

ROY M. MACLEOD
Public Science and Public Policy in Victorian England

JOHN HEDLEY BROOKE
Thinking about Matter
Studies in the History of Chemical Philosophy

MAURICE CROSLAND
Studies in the Culture of Science in France and Britain since the Enlightenment

TREVOR H. LEVERE
Chemists and Chemistry in Nature and Society, 1770–1878

P.M. HARMAN
After Newton: Essays on Natural Philosophy

ROBERT FOX
The Culture of Science in France, 1700–1900

VARIORUM COLLECTED STUDIES SERIES

From Atoms to Molecules

Colin A. Russell

Colin A. Russell

From Atoms to Molecules

Studies in the History of Chemistry
from the 19th Century

ASHGATE
VARIORUM

Published in the Variorum Collected Studies Series by

Ashgate Publishing Limited
Wey Court East
Union Road
Farnham, Surrey
GU9 7PT
England

Ashgate Publishing Company
Suite 420
101 Cherry Street
Burlington, VT 05401–4405
USA

Ashgate website: http://www.ashgate.com

ISBN 978–0–7546–6855–8

British Library Cataloguing in Publication Data
From atoms to molecules : studies in the history of chemistry from the 19th century. – (Variorum collected studies series ; 936)
1. Chemistry, Organic – History – 19th century.
2. Chemistry, Organic – History – 20th century.
3. Chemistry, Organic – Research – History – 19th century.
4. Chemistry, Organic – Research – History – 20th century.
I. Series II. Russell, Colin Archibald.
547'.009–dc22

ISBN 978–0–7546–6855–8

Library of Congress Control Number: 2009933507

VARIORUM COLLECTED STUDIES SERIES CS936

FSC
Mixed Sources
Product group from well-managed forests and other controlled sources
www.fsc.org Cert no. SGS-COC-2482
© 1996 Forest Stewardship Council

Printed and bound in Great Britain by
TJ International Ltd, Padstow, Cornwall

CONTENTS

PREFACE

One of the strange aspects of the history of science is the rate of growth of chemical theory. From simple ideas about matter it took much less than two centuries for the emergence of a sophisticated understanding of important and often complex substances. Oddly enough this came about in one of the more abstruse branches of the science, organic chemistry. Within 100 years of Dalton's classic atomic theory it became possible to define a substance, and understand most of its properties, in terms of the arrangement of atoms within what became called its molecule, what was eventually known as its 'structure'. Organic chemistry is hugely important, for helping us to understand life processes, for example, together with its related subjects of biochemistry and medical chemistry. But those are not the reasons for its appearance here. It is rather the centrality of its role in chemical history in the broadest sense possible. Recently, there have been several excellent studies of the development of sub-branches of the subject, and of distinguished chemists associated with its growth. But not since 1920 has a comprehensive history of organic chemistry appeared. In this book we are proceeding 'from atoms to molecules', and many of these are molecules of organic chemistry.

The history of science has not always been popular, and it has even been embarrassing to some modern practitioners, as it exposes the 'mistakes' of their predecessors and adds little (they think) to the modern subject. I believe this view to be profoundly misconceived, for all kinds of reasons, and the first chapter explores a version of it that was rife in the 19th century. The 'rude and disgraceful beginnings' of chemistry were explored and exploded by men who were practising chemists so long ago. Unfortunately it is still occasionally heard today.

After Dalton's great atomic theory early in the 19th century, a series of spectacular discoveries, including the electrolysis of fused alkalis, led some to think of a permanent electrical polarisation of matter. Among these were Humphry Davy (who never accepted Dalton) and Berzelius (who did). The next three chapters (II–IV) expound the electrochemical theory proposed by Davy, and the following five (V–IX) deal with the Swedish chemist Berzelius who developed it and in so doing became for a time the most influential chemist in Europe.

The electrochemical theory led to exciting experiments, much progress and many arguments. It was a rival to the alternative 'type theory'. This regarded molecules as entities with many family resemblances to molecules of kindred substances, and not just made of electrically positive and negative constituents. A reconciliation of type and radical (electrochemical) theories was achieved by someone who became, in Victorian times at least, the leading chemist of Britain. He was Edward Frankland,[1] whose union of these theories led to the critically important theory of valency (or valence).[2] He has been relatively unknown today, partly because few of his private papers were available. The recent discovery of an immense collection of documents relating to him is charted in Chapter X. The extraordinarily difficult techniques of this man have been further discussed in Chapter XI. As a matter of very recent history it is worth noting that the whole of the Frankland archives referred to above have now been safely deposited in the John Rylands Library of the University of Manchester.

Frankland became famous also for his work on water analysis, but as a theoretical chemist his *forte* was organic chemistry, so we are back to that subject again. One of his rivals was the German chemist August Kekulé, who for some strange reason pursued a course entirely independent of Frankland. Their relationship is explored in Chapter XII, while Chapter XIII deals with the subject for which Kekulé was justly famous, the structure and reactions of benzene. Another nearly contemporary chemist was A.W. Hofmann, who in 1845 forsook his native Germany to head London's Royal College of Chemistry. Aspects of his work are briefly recounted in Chapter XIV.

With the development of a coherent theory of atoms and molecules there slowly came the recognition of their potential complexity. As methods for creating new or existing molecules were explored it became clear that to the number of possible organic compounds there was virtually no end. This was largely due to the capacity of carbon atoms to join with each other, what we call catenation. As the 19th century progressed the building up, or synthesis, of new organic compounds proceeded apace so that the great catalogue of Beilstein now records literally millions of molecules. However it was not only a desire for 'more of the same' that goaded chemists into syntheses, but a number of other factors as well. Some of these appear in Chapter XV.

As organic theory progressed it soon (1874) became obvious that atoms were not necessarily the planar entities seen in printed textbooks. They occupied three dimensions of space, not two. In particular the carbon atom was (often) to be regarded as though its valency bonds pointed to the corners of a regular tetrahedron, with the atom in the centre. Thus arose the science of

---

[1] C.A. Russell, *Lancastrian Chemist: The Early Years of Sir Edward Frankland*, Open University Press, Milton Keynes, 1987; also *Edward Frankland: Chemistry, Controversy and Conspiracy in Victorian England*, Cambridge University Press, 1996.

[2] C.A. Russell, *The History of Valency*, Leicester University Press, 1971.

stereochemistry, and at first it was concerned with problems associated with optical activity and the like. Later, geometrical isomerism, as it became known, concerned different atoms attached to carbon atoms that were doubly bound to each other, in different configurations. At the very end of the century it became clear that certain rings (as in cyclohexane) were not planar but could switch into each other with surprising ease, giving different 'conformations'. Factors were discovered which either facilitated or inhibited such behaviour. The whole science of conformational analysis was born. Though not yet studied in many historical analyses, it probably deserves the exploration of Chapter XVI.

Finally, it is well to realise that none of these advances came in a kind of cultural vacuum. In fact a chemical community had been growing visibly in Britain with the birth and progress of the Chemical Society of London, the oldest in the world of such national institutions. 150 years of cultural change is chronicled in Chapter XVIII. One of its activities has been the publication of *Annual Reports on the Progress of Chemistry*, and the centenary of this invaluable compilation was celebrated in 2004, when an article appeared summarising advances in organic chemistry over that long period. That article has been reproduced here as Chapter XVII.

COLIN A. RUSSELL

Bedford
July 2009

ACKNOWLEDGEMENTS

Grateful acknowledgment is made to the following for permission to reproduce material originally published by them: Prof. S.J. Schaffer, editor of *The British Journal for the History of Science*, and Cambridge University Press (for Chapters I and X); Taylor and Francis, Abingdon (II, III, IV, V, VI); Manchester University Press (VII); the Johnson Reprint Corporation, New York (VIII); Cambridge University Press (IX); The MIT Press, Cambridge, MA (XI); Cache River Press, Vienna, IL (XII, XIII); VCH Verlagsgesellschaft mbH, Weinheim (XIV); Ms. Lisa Johnstone, Managing Editor, Maney Publishing, Leeds (www.maney.co.uk/journals/ambix and http://www.ingentaconnect.com/content/maney/amb)(XV); Mr E.S. Slater, American Chemical Society, Washington, DC (XVI); Ms Gill Cockhead, The Royal Society of Chemistry, Cambridge (XVII, XVIII).

I am also grateful for help and encouragement from colleagues in the History of Science Department at the Open University, and before that from the Chemistry Department at the [then] Harris College in Preston. My deepest thanks must also go to my wife Shirley (who collaborated in Chapter X), and my children Caroline, Jeremy, Kate and Helena who, in their early years, especially, coped in a splendid fashion with my preoccupation with these matters.

I

'Rude and Disgraceful Beginnings': A View of History of Chemistry from the Nineteenth Century

In a conference devoted to the history of science and its publics it might momentarily be supposed that this phrase has been intended to refer to our subject in its earlier, less enlightened period. That a few historians of science today do indeed regard their predecessors in a less than flattering light is, perhaps unfortunately, all too true. Moreover, this chauvinistic impression would be reinforced if I were to declare (as I now do) that it *is* an earlier phase of the subject that constitutes my theme for this paper. However, the intention is not to pillory, but to understand, earlier historians of science—possibly to learn from them. And the pejorative words of this title are not mine, but theirs; and they apply not to even earlier writers of history but to those whose work many centuries ago may be said to have marked one of the origins of a modern science. They refer, of course, to the alchemists whose famous activities have engendered a mixture of shame, embarassment and curiosity in those who may be considered their intellectual descendants, the chemists of today.

It would be possible, in the present day, to examine the way in which alchemy has been treated in historiography just as others have discussed it in the context of literature[1] and iconography.[2] This would, at the very least, have something to tell us about changing attitudes in history of science. But it is a shade too specialized for an Address like this and also limiting for another reason. Partly because of the 'rude and disgraceful beginnings' of the subject writers of chemical history often commence their labours in the post-alchemical field, thereby excluding themselves from such a discussion. On the other hand, to cast our net so widely as to include all former historians of science would be a self-defeating exercise in compression, even assuming it were possible (which in my case it is emphatically not). As a middle way I propose to focus on one branch of science (chemistry) and one period (the nineteenth century). In posing the question

1 J. Read, 'Alchemy and literature', *Proc. Chem. Soc.*, (1957), pp. 138–143, and many other sources.

2 J. Read, 'Alchemy and art', *Proc. Chem. Soc.*, (1958), pp. 162–166; C.R. Hill, 'The iconography of the laboratory', *Ambix*, (1975), **22**, 102–110.

'how did nineteenth century writers on chemical history see their task?' we may glean a few insights into Victorian chemistry on the way and perhaps see our own endeavours in a larger context.

It may, of course, be objected that chemistry is not typical of science as a whole (which is true in some respects) and therefore not relevant to a general view of science (which would not appear to follow from the premiss). With more truth it could now be argued that historical studies in chemistry are less important than they once were. This, however, needs to be seen in the context of specialist studies as a whole, and depends to some extent on which indicators are used: numbers and percentages of papers, or of doctoral students. Although comparable figures are not available for the previous century a qualitative impression is one of very considerable interest and activity in the history of chemical science, and that is my chief justification.[3]

In the nineteenth century and well into our own it used to be the habit to introduce chemical compounds to students under the headings: Occurrence—Preparation—Properties—Uses. Not all occurred naturally, nor did all have known uses, but it was a valuable framework for individuals and whole classes of substances. For more than symbolic reasons it will also be appropriate for the present survey of historico-chemical literature.

1. OCCURRENCE

What nineteenth century histories of chemistry can be identified and who were their authors? Professional historians of chemistry or any other science did not exist and these people are often called the 'chemist-historians'. On any basis, the first off the mark was that notable breeder of research chemists and university reformer, Thomas Thomson, Professor of Chemistry in Glasgow, defender of John Dalton and author of *The History of Chemistry* of 1830–1831.[4] Equally important was his textbook *A System of Chemistry,*[5] which first appeared in 1802 and was shot through with historical insights and emphases. His influence in Britain was considerable, though as a historian he was overshadowed on the continent by two other figures. First, the little-known Ferdinand Hoefer published his two-volume *Histoire de la Chimie* in 1842–1843 (thereby inhibiting the notable French chemist Chevreul from producing his own *magnum opus* on the subject but stimulating him to develop and expand his own historical insight in a series of articles).[6] This was immediately followed by the massive four

3 History of chemistry did not start in the nineteenth century, though that period saw a major efflorescence of the subject. See A.G. Debus, 'An Elizabethan history of medical chemistry', *Ann. Sci.,* (1962) **18**, 1–29; 'The history of chemistry and the history of science', *Ambix,* (1971), **18**, 169–177. A paper on 'Heroes and hero-worship: history of chemistry in the nineteenth century' was given by Dr Nicholas Fisher at a BSHS Symposium on problems and perspectives in the history of chemistry in April 1979 at Leicester. A post-circulated typescript included a specific request that the paper be neither cited nor quoted; it has not been used in the preparation of this Address.

4 T. Thomson, *The History of Chemistry,* London, 1830.

5 T. Thomson, *A System of Chemistry,* Edinburgh, 1802 (4 vols), 1804 (4 vols), 1807 (5 vols) and later editions.

6 J.C.F. Hoefer, *Histoire de la Chimie,* Paris, 1842 (2 vols). See G. Sarton, 'Hoefer and Chevreul', *Bull. Hist. Med.,* (1940), **8**, 419–445; H. Metzger, 'Eugène Chevreul, historien de la chimie', *Archeion,* (1932), **14**, 6–11.

Figure 1. H. Kopp. (Reproduced by kind courtesy of The Royal Society of Chemistry.)

volume *Geschichte der Chemie*[7] of Hermann Kopp of 1843–1847,[7] then a youthful professor at Giessen. In the next forty years three other major historical works flowed from his pen.[8] Many a subsequent historian of chemistry has confessed that Kopp was his starting-point.

The 1860s witnessed a relative flood of historico-chemical writing, perhaps reflecting the new spirit of optimism abroad in the post-Karlsruhe era of agreed atomic weights,

7 H. Kopp, *Geschichte der Chemie*, Brunswick, 1844–1847 (4 vols).

8 H. Kopp, *Beiträge zur Geschichte der Chemie*, Brunswick, 1869–1875; *Die Entwickelung der Chemie in der neueren Zeit*, Munich, 1873; *Die Alchemie in älterer und neuerer Zeit*, Heidelberg, 1886.

Figure 2. A. Ladenburg. (Reproduced by kind courtesy of The Royal Society of Chemistry.)

valency and structure. In 1866–1869 a second edition of Hoefer's history appeared (two volumes). Adolph Wurtz, a founder of the Societé Chimique de Paris and Professor of Chemistry in the Sorbonne, achieved a new kind of fame by his *Histoire des Doctrines Chimiques depuis Lavoisier jusqu' à nos Jours* of 1869, with an almost instantaneous translation in English.[9] Meanwhile, a young German chemist in Heidelberg, Albert

9 A. Wurtz, *Histoire des Doctrines Chimiques depuis Lavoisier jusqu'à nos Jours*, Paris, 1869; *A History of Chemical Theory from the Age of Lavoisier to the present time*, tr. H. Watts, London, 1869.

Figure 3. C. Schorlemmer. (Reproduced by kind courtesy of The Royal Society of Chemistry.)

Ladenburg, made 1869 a bumper year with his *Vorträge über die Entwicklungsgeschichte der Chemie*; this went through several editions and appeared in English translation in 1905.[10] Ladenburg was to make important contributions to organic chemistry, not least in the synthesis of coniine alkaloids and in his famous 'prism' structure for

10 A. Ladenburg, *Vorträge über die Entwicklungsgeschichte der Chemie von Lavoisier bis zur Gegenwart*, Brunswick, 1879; 2nd edn, 1907; *Lectures on the History of the Development of Chemistry since the Time of Lavoisier,* tr. L. Dobbin, Edinburgh, 1905.

Figure 4. M. Berthelot. (Reproduced by kind courtesy of The Royal Society of Chemistry.)

benzene that has only recently been established for a transient valence isomer. He was working on benzene and the history of chemistry at the same time.

In 1879 the first history of organic chemistry appeared, written by the first incumbent of a British chair in that subject, Carl Schorlemmer, of Owens College, Manchester. It was he who first clarified the nature of the alkane series of hydrocarbons [paraffins]. His predecessor, Edward Frankland, had incorrectly inferred two different species with (for instance) the formula C_2H_6; Schorlemmer's end to the confusion removed a great hindrance to the development of organic chemical theory in the 1860s. A friend of Engels and Marx, he belonged to the German Social-Democratic Party, and that would have precluded tenure of any chair in Germany in the later years of the century. His

Figure 5. T.E. Thorpe. (Reproduced by kind courtesy of The Royal Society of Chemistry.)

book *The Rise and Development of Organic Chemistry* was translated into German and appeared as a second English edition in 1894.[11] An unpublished manuscript history of chemistry as a whole was discovered after his death.

As the century drew to a close other histories appeared, some continuing the more

11 C. Schorlemmer, *The Rise and Development of Organic Chemistry*, Manchester, 1879; 2nd edn, London, 1894; *Der Ursprung und die Entwicklung der organischen Chemie*, Brunswick, 1889.

specialist emphasis of Schorlemmer. In 1885 Marcellin Berthelot produced *Les Origines de l'Alchimie*, in 1887–1888 his *Collection des anciens Alchimistes Grecs*, in 1889 his *Introduction à l'Étude de la Chimie des Anciens et du Moyen Age*, in 1890 *La Revolution Chimique, Lavoisier*, and three years later his three-volume *Histoire des Sciences. La Chimie au Moyen Age*.[12] Berthelot was one of the great chemists of his day, as was Friedrich Wilhelm Ostwald who shared some of his philosophical attitudes. Ostwald wrote his *Elektrochemie, ihre Geschichte und Lehre* in 1896,[13] although his book on the history of chemistry, *Der Werdegang einer Wissenschaft*, had to wait until the new century (1909).[14] Nor should one forget Ernst von Meyer, of Leipzig and later Dresden, whose *Geschichte der Chemie* first appeared in 1889.[15] In Britain, Pattison Muir, Praelector in Chemistry at Gonville and Caius College, Cambridge, added *Chemists* to the gallery of 'Heroes of Science' in 1883, supplementing this book with *The Story of Chemical Elements* in 1897.[16] William Tilden's *Short History of Chemistry* appeared in 1899,[17] with more substantial works in the new century from both Tilden and Muir.[18] T. E. Thorpe entitled his Presidential Address to the Chemical Society 'On the progress of chemistry in Great Britain and Ireland in the nineteenth century',[19] reaching a wider public with his *Essays in Historical Chemistry*[20] and other incursions into chemical biography, together with a small two-volume *History of Chemistry* in the new century.[21]

2. PREPARATION

Even that does not quite exhaust the list, but it is long enough. Fortunately, it is possible to go beyond a mere catalogue of names and to ask what kind of people these men were and whether any general statements can be made about them. Indeed, one generalization is already in common circulation, and has been widely repeated. It is that historians of chemistry of previous generations, including the nineteenth century, were often retired chemists spending their declining years rewriting the history of their subject so as to paint their own endeavours in the best possible light. They knew nothing of historical analysis, were chiefly concerned with compilation of lists of curious facts,

12 M. Berthelot, *Les Origines de l'Alchimie*, Paris, 1885; *Collection des anciens Alchimistes Grecs*, Paris, 1887–1888 (3 vols); *Introduction à l'Étude de la Chimie des Anciens et du Moyen Age*, Paris, 1889; *La Revolution Chimique, Lavoisier*, Paris, 1890; *Histoire des Sciences. La Chimie au Moyen Age*, Paris, 1893 (3 vols).

13 W. Ostwald, *Elektrochemie, ihre Geschichte und Lehre*, Leipzig, 1896.

14 W. Ostwald, *Der Werdegang einer Wissenschaft*, Leipzig, 1909; *L'Évolution d'une Science, la Chimie*, tr. M. Dufour, Paris, 1919.

15 E. von Meyer, *Geschichte der Chemie, von den ältesten Zeiten bis zur Gegenwart*, Leipzig, 1889, 1891 and 1905; *A History of Chemistry from Earliest Times to the Present Day*, tr. G. McGowan, London, 1891, 1898 and 1906.

16 M.M. Pattison Muir, *Heroes of Science: Chemists*, London, 1883; *The Story of Chemical Elements*, London, 1897.

17 W. Tilden, *A Short History of the Progress of Scientific Chemistry in our own Times*, London, 1899.

18 W. Tilden, *The Progress of Scientific Chemistry in our own Times*, London, 1913; *Chemical Discovery and Invention in the Twentieth Century*, London, 1916; *Famous Chemists, the Men and their Work*, London, 1921.

19 T.E. Thorpe, Presidential Address, *J. Chem. Soc.*, (1900), 77, pp. 555–586 (562–586).

20 T.E. Thorpe, *Essays in Historical Chemistry*, London, 1894.

21 T.E. Thorpe, *History of Chemistry*, London, 1909–1910 (2 vols).

often failed to distinguish between good and bad evidence, were only concerned with what their generation regarded as successful science, ignored the blind alleys and cul-de-sacs of chemical enquiry, and (in short) represented the very quintessence of Whiggism. Such a view I call 'the dotage theory'. Let us see if it bears any relation to the facts.

Author	Dates	Book dates	Age
T. Thomson	1773–1852	1830	57
J.C.F. Hoefer	1811–1878	1842+	31+
A. Wurtz	1817–1884	1868	51
H. Kopp	1817–1892	1843–1847	26+
M. Berthelot	1827–1907	1885 (Alchemy)	58
		1892 (Middle Ages)	65
C. Schorlemmer	1834–1892	1879	45
C. Graebe	1841–1927	1920	79
A. Ladenburg	1842–1911	1869	27
W.A. Tilden	1842–1926	1899 (Short Hist.)	57
		1913 (Prog. Sci. Ch.)	71
T.E. Thorpe	1845–1925	1894 (Essays)	49
E. von Meyer	1847–1916	1889	42
M.M.P. Muir	1848–1931	1883 (Heroes)	35
		1897 (Elements)	49
W. Ostwald	1853–1932	1896 (Electrochem.)	43
		1909 (Hist. Chem.)	56
G.W.A. Kahlbaum	1853–1905	1897+	44+
P. Walden	1863–1957	1941	78
T.M. Lowry	1874–1936	1911	37
P. Diergart	1875–1943	1909	34
J.R. Partington	1886–1965	1935 (App. Chem.)	49
		1937 (Short Hist.)	51
		1961+ (4 vol.)	75+
E. Farber	1892–1969	1931 (Hist. Dev.)	39
		1952 (Evol. Chem.)	60

Figure 6. The Dotage Theory.

This chart lists the main figures just mentioned, together with their successors in the present century. Several things are immediately apparent. First, there are several authors well into old age, of whom the organic chemists Graebe and Walden appear to have started their historical endeavours well after their distinguished chemical careers were over. Though not completely true (for Walden had written on the history of solution when he was a mere forty-seven[22]) it is largely correct, and Graebe admits that only after he retired from teaching did he develop his predilection for history of chemistry.[23] Perhaps here is the origin of the dotage legend, for the rest of the data

22 P. Walden, 'Die Lösungstheorie in ihrer geschichtlichen Aufeinanderfolge', *Samml. chem. u. chem.-techn. Vortrage*, (1910), **15**, 277.

23 C. Graebe, *Geschichte der organischen Chemie*, Berlin, 1920, p. iii.

indicates that that is exactly what it is. To be sure Partington was long retired when his monumental four-volume *History of Chemistry* began to appear, but his *Origins and Development of Applied Chemistry*[24] had appeared over a quarter of a century earlier. Similarly, neither Tilden nor Farber waited for retirement to produce their first historical work. Nor did Berthelot, whose magisterial work on alchemy was conducted while he was at the height of his powers and actually preceded his work on thermochemistry, his discovery of iron carbonyl and much else. Thomson had been writing historically for over twenty years before his *History* was published and Hoefer, although fifty-five when the second edition of his book appeared, was only thirty-one for the first edition. Of the rest most were youngish men, while Kopp, perhaps the most influential of them all, was a mere stripling of twenty-six when he first went into print.

That most of the major chemist-historians of the last two centuries wrote important historical work well before their dotage seems incontrovertible. But age, by itself, is only one consideration. Another is the provenance of their studies, and here it seems likely that many began in the same kind of way as historians of other sciences in the last century. We are told, for instance, that J. C. Poggendorf had 'for many years' delivered lectures on the history of physics at the University of Berlin, and out of these his *Geschichte der Physik* was posthumously produced for publication.[25] Many years later a similar course of lectures at Northwestern University, Illinois, led to the publication of Henry Crew's *The Rise of Modern Physics,* just as his colleague William Locy had previously done with his lectures on the history of biology.[26] In chemistry such lectures may have been popular expositions, like those of T. E. Thorpe to working men in Manchester[27] or of Tilden to a similar audience in London.[28] Or they may have been delivered to university classes, either as part of an undergraduate course or as an optional extra—sometimes it is hard to tell which. Two examples must suffice, both around 1900.

James Campbell Brown (1843–1910), a pupil of Tyndall and Hofmann, was Professor of Chemistry at Liverpool at the turn of the century. The Public Analyst for the city, he was a founding member of both the Society of Chemical Industry and the Society of Public Analysts. His *History of Chemistry from the Earliest Times to the Present Day* was published posthumously[29] on the basis of lectures delivered as 'a labour of love' and illustrated with 'lantern slides' as part of his chemical course.[30] At that time the University College of Liverpool was plagued by low numbers and low entry standards[31] and historical lectures may have been seen as an alternative to laboratory work. The other example [for which I am indebted to Dr W. A. Smeaton] is

24 J.R. Partington, *Origins and Development of Applied Chemistry,* London, 1935.

25 W. Barentin, Foreword to J. C. Poggendorf, *Geschichte der Physik,* Leipzig, 1879.

26 H. Crew, *The Rise of Modern Physics,* 2nd edn, London, 1935, p. ix.

27 E.g., T.E. Thorpe, 'Cavendish and his discoveries', in *Science Lectures for the People,* Manchester, 1875–1876, pp. 89–103.

28 W. Tilden (1899), p. v.

29 J.C. Brown, *History of Chemistry from the Earliest Times to the Present Day,* London, 1913.

30 Ibid., p. v.

31 Only thirty chemistry graduates had been produced in the ten years since its foundation in 1890: G.W. Roderick and M.D. Stephens, *Scientific and Technical Education in the Nineteenth Century,* Newton Abbot, 1972, p. 68.

Figure 7. J.C. Brown.

that of William Ramsay who, in 1900–1901, gave a course on history of chemistry to third-year students at University College, London. However, it was only in the final term, and at 9 a.m. on Saturdays![32] Perhaps it was poor attendance that caused the course to be withdrawn shortly afterwards. If any lesson is to be drawn from these melancholy events it is possibly that history of science is no substitute for science teaching for less able students, and that, like anything else, it suffers from exposure at

32 Information from Dr W.A. Smeaton; these lectures were probably one ingredient of Ramsay's *Essays Biographical and Chemical,* London, 1908.

unsocial hours. Of other, more successful, attempts to include history in chemistry courses we shall take notice later.

Not all the great histories of chemistry originated as lecture courses. Many were the outcome of research conducted between spells of experimental work, others as an essentially spare time activity, and a few as a long-anticipated luxury of retirement years. In the last category were doubtless several that never saw the light of day, such as the sadly incomplete manuscript of Schorlemmer's history of the whole science.[33] Others again must be recognized as the chief literary activity of authors whose experimental chemical activity was negligible. Men like Hoefer were to produce considerable works of historical scholarship long before history of science was professionalized in anything like the modern sense.

And we must not forget another class of writers. These were the many chemists who did not aspire to writing whole histories of their subject but nevertheless wrote impressively on either individual biography or on special episodes in chemical history. A notable example of the latter authors is Henry Roscoe, who wrote both a biography of John Dalton and (with Harden) *A New View of the Origin of Dalton's Atomic Theory*.[34] Nor must the multitude of lesser men be forgotten whose textbooks of chemistry were liberally sprinkled with allusions to past events in the subject, real or supposed. From such diverse origins did the nineteenth century treatments of chemical history arise. Equally diverse were the characteristics they displayed.

3. PROPERTIES

Far from being monolithic expressions of Whiggism or of anything else the nineteenth century histories of chemistry are at least as different from one another as from any of our own time. They differ in the extent to which they represent original research or rely on other secondary sources. Many acknowledge their indebtedness to Kopp, especially, and some are patently derivative for almost all their material. Others, like the works of Thomson and Berthelot, display a familiarity with original sources that many a modern might profitably emulate. However, Thomson emerged from his alchemical studies to remark ruefully 'that the information gained by such a perusal will very seldom repay the trouble', and regretted time spent with Paracelsus, 'this prince of quacks and imposters'.[35] Most obviously there are differences in the extent of historical coverage, many attempting comprehensive coverage but others focusing on specific areas of chemistry (as Schorlemmer or Roscoe and Harden) or more recent times (as the spate of studies on the departing century around 1900). In all cases there are exact parallels in other sciences.[36] Yet without exception all the works contain three elements, and it is the ratio of these to one another as well as their separate treatment that

33 On this see *DNB* and *DSB*.

34 H.E. Roscoe, *John Dalton and the Rise of Modern Chemistry*, London, 1895; H.E. Roscoe and P. Harden, *A New View of the Origin of Dalton's Atomic Theory*, London, 1896.

35 Thomson (1830), pp. ii and iv.

36 Thus in astronomy there were specialist studies such as R. Grant, *History of Physical Astronomy*, London, 1852; and Agnes M. Clerke, *A Popular History of Astronomy during the Nineteenth Century*, London, 1893.

constitutes some of the most revealing characteristics of both the author and his work. These are discoveries, concepts and people. Some of the most useful histories of chemistry have these so well blended together that it is impossible to argue for a special emphasis on one rather than another. Outstanding examples of such eirenic treatments are furnished by Thomson, Meyer and Ladenburg. Other accounts, still of great value, are distinctively slanted in one of the three directions.

3.1 Discoveries

The pragmatic, empirical spirit of chemistry is well displayed in the title of a book appearing well into the twentieth century but redolent of the values of Victorian chemical history: *Chemical Discovery and Invention in the Twentieth Century* by Sir William Tilden (1917). Relatively light on theory, it is dedicated to showing that chemistry is a useful science for which industry ought to be prepared to make better provision.[37] (This was also a time when Government was glad enough for chemists to run munition factories but otherwise treating them with scandalous neglect.[38]) Introducing an earlier book in 1899, Tilden declared his intention 'to trace the course of discovery in chemistry' in the Victorian period.[39]

A few other histories centred on discoveries but the tendency is far more common in chemistry textbooks. One of the most famous textbooks in English was *Inorganic Chemistry* by Frankland and Japp,[40] in which all major compounds were introduced by a section on 'history', thus adding a new category to occurrence, preparation, properties and uses. They set a pattern for generations of textbooks on elementary chemistry where such historical synopses, often in very small print, were quite sufficient to minimize historical interest in those who read them. This formalization of history did not appear in Frankland's earlier books and appears to reflect a growing interest in chemistry's past as well as Frankland's inveterate tendency to systematize for didactic purposes. However, 'discovery' was not the major *Leitmotiv* of Victorian chemical historiography.

3.2 Concepts

The history of chemical ideas was a favourite theme of these chemist-historians. But it must not be imagined that most of them wished merely to chronicle minute fluctuations of ideas within their own limited sphere of chemical understanding, still less that they wanted to mobilize contemporary opinion in breathless admiration at their own latest intellectual achievement. These were often men of wide and general culture, and they saw the history of chemistry as part of a much wider history of ideas. A good example is Justus von Liebig.

37 W.A. Tilden, *Chemical Discovery and Invention in the Twentieth Century*, London, 1917.

38 G.K. Roberts in C.A. Russell, N.G. Coley and G.K. Roberts, *Chemists by Profession*, Milton Keynes, 1977, pp. 212–213.

39 W.A. Tilden (1899), p. 16.

40 E. Frankland and F.R. Japp, *Inorganic Chemistry*, London, 1884.

The third edition of Liebig's *Familiar Letters on Chemistry* (1851) contains several chapters on the history of the subject. He confesses a debt to Kopp but also to several general historians and observes 'the history of science fills a page in the history of the human mind'. Parting company from those who despise chemistry's humble origins, he complains of 'the very ludicrous and excessive estimation of ourselves' which leads us to despise the age of alchemy and which springs from a 'prevailing ignorance of chemistry, and *especially of its history*'. If that were not enough he goes on to ascribe a contempt for the theory of phlogiston as a result of 'ignorance of the history of science'.[41] In other words he wants to use history to exalt the past, not the present, so turning the conventional caricature of a chemist-historian on its head.

In a similar way Ladenburg argues the value of a history that 'is no longer a mere enumeration of isolated facts in chronological order' but one that 'embraces the development of the human mind and of human civilisation'. Darwinians and even their opponents must surely see the importance of historical studies. But it is history of science, even more than history of philosophy, that can help us to understand the history of the development of the human intellect. He instances several specific advantages of such an enterprise: it can help us to understand the present; it displays 'the mutability of opinions'; it undermines a blind faith in authority; and it stresses the importance of theories as compared with mere observations.[42]

Examples can be multiplied of this high valuation of the history of science, and thus of the treatment of chemical concepts (as opposed to mere discoveries). It was said of Ostwald's historico-chemical studies in the late nineteenth century that 'he became more and more convinced that in the history of science the general laws of historical phenomena can be more readily grasped than in the entangled web of universal history'.[43]

This general dedication to intellectual history was sometimes associated with a clearly displayed philosophy of life. The values of positivism are not far below the surface of Ostwald's exposition, for instance. He deals with Berzelius' electrochemical theory largely in terms of equivalents, announcing that 'equivalent and combining weights are nothing more than the magnitudes of the capacity for chemical energy', and complains that in the development of isomerism theories 'chemists have got themselves more and more into the habit of regarding the atoms as real'.[44] Even more pronounced was the ideological stance of Schorlemmer.

Schorlemmer's history of chemistry was deeply indebted to Kopp, as he gladly admitted, but he went far beyond his teacher in developing a history of science based on Marxism. Indeed his recent biographer, Karl Heinig, asserts that here was the beginning of Marxist historiography of science. Thoroughly disenchanted with the empiricist legacy of *Naturphilosophie,* Schorlemmer adopted the standpoint of dialectical materialism, concerning himself with 'the basic laws of scientific development, the

41 J. von Liebig, *Familiar Letters on Chemistry,* London, 1851, pp. 25, 42 and 46 (italics mine).

42 A. Ladenburg (1905), pp. 1–4.

43 M. Dufour in Preface to W. Ostwald, *L'Évolution d'une Science: la Chimie,* tr. M. Dufour, Paris, 1919, p. 2.

44 Ibid., p. 85.

context in which the individual sciences relate to one another, theoretical questions about the role of hypotheses and theories, and the connection between Theory and Praxis'.[45] He held a dynamic view of chemical change, holding that even for molecules 'everything is in an eternal flux' as Heraclitus had said.[46] He was fortunate in finding organic chemistry just then embracing dynamic concepts (as in tautomerism) and therefore ideal for his exegesis. And the Hegelian notion of quality emerging from quantity was able to account for the incremental effect of successive CH_2 additions in building up homologous series,[47] an argument later used by Engels.[48] In 1892 Engels wrote a highly appreciative obituary notice for him.[49]

Schorlemmer's history of organic chemistry is undoubtedly a major achievement, and it does highlight the importance of conceptual analysis in history as in science. Yet if it be compared with its predecessors—even with Thomson half a century earlier—its methodology does not seem to be so very different. Nor, in fact, does it have much to say about economic or political factors. What it does accomplish is the establishment of a historiography that takes into account the dynamic processes and distinctive features of the new organic chemistry. And it was a means to an end. 'For Schorlemmer the history of science was an unconditional necessity for understanding science itself'.[50]

This conviction, together with his high evaluation of a conceptual history of chemistry, was shared by many of his successors. Pattison Muir is chiefly remembered for his historical writing, and that of a conceptual kind. It was said of him that 'as behoved his Scottish birth [he] was more a metaphysician than a typical man of test tubes and reagent bottles'.[51] When his *magnum opus* did appear, early in the new century, it was appropriately called *A History of Chemical Laws and Theories*.[52] Not for nothing had he been secretary of the Cambridge Philosophical Society. But there still remains a third emphasis in nineteenth century chemical history.

3.3 People

When T. E. Thorpe wrote the Memorial Lecture for his former master Kopp he announced 'The history of an epoch is the history of its leading men'.[53] Such opinions may not have been held by Kopp but they were firmly believed by his obituarist. Thorpe excelled at chemical biography, and made it the basis of his later *History of Chemistry*.[21] His *Essays in Historical Chemistry*[20] consisted of a selective series of biographical studies, culled from a variety of sources. Its approach is typical of much history of chemistry at the end of the Victorian era including, of course, many single-volume

45 K. Heinig, *Carl Schorlemmer, Chemiker und Kommunist ersten ranges*, Leipzig, 1979, p. 53.
46 Schorlemmer (1894), p. 184.
47 Schorlemmer (1894), p. 142.
48 F. Engels, *Herrn Eugen Dühring's Umwälzung der Wissenschaft*, Zurich, 1886, 1878, p. 104.
49 *Vorwärts Berliner Volksblatt*, 3 July, 1892.
50 Heinig, p. 55.
51 R.S. Morrell, *J. Chem. Soc.*, (1932), p. 1334.
52 M.M. Pattison Muir, *A History of Chemical Laws and Theories*, London, 1907.
53 T.E. Thorpe, *Memorial Lectures of the Chemical Society*, 1893–1900, London, 1901, p. 776.

biographies of great chemists. This tendency was not unique to Britain. Of the continental writers one of the most prolific was the G. W. A. Kahlbaum of Basle. In addition to his eight-volume *Monographien* he wrote about Paracelsus, Schönbein, Wöhler and others and edited several volumes of their letters.[54]

4. USES

In asking for what uses the history of chemistry was intended we inevitably raise the most important underlying question of the dominant ideologies. At the risk of oversimplification we can perhaps identify three separate approaches.

4.1 Patriotic

At both a national and local level patriotism was a powerful force in European culture, and even history of science was pressed into its service. Chemistry recognized its international bonds, as the Memorial Lectures of the Chemical Society testify, but at moments of stress or provocation nationalism may be seen to have determined historical writing to a considerable extent. Writing in the aftermath of the 1848 revolutions Liebig complained of the fallacious view that chemical knowledge began with 'the French school' of Lavoisier.[55] When, in 1869, Franco-German relations reached a new low point, Wurtz introduced his history of chemistry with the words 'Chemistry is a French science; it was founded by Lavoisier of immortal memory',[56] a remark that called from Thorpe the riposte 'Chemistry is an English Science, its founder was Cavendish of immortal memory'![57] One is reminded of the comments of Berzelius on Turner's *Elements of Chemistry* in 1827: 'What damned nonsense, written in an impertinent style, oblivious to everything which has not been done in England by Englishmen . . . in spite of the fact that the man has been studying under Stromeyer!'[58] But then, he was suffering badly from gout.

Nor must such partisan uses of the history of science be regarded solely as the propensity of historians. A history of science in Germany was actually commissioned by King Maximilian II of Bavaria to exalt his country's honour.[59] And patriotism is not restricted to German monarchs: at a much more local level the civic pride of Manchester may be seen in the intense veneration of Dalton's memory, as shown by Roscoe's biography of him.[34] In conjunction with Schorlemmer, Roscoe wrote a textbook of inorganic chemistry with a frontispiece portrait of Dalton and a long

54 G.W.A. Kahlbaum, *Monographien aus der Geschichte der Chemie*, Leipzig, 1897–1904, (8 vols); *Theophrastus Paracelsus*, Basle, 1894; *Christian Friedrich Schönbein, 1799–1868*, Leipzig, 1879–1901 (3 vols); (ed. with Darbyshire) *The Letters of Faraday and Schönbein*, London, 1899; (ed. with E. Thon) *Justus Liebig und Christian Friedrich Schönbein. Briefwechsel 1853–1868*, Leipzig, 1900, etc.

55 Liebig, p. 26.

56 A. Wurtz, *Histoire des Doctrines Chimiques depuis Lavoisier jusqu'à nos Jours*, Paris, 1869, p. xx.

57 T.E. Thorpe, (1875/1876), p. 103.

58 J.J. Berzelius, Letter to Wöhler, 10 July, 1829, in O. Wallach (ed.), *Briefwechsel zwischen J. Berzelius und F. Wöhler*, Leipzig, 1901, vol. i, p. 267.

59 J. von Sachs, *Histoire de la Botanique du XVIe Siècle à 1860*, tr. H. de Varigny, Paris, 1892, p. xi.

historical introduction that specially relates the rise of Daltonian atomism.[60] It is not hard to think of modern histories of chemistry with an overtly nationalistic function. Looking back on the previous century one may well be grateful for such endeavours, and perhaps think more kindly (though no less critically) of their latter-day equivalents.

4.2 Didactic

That history of chemistry has been used as a vehicle for ideological purposes is obvious for the cases of Ostwald (positivism) and Schorlemmer (dialectical materialism). The latter's biographer remarked that his book was 'really much more than a history of organic chemistry. It is a history of human thought isolated in concrete form in the example of the development of one science'.[61] It might also be inferred that history of chemistry was considered to be an agent of humanism or atheism since Thorpe's *History of Chemistry* was 'issued for the Rationalist Press Association, Limited'.[54] But there is the counter-example of the 'Heroes of Science' series, which included Pattison Muir's essay on chemists, which was issued by the Society for the Promotion of Christian Knowledge (who also published *Our Secret Friends and Foes* by the agnostic Percy Frankland![62]). All that this seems to show is that history of chemistry was seen as a marketable commodity at a time when science was enjoying greater popularity than ever before.

There was, however, another form of didacticism that pervaded late Victorian thought and that is endemic in some (but by no means all) histories of chemistry at the turn of the century. This was the triumphalist belief in progress unlimited. In chemical historiography it is visible at an almost ludicrous level in *The Story of Chemistry,* by a young research worker in Sir William Ramsay's laboratory, Harold W. Picton. From mysticism chemistry has progressed from 'the childhood of truth' to 'the conflict with error' and thence to 'the triumph of truth' and the broad daylight of modern science.[63] Even Roscoe commended 'this little book' in a Preface. In the same spirit 'heroes of science' are paraded as in a Roman triumph, 'progress' is on everyone's lips and all is well with the chemical world. It is hard to avoid the impression that this is a particularly English phenomenon. If so, Herbert Spencer must bear some of the responsibility, but so also must Queen Victoria. Her Diamond Jubilee, coming so near the end of a century of unparalleled industrial and imperial progress, cried out for celebration and the chemists were not slow to respond. It was for that very reason that Roscoe launched his 'Century of Science' series that included his own life of Dalton, and biographies of other scientists including the chemists Liebig and Davy.[64]

Not all of this plunged to Pictonian depths of Whiggishness but there is enough to provide a nucleus for the dotage myth referred to at the beginning. However, given a predisposition for Whig-hunting (if I may be allowed the term), modern historians can

60 H.E. Roscoe and C. Schorlemmer, *A Treatise on Chemistry*, vol. i, 'The non-metallic elements', 2nd edn, London, 1878, pp. 3–40.

61 Heinig, p. 56.

62 P.F. Frankland, *Our Secret Friends and Foes*, London, 1894.

63 H.W. Picton, *The Story of Chemistry*, London, 1889.

64 W.A. Shenstone, *Justus von Liebig: His Life and Work*; T.E. Thorpe, *Humphry Davy, Poet and Philosopher*, London, 1901 and 1896, respectively.

CONTENTS.

Figure 8. Contents pages of Picton's *The Story of Chemistry.*

find plenty more evidence elsewhere if they so desire. Much of this 'evidence' I believe to be misunderstood and capable of an entirely different interpretation. Let me offer two examples. At one point in his *Familiar Letters* Liebig compares scientific and political progress, seeing both as a three-phase sequence. An apparent gift to the Whig-hunters this analogy turns out to imply not that we are living in the last days of all-conquering science, but simply that we must soon address ourselves to the third and most sophisticated method of analysis, the quantitative. After concentrating on *substance,* and then on *ideas* we call in *mathematics* which, 'with aid of numbers, completes the work'.[65] My second example is from Thomson, whose description of alchemy provides the title of this Address: 'rude and disgraceful beginnings'.[66] Yet repeatedly the term 'disgrace' is applied not to ignorance and credulity but to duplicity and deception. It is not intellectual but moral (or at least ethical) categories that are involved in his denunciations. Neither Liebig nor Thomson, nor for that matter many of their contemporaries and successors, held to an inexorability of chemical progress and neither can be accused of that modern heresy, Whig historiography.

65 Liebig, p. 54.
66 Thomson (1830), p. 2.

The one conviction shared by every one of the nineteenth-century historians of chemistry is that (like every other science) it has progressed. Its progress has sometimes been in an almost linear fashion, sometimes by fits and starts. That they wrote from the standpoint of their own day is no less true of them than it is of us. As some unlikely defenders of a (modified) Whig historiography have recently written there *is* sometimes a case to be made for writing about the past from the standpoint of the present. Only then can we communicate effectively to our scientific contemporaries, and only then can we demonstrate the continuity of ideas. 'Our purpose in doing so is not to demonstrate our superiority over our predecessors, but to learn from their mistakes and successes'.[67] And, of course, to condemn our predecessors as being more Whiggish, and thus less enlightened, than ourselves is to adopt the most Whiggish stance of all. It may be that some modern historians are doing unconsciously exactly what they imagine these chemist-historians were up to: highlighting the achievements of the present generation with its new professionalism at the expense of their forebears.

4.3 Pedagogic

One final application of the history of chemistry in the previous century remains to be mentioned. So far as can be ascertained every one of our chemist-historians felt passionately that history was part of chemistry, and that it was of value in communicating the rest of the science. In fact the nineteenth century witnessed two opposing tendencies on the part of the chemist-historians.

The older tendency, long predating 1800, was to treat modern and earlier chemistry as an integrated whole, much as Priestley had done in his *History and Present State of Electricity*[68] or his *History and Present State of Discoveries relating to Vision, Light and Colours.*[69] For chemistry, Booerhaave, Black and others had done the same.[3] As Debus reminds us, when William Henry proposed to omit this standard feature from a new textbook in 1819[70] he felt it necessary to explain why. Thomas Thomson had no such scruples and made clear why the history was included:

> The object of this work is to exhibit as complete a view as possible of the present state of chemistry; and to trace at the same time its gradual progress from its first rude dawnings as a science, to the improved state which it has now attained. By thus blending the history with the science, the facts will be more easily remembered, as well as better understood; and we shall at the same time pay that tribute of respect to which the illustrious improvers of it are justly entitled.[71]

In the century ahead others were made to emulate Thomson in this objective. At the same time the history of chemistry began to be administered to chemical students in a rather different way, as a separate course of lectures or in separate books (as Thomson's own *History of Chemistry*). This differentiation of the subject may reflect

67 J.D. Barrow and F.J. Tipler, *The Anthropic Cosmological Principle*, Oxford, 1986, pp. 10–11.

68 J. Priestley, *The History and Present State of Electricity*, London, 1767.

69 J. Priestley, *The History and Present State of Discoveries relating to Vision, Light and Colours*, London, 1772.

70 W. Henry, *The Elements of Experimental Chemistry*, Philadelphia, 1819, vol. i, p. ix.

71 T. Thomson (1820), vol. i, pp. 9–10.

the growing interest in history generally and a wish to give it a more prominent role still in scientific education. Thus we find Kopp adding the subject to his list of lecture courses at Giessen in about 1841,[72] and Ladenburg making it the topic of his first course of lectures at Heidelberg in about 1868.[73] By 1904 E. von Meyer could write of 'frequent courses of lectures on the history of chemistry' then being delivered in German colleges and universities.[74] Six years later a course on the history of organic chemistry was given by Norman Collie to research students in that subject at University College, London. And, as we have seen, many of these courses were eventually transformed into books.

Meanwhile that other tradition of an integrated historico-chemical approach showed unmistakeable signs of withering, though it continued in an attenuated form until World War II. The famous textbook of Frankland and Japp,[40] though setting a trend by formally installing 'history' as a desirable category of knowledge about each individual substance, seems to have accelerated the dehistoricisation of chemistry. By screening off the history, so to speak, their treatment effectively questioned its relevance. By greatly compressing it they reduced its interest and appeal. And by failing to relate one discovery to any other they evacuated it of any real significance. It is interesting to recall that their index does not include any historical names except the very few who are noticed in connection with the development of theory. Thus Dalton and Mendeléef may be located, but not Priestley, Cavendish or Davy. Many years later this very approach, with its 'snippets' of history, was to receive severe treatment in the context of school teaching.[75]

The demise of chemical history in chemistry textbooks doubtless owed much to two other considerations. One was the rapidly increasing volume of chemical facts and theories to be absorbed at every academic level. The other was the growth of faculty psychology and the attendant rise of heurism, when chemistry by discovery in the here and now was so much more to be desired than gaining second-hand knowledge about other people's discoveries in the remote past.[76] It is only in our own day that the value of the integrated historico-chemical approach urged so long ago by Thomson is reappearing as a viable possibility in certain educational situations. It is not hard to speculate on the reasons. In his day, as in ours, science was far less popular than it was to become in late Victorian Britain; both early Romantic and latter-day environmentalist lobbies have rebelled against the soullessness of a dehumanized science; and a separation of science from its history was not demanded (as then) by an explosive growth of science or (as now) by a professionalization of its history. For all these reasons, and doubtless for others, the opportunities for integrating historical insights and material within a scientific curriculum are greater today than for many years past. To take the case of physics, it has been possible for a recent writer to assert that 'we now have enough concrete evidence showing that the historical approach does have

72 T.E. Thorpe (1901), p. 778.
73 F.S. Kipping, Ladenburg Memorial Lecture, *J. Chem. Soc.* (1913), **103**, 1871–1895 (1874).
74 E. von Meyer (1906), p. xiv [dated October, 1904].
75 E.W. Jenkins, *From Armstrong to Nuffield*, London, 1979, p. 70.
76 W.H. Brock (ed.), *H.E. Armstrong and the Teaching of Science 1880–1930*, Cambridge, 1973.

Figure 9. At the British Association Meeting, Manchester, 1887: back row (l. to r.): J.A. Wislicenus, G. Quincke, H.E. Schunck, C. Schorlemmer, J.P. Joule; front row (l. to r.): L. Meyer, D.I. Mendeléev, H.E. Roscoe.

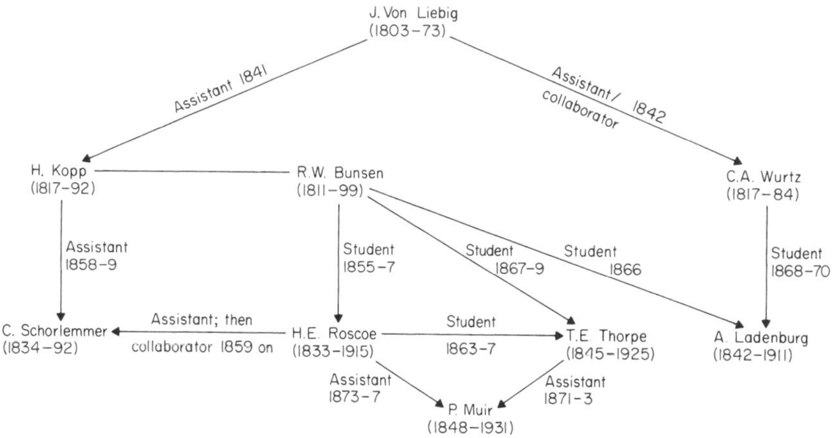

Figure 10. History of Chemistry Networks.

an impact on students, without interfering with their learning of the technical content of science'.[77] And if the opportunities are immense then the resources are greater than ever before. Let me therefore comment finally on one other aspect of the work of the nineteenth century chemist-historians.

Hardly any of the major figures in nineteenth century chemical historiography worked in isolation from others sympathetic to his enterprise. There was a network of communication in nineteenth century science about which we still have much to learn. But is it certain that it included the chemist-historians. A few of the more obvious links are sketched in Fig. 10. One example must suffice, Henry Roscoe. He had been, like Kopp, a pupil of Bunsen; his assistant Schorlemmer (a former pupil of Kopp) became his colleague and co-author; one of his pupils was Thorpe, under whom Pattison Muir had worked before studying under Roscoe. And Thorpe himself had encountered Kopp at Heidelberg and was to write his Memorial Lecture.

Among the immense resources available to us today is the modern history of science network, of which this Society is just one constituent. In an era of unparalleled opportunity it is our responsibility and privilege to serve the community of science at least as well as did the chemist-historians of the nineteenth century.

77 S.G. Brush, 'History of science and science education', *Scientific Literacy Papers* (1987), **1**, 75–85 (82).

II

THE ELECTROCHEMICAL THEORY OF SIR HUMPHRY DAVY

PART I : THE VOLTAIC PILE AND ELECTROLYSIS *

SIR HUMPHRY DAVY has been widely acknowledged as the founder of the science of electrochemistry[1]. The discoveries of the Voltaic pile[2] and electrolysis[3] announced in 1800 aroused Davy's immediate interest and caused him to begin a series of experiments on the new " galvanic phenomena "[4]. He pursued these with such success that within a year or two public interest was to be demonstrated by the provision of courses of lectures at the Royal Institution[5].

This early work was largely concerned with the acquisition of experimental data rather than the formulation of theoretical views. There followed a break of four years during which time, no doubt, the new electrical phenomena must often have been in Davy's mind and when various relationships were suggesting themselves to him. The results of his meditations were made public in the celebrated Bakerian Lecture to the Royal Society on 20 November 1806 : " Some chemical agencies of electricity "[6]. A second Bakerian Lecture the next year[7] surpassed

[1] E.g., A. Tiselius, *Purity and Purification of Chemical Substances*, London, 1958, p. 1.

[2] A. Volta, *Phil. Trans.*, 1800, **90**, 403.

[3] W. Nicholson, *Nic. J.*, 1800, **4**, 179.

[4] *Ibid.*, 1800–1, **4**, 275, 326, 337, 380, 394, 527 ; *Phil. Trans.*, 1801, [**91**], 397.

[5] *J. Roy. Inst.*, 1802, **1**, 49, 79, 165, 209, 284 ; also *A Syllabus of a course of lectures on Chemistry delivered at the Royal Institution of Great Britain*, London, 1802.

[6] *Phil. Trans.*, 1807, [**97**], 1. The volumes of *Phil. Trans.* from 1801 to 1853 were not numbered on the title-page, but were sometimes numbered on the spine by librarians.

[7] *Ibid.*, 1808, [**98**], 1.

---

* The material of Parts I and II of this paper is based on sections of a Dissertation submitted at University College, London, in 1958 for the degree of M.Sc. (History and Philosophy of Science) in the University of London. The author's thanks are expressed to Professor D. McKie for his advice during its preparation.

Annals of Science 15, 1959, 1–13. Reprinted by permission of Taylor and Francis Ltd (http://www.tandf.co.uk/journals).

2

its predecessor in importance, and was followed by addresses to the Royal Institution[8] and papers to the Royal Society[9]. By 1809 a vast mass of information had been made available by Davy, including details of his electrolytic isolation of potassium, sodium and other elements. His theoretical interpretations of the facts that came to light may be found in some of these papers of 1806 to 1809, and are usually met in fairly scattered references rather than in one or two systematic expositions. The nearest he came to giving the latter seems to have been in the first (and only) volume of his *Elements of Chemical Philosophy* (London, 1812). By this time his views had suffered some modification, as indeed they appeared to do frequently, often for the good reason that further facts had emerged, but sometimes, one feels, because of his well-known mistrust of theory in general except as a guide to further experiment and his consequent reluctance to attach himself firmly to any hypothesis.

The gap between Davy's second and third periods of electrochemical research was nearly fourteen years. Various reasons suggest themselves for this break, the most obvious being his preoccupation with other matters ; during the period in question (1810 to 1824), he was busy with his investigations on the nature of the halogens and of acids, together with researches on boron, phosphorus compounds and much else. He was becoming increasingly a public figure, being a Secretary of the Royal Society from 1807 to 1812 and President from 1819 to 1826. Several tours made more incursions into his time, as when he went to Dublin (1810 and 1811) and the Continent (1813–5). He had probably obtained as many facts in electrochemistry as he desired for the time being, and others were left to add to them and to produce improved theories. A perusal of the *Philosophical Magazine*, for example, will produce enough evidence for this. It may well have been that Davy felt that others were taking things too far in their theoretical speculations, and possibly the figure of Berzelius was beginning to dominate the electrochemical scene a little too much for Davy's liking.

However, in 1824, a series of publications[10] culminating in Davy's last Bakerian Lecture, " On the relations of electrical and chemical changes "[10c], marked a resumption of his interest in matters electrochemical. This seems to have been occasioned mainly by a problem from the Admiralty about the prevention of corrosion of copper plates attached to ships. Davy's solution (attachment of less noble metals)

[8] A course of lectures entitled " Introductory to Electrochemical Science ", commencing 12 March 1808 : printed in Davy's *Works*, London, 1840, vol. viii, p. 275.

[9] *Phil. Trans.*, 1808, [**98**], 333 ; *ibid.*, 1810, [**100**], 16.

[10] (*a*) *Phil. Trans.*, 1824, [**114**], 151, 242; (*b*) *ibid.*, 1825, [**115**], 328; (*c*) *ibid.*, 1826, [**116**], 383.

did not in fact prove workable, but concentration on the problem did give rise to the further development of his electrochemical theories.

In these present papers, an attempt is made to describe the main points of Davy's views on the Voltaic pile, on electrolysis and on the nature of chemical reaction. Part I is concerned with the first two of these.

1. *Davy's Theory of the Voltaic Pile.*

At the beginning of the nineteenth century there were two main views on the cause of Voltaic electricity. The " contact theory " of Volta[11] implied that the electricity originated in a bimetallic contact, for the condensing electroscope had indeed shown the existence of minute charges at such junctions ; the function of the liquid was simply to act as a conductor. The fact that the liquid was always decomposed chemically led to the alternative " chemical theory " of Fabroni and others[12], which tended to discount the facts of contact electrification and to stress the necessity of chemical action which was regarded as the ultimate source of the electricity. Both theories coloured Davy's early thinking, but at first he inclined to the latter view. Since the power of a pile using zinc and silver plates was related to the ability of the intervening solution to oxidize the zinc, he concluded " that the oxydation of the zinc in the pile, and the chemical changes connected with it are *somehow* the cause of the electrical effects it produces "[13].

This position, however, he was soon to abandon as further facts came to light and rendered it untenable. The above statement was published in November 1800, five months after the publication of the full text of Volta's paper ; indeed it occurs in a paper where specific reference is made to the " Galvanic pile of Signor Volta ". It is therefore surprising to read in Davy's first Bakerian Lecture (1806) that " Before the experiments of M. Volta on the electricity excited by the mere contact of metals were published I had to a certain extent adopted this opinion [chemical theory of galvanism] ; but the new facts immediately proved that another power must necessarily be concerned "[14]. There must have been a time-lag between his hearing the facts and changing his mind ; " immediately " must have a relative significance.

Whatever first discredited the " chemical theory " in his eyes, he was to give ample grounds for its rejection, and the following are his most cogent objections :

[11] *Ibid.*, 1800, **90**, 403.
[12] *Nic. J.*, 1799, **3**, 308, etc.
[13] *Ibid.*, 1800, **4**, 341.
[14] *Phil. Trans.*, 1807, [**97**], 49.

4

(*a*) *Contact electrification without chemical change.*

Electrical effects occur " when not the slightest chemical change can be observed "[15]. Moreover, they carry furthest in just those conditions generally opposed to chemical action on metals, *viz.*, a dry atmosphere[16].

(*b*) *Experiments with the electrical machine.*

In 1801 Wollaston carried out some experiments with an electrical machine, in the course of which he was able to prove the identity of the electricity from this source with that from a Voltaic pile. He also showed that when the cylinder of the machine was rotated in carbonic acid gas after previous rotation in air " the excitement was immediately destroyed, but that it returned on the readmission of ordinary air "[17]. He concluded that the cause was oxidation of the amalgam that adhered to the cushion of the machine ; he assumed a *chemical* cause for the electricity from the machine. It was probably this idea that led Davy to repeat the experiment ; for, if frictional and Voltaic electricities are the same, a chemical cause for the one could be taken as evidence for a chemical cause for the other. Davy, however, found that his machine " was active in hydrogene gas, and more active in carbonic acid gas, than in the atmosphere (probably owing to its greater density) "[18]. Thus another " chemical " explanation for electricity was discredited.

(*c*) *Variation in the sign of the charge.*

In 1807 Davy referred to the two cells :—

	zinc	diluted " nitrous acid "	copper
and,	water	zinc	diluted " nitric acid ".

It seems the same acid is intended here, as the point of the argument is lost if this is not so. Assuming we may identify " nitrous " and " nitric " acids, we should expect the reactions at the zinc/acid interface to be the same ; yet in the first cell the side of the zinc opposite the acid is positive, and in the second it is negative. Davy concluded that if chemical action is the cause of these electrical effects, then these charges should be the same. As experiment showed them different, the " chemical theory " was again rendered improbable[19].

[15] *Elements of Chemical Philosophy*, London, 1812, vol. i, p. 163.
[16] *Phil. Trans.*, 1807, [**97**], 49.
[17] W. H. Wollaston, *Phil. Trans.*, 1801, [**91**], 427.
[18] *Elements of Chemical Philosophy*, London, 1812, vol. i, pp. 163–4.
[19] *Phil. Trans.*, 1807, [**97**], 49.

(*d*) *Occurrence of chemical changes without electrical changes.*

Although a number of cases of electrical change following chemical action had been reported, Davy regarded evaporation (a *physical* cause) as responsible. As a corollary, he was able to show that, where many chemical reactions did occur, electrification was undetectable, e.g., between iron and oxygen, nitre and charcoal, zinc and mercury, etc.[19].

For these reasons the " chemical theory " was found to be unsatisfactory, and was rejected. But the alternative " contact theory " was equally unsatisfactory on its own. It could perhaps account for the behaviour of an insulated pile, but was inadequate when applied to a completed circuit. The problems remained of the almost unlimited quantities of electricity obtainable merely by attraction from one plate to another, and the manner in which a plate could both give and receive electricity at the same time[20]. Many years later (in 1826) an experiment was described by which the " contact theory " received a further severe blow[21]. Davy set up an apparatus of the kind represented schematically in Fig. 1 :

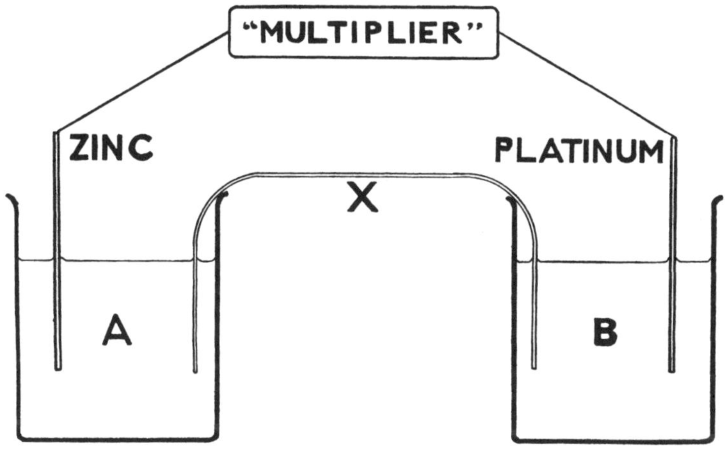

Fig. 1.

The vessels A and B contained solutions of nitrate of potassa, and the part X was a conductor whose nature was varied in different experiments. The following observations were made :

(1) X = asbestos moistened with the same solution as in A and B : definite electrical action resulted.

[20] *Elements of Chemical Philosophy*, London, 1812, vol. i, p. 174.
[21] *Phil. Trans.*, 1826, [**116**], 410.

(2) X = a bimetallic strip of platinum (in A) and zinc (in B) : the electrical action was increased.

(3) X = zinc : the effect was less than (2) but more than (1), the end of X in B being oxidized, with evolution of hydrogen at the end in A.

(4) X = other metals : electrical effects became progressively smaller as the oxidizability of the metals decreased.

Only in the case of (2) was there a metallic junction within the apparatus, and so Davy concluded :

> " If contact of the metals only was necessary for the continued electromotion, these results, in which a simple homogeneous chain is interspersed between the fluids, would be impossible."[21]

Thus, just as the occurrence of chemical change *by itself* is no explanation of voltaic phenomena, so also mere contact *by itself* is similarly insufficient, and " there can be no accumulation of electricity in voltaic combinations unless the same or similar conditions of chemical change exist in the elements or single circles composing them ".[22] What explanation could be offered? The genius of Davy lay in that he took *both* opposing theories, welded them together and showed how each was necessary, with some alteration, and each was complementary to the other. This reconciliation he describes particularly in his first Bakerian Lecture[6], his *Elements of Chemical Philosophy*, and his last Bakerian Lecture[10c]. Each of these represents a clarification rather than a modification relative to the one before it.

Davy's basic assertion was that a distinction must be made between two aspects of the operation of a pile. In his *Elements*, though not earlier, he calls these " the power of action " and " permanent action ", and these two aspects correspond to the contact and chemical theories respectively :

> " The power of action of the Voltaic apparatus seems to depend upon causes similar to those which produce the accumulation in the Leyden battery, namely, the property of non-conductors and imperfect conductors to receive electrical polarities from, and to communicate them to conductors."[23]

On the cause of this property he did not speculate, but accepted as axiomatic this ability of, for example, zinc to become positive when in contact with copper. He regarded this intermetallic contact as " destroying " or at least " disturbing " the electrical equilibrium :

> " The electrical energies of the metals with regard to each other, or the substances dissolved in water, in the Voltaic and analogous instruments seem to be the causes that disturb the equilibrium."[24]

[22] *Ibid.*, p. 411.

[23] *Elements of Chemical Philosophy*, London, 1812, vol. i, p. 166.

[24] *Phil. Trans.*, 1807, [**97**], 45.

It may be necessary in the light of modern scientific thought to emphasize that for Davy electrical equilibrium is roughly the same at electrical neutrality ; copper and zinc in contact are not therefore in this kind of equilibrium.

In a battery consisting of several cells the cumulative effect was considered to arise from induction :

> " With regard to electricities of such very low intensity, water is an insulating body : every copper plate consequently produces by induction an increase of positive electricity upon the opposite zinc plate ; and every zinc plate an increase of negative electricity on the opposite copper plate ; and the intensity increases with the number and the quantity with the extent of the series."[25]

However, the *continuous* action of the pile cannot be explained in terms of this kind. To explain the " permanent action " Davy invoked the " chemical theory " : " Permanent action is connected with the decomposition of the chemical menstrua between the plates "[26]. His later experiments[27] were to show that this hypothesis is at least credible, but some much earlier work[28] demonstrated that chemical decomposition was essential for production of an electric flow and with it a rise in temperature.

At this point we must notice that Davy spoke of " restoring the electrical equilibrium " in two slightly different senses, particularly in the Bakerian Lecture of 1806. He employed it, firstly, of the state of affairs arising when connexion is made between the two ends of a pile containing a fluid incapable of chemical action on the metals (e.g., pure water) ; such a pile would give single shocks only, and the equilibrium would be " permanently restored ". On the other hand, these words are also applied to a temporary restoration, a result of chemical changes inside the cell making it possible for the equilibrium to be re-destroyed immediately by contact, thus enabling " the process of electromotion " to continue. An example of a cell where this " temporary restoration " occurred is the following :

zinc ⋮ " muriate of soda " ⋮ copper.

When this cell is working hydrogen and alkali appear at the copper, and oxygen and acid at the zinc :

> "A solution of zinc is formed, and the hydrogene disengaged. The negative energy of the copper and the positive energy of the zinc are consequently again exerted, enfeebled only by the opposing energy of the soda in contact with the copper, and the process of electromotion continues as long as the chemical changes are capable of being carried out."[29]

[25] *Ibid.*, 1807, [**97**], 45.

[26] *Elements of Chemical Philosophy*, London, 1812, vol. i, p. 166.

[27] *Phil. Trans.*, 1826, [**116**], 410.

[28] *Ibid.*, 1807 [**97**], 48.

[29] *Ibid.*, 1807, [**97**], 45–6.

It is this latter conception that was to prove so fruitful. In the *Elements of Chemical Philosophy* he spoke of the restoration of electrical equilibrium only in terms of this temporary effect, and gave a further statement of the nature of the cyclic process envisaged :

> " The action of the chemical menstrua exposes continually new surfaces of metal ; and the electrical equilibrium may be conceived in consequence to be alternately destroyed and restored, the changes taking place in imperceptible portions of time."[30]

The same view was reaffirmed in 1826[31].

Thus Davy considered that a Voltaic cell operated by means of a cycle of events, and these can now be considered in more detail. The clearest statement in his own words appeared in 1826 :

> " It is a fact that zinc and platinum become electrical by contact, the zinc positive, the platinum negative ; and the two kinds of electricity are apparently most intense at the surfaces where they are in contact with the fluid, which is too imperfect a conductor to allow them to neutralise or destroy each other : they consequently assert their attractive and repellent powers upon the elements of the menstruum ; acids and oxygen circulate to the surface of the zinc, which in consequence is dissolved, and alkali and hydrogen to the surface of platinum, of which the hydrogen is disengaged, and the equilibrium broken by the contact of the metals is restored by the chemical changes ; so that a constant circulation, or a current of electricity, takes place, the power of the combination becoming feebler in proportion as the solution is decomposed, and acid accumulated round its positive, and alkali round its negative surface."[32]

Finally, it is necessary to refer to the rôle of water in a pile. In 1812 Davy expressly claimed that " there are no fluids known, except such as contain water, which are capable of being made the medium of connection between the metals, or metal of the Voltaic apparatus ".[33] Claims to the contrary there certainly had been, the most noteworthy being that of J. A. de Luc, whose column of zinc, Dutch leaf and paper seemed to operate very well without any moisture[34]. Davy was aware of this : he had met de Luc and seen his column in July 1808, but was convinced that traces of water were probably present and responsible for its action. Davy's insistence on this point was to cause a good deal of trouble later ; Faraday[35] attributed the idea that water was always necessary in *electrolysis* to an extrapolation from Davy's earlier remark about the Voltaic cell. In 1826 Davy stated that water was *not* always essential[36],

[30] *Elements of Chemical Philosophy*, London, 1812, vol. i, p. 170.
[31] *Phil. Trans.*, 1826, [**116**], 406.
[32] *Ibid.*, 1826, [**116**], 407.
[33] *Elements of Chemical Philosophy*, London, 1812, vol. i, p. 169.
[34] J. A. de Luc, *Nic. J.*, 1810, **26**, 69, 113, 241.
[35] M. Faraday, *Experimental Researches in Electricity*, London, 1849, vol. i, p. 133.
[36] *Phil. Trans.*, 1826, [**116**], 407.

though without reference to his former statement. He quoted the cell :

zinc ⋮ fused litharge or oxychlorate of potassa ⋮ platinum.

His brother John Davy, who edited his works, faced with the difficulty of two contradictory statements, said of the earlier : " it appears from the context that he was speaking of the apparatus as it was commonly used "[37]. This, however, hardly agrees with the forthright tone of the original pronouncement and perhaps it is unlikely that Davy changed his mind on new experimental evidence or his brother would have said so. Whatever the cause, there is no doubt of an altered point of view in 1826.

The fact remained, however, that water was *nearly* always necessary, though always with an impurity or other added substance that could attack one metal. The reason for this necessity Davy considered to lie in the " dipolarity " of water. This concept involved the ability of the fluid to have opposite charges at opposite surfaces, and to conduct a current in either direction. Substances capable of only unidirectional conductance were " unipolar ". In the case of a liquid like water, Davy seemed to visualize an arrangement of doubly polarized water particles like that of floating wires that he investigated[38]. When these were placed between two charged wires, they became electrically polarized and tended to be orientated between the electrodes as in Fig. 2 :

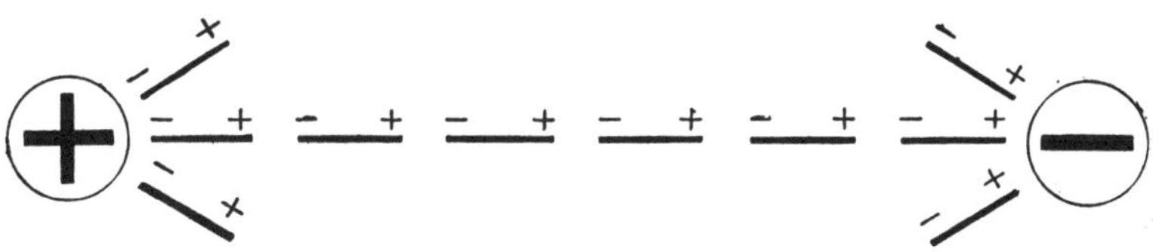

Fig. 2.

With the aid of the theory that he developed, Davy was able to make a number of successful predictions in the field of electrolysis. He showed that the action of a pile was reversible and that, when one was connected to a stronger pile, electrolysis took place. An example of this was the system :

zinc ⋮ solution of nitre ⋮ copper.

When this was connected " in a reverse or unconformable manner " with a 50-pair Voltaic battery, hydrogen was evolved at the zinc, and the copper became oxidized[39]. Other predictions were suggested by his own theoretical views on the Voltaic pile, and as these were so closely

[37] Editor's foot-note in H. Davy's *Works*, London, 1840, vol. vi, pp. 328–9.
[38] *Elements of Chemical Philosophy*, London, 1812, vol. i, p. 170.
[39] *Phil. Trans.*, 1826 [**116**], 414.

related to those on electrolysis, some attention to the latter will now be given.

2. *Davy's Theory of Electrolysis.*

Although the practical achievements of Davy in this field are far better known than his experiments with the Voltaic pile, his interpretations of the facts of electrolysis were less clearly expounded than his views on other aspects of electrochemistry. Indeed, his first feelings were those of great perplexity ; he wrote in 1802 :

> "The general chemical agency of galvanism is at present involved in obscurity. The facts relating to the separate production of oxygen and hydrogen, acid and alkali in water, are totally incommensurable with the usually received theory of chemistry ; and even supposing that the appearance of the two last of these bodies is connected with the presence of atmospheric air, it is still extremely difficult to conceive that either oxygene or hydrogene can pass in an invisible form through fluids or organic bodies."[40]

Nevertheless, in the few years between his first experiments and his Bakerian Lectures of 1806 and 1807 he was able to come to certain conclusions. Thus in the first place, he was convinced of the presence of an electrical attraction between the charged surfaces and parts of the medium[41]. Although only unsatisfactory results were obtained when he sought for evidence of contact electrification between metals and solutions, he did obtain definite indications of charges after *dry* solids had been in contact with each other[42]. Perhaps because of the inconclusive results with solutions, he did not commit himself definitely to the view that attraction arises from a preliminary contact electrification of particles of solute by the electrode metals. Probably he thought that the alkali received its positive electricity from the positive plate, but he does not appear to have said as much explicitly.

In the electrolysis of salts, he assumed that alkalis tend to become positive and acids negative. They would then be incapable " of remaining in combination when placed in a state of electricity different from their natural order. Thus, as we have seen, the acids in the positive part of the circuit separate themselves from alkalies, oxygene from hydrogene, and so on ". This early statement[43] contains an allusion to what was to become later[44] an important part of his theory, namely, that the solution was to be regarded as being "divided into two zones, directly opposite in their powers to the poles of a battery ". Presumably the

[40] *J. Roy. Inst.*, 1802, **1**, 64.
[41] *Phil. Trans.*, 1807, [**97**], 30.
[42] This work will be further considered in Part II of this paper.
[43] *Ibid.*, 1807, [**97**], 38.
[44] *Ibid.*, 1826, [**116**], 416–7.

zones would become charged by induction from the plates, and the intensity of their charges would vary with distance from these surfaces. Thus at one point in the solution the two zones would meet in a " neutral point ".

This gave a general way of looking at the causes of decompositions, but it did not *per se* suggest any detailed steps. It left unanswered the questions as to why the products appeared at the plates only, and what were the steps leading to their production there. After a little hesitation Davy concluded that in fact a " series of decompositions and recompositions " took place, and that it was only the " extreme parts " that were active. His own experiments on electrolytic transfer convinced him on

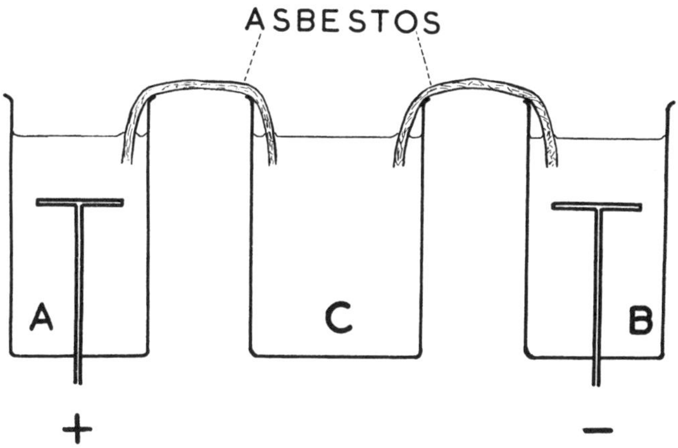

Fig. 3.

this matter. Using apparatus of the type in Fig. 3, he varied the nature of the solutions in A, B and C, and found in many cases that materials in A could be successfully transferred to B without affecting the contents of C. If, however, B contained sulphuric acid and C contained baryta, no sulphuric acid reached A, but a precipitate of sulphate of baryta appeared in C. He interpreted this as an interference by the baryta with the chain of particles, precipitating them out of the sphere of electrical action.

The initiating stage of the process was considered by Davy (unlike Grotthus[45], to whom he has been compared[46]) to take place, not at the two ends of a chain of molecules orientated between the electrodes, but in the one particle *situated in the neutral zone* :

> " If it be supposed that the fluid is divided into two zones, directly opposite in their powers to the poles of a battery, the virtual change

[45] T. von Grotthus, *Ann. de Chim.*, 1806, **58,** 54.

[46] E.g., M. Faraday, *Experimental Researches in Electricity*, London, 1839, vol. i, p. 137.

may be regarded as taking place in the two extremities of these zones nearest the neutral point ; so that by a series of decompositions and recompositions the alkaline matters and hydrogene separate at one side, and oxygene, pure or in union, at the other."[47]

It is at first difficult to see what should have led Davy to propose the first step to occur where the electrical fields are at their *weakest*. Although his reason is nowhere stated explicitly, it might well have been that only at or very near the neutral point could both parts of a " particle " be in oppositely charged fields *at the same time*. For decomposition the basic part must be in the negative zone (i.e., that around the positive plate), and the acidic part in the positive zone. It was then assumed that the liberated parts of the salt would attack other parts around them until at length charged parts arrived at the plates and discharge took place. However, Davy did not give a precise account of his detailed views, and it was of this aspect (electrolysis) of his theory that Faraday wrote " probably a dozen precise schemes of electrochemical action might be drawn up, differing essentially from each other, yet all agreeing with the statement there given "[48].

Conclusion.

It is an irony of history that of all the ways in which Davy's ideas could have been transmitted it should have been by Michael Faraday that this process was to be accomplished, at least so far as the Voltaic cell and electrolysis were concerned. Despite the well known personal difficulties that arose between these men, Faraday was the one to influence the world to look favourably on his master's theories. And he did this by enshrining them in his own.

The influence of Davy on Faraday's views of the Voltaic pile may be discerned. Although the early years of the century saw a rather indecisive conflict between advocates of the " contact " and " chemical " theories of the pile, it is hardly surprising that Faraday leant towards " that one to which all his antecedents and connections predisposed him "[49]. He approved of Davy's work showing that chemical action is not essential to electrification[50], specifically identified Davy's " electrification by contact " as one stage in the production of a current[51], and was " fully of the opinion with Davy that it is at least continued by chemical action, and that the supply constituting the current is almost entirely from that source "[52].

[47] *Phil. Trans.*, 1826, [**116**], 416–7.

[48] M. Faraday, *Experimental Researches in Electricity*, London, 1839, vol. i, p. 136.

[49] E. Whittaker, *History of the Theories of Aether and Electricity*, London, 1951, vol. i, p. 180.

[50] M. Faraday, *Experimental Researches in Electricity*, London. 1839, vol. i, p. 555.

[51] *Ibid.*, pp. 554–55.

[52] *Ibid.*, p. 252.

In the field of electrolysis, as has been mentioned, the vagueness of Davy's ideas seemed rather to irritate Faraday[48]. Rejecting any idea of a neutral zone, Faraday agreed that a series of decompositions and recompositions was probable, and put forward a theory that accounted for a good deal that Davy had left unexplained[53]. He denied that the attractive power of the poles could cause decompositions, and supplied experimental evidence to support this by effecting some decompositions at liquid/liquid interfaces[54].

Other workers of the time owed something to Davy's speculations in these fields. Those who held to the belief in decompositions and recompositions throughout the solution, as Riffault and Chompré[55], may have been indebted to Davy as well as to Grotthus. The same idea appears to be present in the writings of Berzelius[56], though is not explicitly mentioned. The latter also agreed with Davy's doctrine of attracting poles, and it is interesting to read that Berzelius was first led to doubt the " chemical theory " of the pile by the work of Davy[57].

[53] *Ibid.*, pp. 141–151, 157.

[54] *Ibid.*, p. 143.

[55] *Ann. de Chim.*, 1807, **63,** 83.

[56] J. J. Berzelius, *J. de Phys.*, 1811, **73,** 277–79 ; *Traité de chimie,* Paris, 1829, vol. i, pp. 155–56 ; *ibid.*, 2nd edn., Paris, 1845, vol. i, pp. 97–99.

[57] *Idem.*, *J. de Phys.*, 1811, **73,** 256.

III

THE ELECTROCHEMICAL THEORY OF SIR HUMPHRY DAVY.

PART II : ELECTRICAL INTERPRETATIONS OF CHEMISTRY.

IN the previous part of this paper[1] the electrochemical theories of Sir Humphry Davy were described in so far as they related to the Voltaic pile and electrolysis. These two phenomena are obvious cases of the interaction of chemical and electrical effects, and it is right to consider Davy's ideas about them. It is, however, the contention of the present paper that his greatest theoretical contribution came essentially as an extension of these views ; this was to regard all chemical reactions as further manifestations of electrical effects, or, later, to ascribe to both a common cause or basis. It is this aspect of his electrochemical theories that will now be considered.

Davy was, of course, aware of a variety of chemical phenomena induced by electrical effects. Apart from electrolysis he had a first-hand acquaintance with the phenomena of "electrical ignition" produced by the discharge of large Voltaic cells through metals and between carbon points[2], and also through liquids[3]. He knew much about the physiological action of electricity[4], and was specially interested in the behaviour of substances in an "unnatural electrical state". Matter in this condition has the opposite charge to that which it will acquire on contact electrification, and was found to show anomalous chemical behaviour ; thus negatively electrified zinc is not oxidized, but oxidation of positively electrified silver becomes possible[5]. On the other hand, Davy found that when "natural" states are enhanced (e.g., by contact between dissimilar metals), so also are the reactivities[6].

Davy was thus in possession of many experimental facts connecting chemical and electrical action. In addition, he greatly increased knowledge of contact electrification, extending Volta's work to contacts between metals and non-metallic substances. He could not prove

[1] *Ann. Sci.*, 1959, **15**, 1–13.
[2] *Elements of Chemical Philosophy*, London, 1812, vol. i, pp. 137–8.
[3] *Ibid.*, pp. 151–54.
[4] *Ibid.*, p. 142 ; *J. Roy. Inst.*, 1802, **1**, 49.
[5] *Phil. Trans.*, 1807, [**97**], 39.
[6] *Elements of Chemical Philosophy*, London, 1812, vol. i, pp. 142–3.

Annals of Science 15, 1959, 15–25. Reprinted by permission of Taylor and Francis Ltd (http://www.tandf.co.uk/journals).

electrification of the fluid in a Voltaic cell[7], but was successful with a number of dry solids, using a gold-leaf electrometer.

All the solid acids he investigated became negative on plates of zinc or copper. This included oxalic, succinic, boracic, benzoic and phosphoric acids. Of the solid bases he tried, lime, strontites and magnesia rendered metals negative ; soda required very drastic exclusion of moisture, and potash gave no effect. Sulphur became positive on contact with metals, agreeing with its separation on the negative surface in electrolysis of sulphuric acid, and with friction experiments. He extended these findings to cases where neither surface was a metal :

> " Bodies possessing opposite electrical energies with regard to one and the same body, we might fairly conclude would likewise possess them with regard to each other."[8]

This he confirmed in the case of lime and oxalic acid which on mutual contact became respectively positive and negative. Nevertheless, electrical effects of this kind disappear on combination. The chief example given of this principle was not, perhaps, the happiest choice ; it was, however, one of the easiest to examine. Davy referred to the process of amalgamation, which is not regarded today as a clearly defined chemical process. He spoke of the positive electrification of zinc in contact with mercury, which vanishes on strong heating, i.e., under the essential condition for speedy amalgamation. On the other hand, iron and mercury form stronger charges, perhaps because of the difficulty of amalgamation in their case. A more truly " chemical " example was the disappearance of charge as copper combined with sulphur[9].

All these facts led Davy to conclude that contact electrification and " chemical affinity " must be identified :

> " In the present state of our knowledge it would be useless to speculate on the remote cause of the electrical energy, or the reason why different bodies, after being brought into contact, should be found differently electrified ; its relation to chemical affinity is, however, sufficiently evident. May it not be identical with it, and an essential property of matter ? "[10]

Shortly after this he asked :

> " Is not what has always been called chemical affinity merely the union or coalescence of particles in naturally opposite electrical states? And are not chemical attractions of particles and electrical attractions of masses owing to one property and governed by one simple law? "[11]

[7] *Phil. Trans.*, 1807, [**97**], 30–3.

[8] *Ibid.*, pp. 33–39.

[9] *Elements of Chemical Philosophy*, London, 1812, vol. i, p. 159.

[10] *Phil. Trans.*, 1807, [**97**], 39.

[11] A course of lectures entitled " Introductory to Electrochemical Science ", commencing on 12 March 1808 : printed in Davy's *Works*, London, 1840, vol. viii, p. 283.

This electrical interpretation of chemistry was unfolded and illustrated in a number of different contexts. For example, when two bodies attract each other Davy believed that the attractive force (i.e., that acting between particles of different substances) would act in the opposite direction to the force of aggregation (i.e., that between particles of the same substance). The resultant of these would determine the intensity of the reaction and the extent to which the properties would be modified. In the case of two bodies that were mutually repellent but attractive to a third, he recognized that " different substances have different degrees of the same electrical energy in relation to the same body "[12]. Thus for lead, sulphuric acid has a more powerful affinity than hydrochloric, while for tin, potash is a stronger reagent than soda[13]. Hence the reagent with the weakest energy will be repelled. If two bodies have different degrees of the same energy with respect to a third, but opposite kinds relative to each other, then complicated situations can arise, such as the formation of triple compounds. Possibly he had in mind sulphur, alkalis and metals.

He attempted to extend the argument to cohesion. In 1777, Guyton de Morveau had stated that metals adhere to mercury with forces proportional to their chemical affinities[14]. Now, Davy thought it probable that the force holding the two metals together would depend upon the degree to which their " electrical states " differ. To some extent he justified this view ; for instance, he found that copper, zinc and tin gave much higher charges to a condensing electroscope than did iron, after these metals had been in contact with mercury and then separated. Thus Guyton's finding could be explained on the simple hypothesis that :

$$\text{chemical affinities} = \text{electrical energies.}$$

However, Davy recognized that this might be a special case, and left open the more general question of cohesion[15].

Thus far in the argument no special difficulties arise. But it was not long before a total identification of electrical and chemical attraction ran into trouble. When electrically charged objects are attracted to

[12] *Phil. Trans.*, 1807, [**97**], 40.

[13] It is doubtful if these statements are wholly correct, or, since conditions were not specified, if they have much meaning. The reactions that were probably being considered can be represented in modern symbolism thus :

$$PbCl_2 + H_2SO_4 = PbSO_4 + 2HCl \; ;$$
$$Sn + 2KOH = K_2SnO_2 + H_2 \; ;$$
$$Sn + 2NaOH = Na_2SnO_2 + H_2.$$

[14] L. B. Guyton de Morveau, H. Maret and J. F. Durande : *Elémens de chymie*, Dijon, 1777, vol. i, p. 61.

[15] *Phil. Trans.*, 1807, [**97**], 51.

each other, their charge remains while attraction continues. At first, Davy supposed that this would also be true where the attraction was manifest as chemical affinity ; " different particles in combining must still be supposed to preserve their peculiar state of energy ".[16] However, the experiments referred to earlier[9], where electrical effects disappeared on combination, caused him to abandon this extension to his theory. One cannot say that chemical forces are just electrical attraction if there are no detectable charges when the chemical binding is strongest, i.e., in the newly formed compound.

Consequently, by the time he was writing his *Elements of Chemical Philosophy* (1812), Davy had slightly shifted his ground. Instead of completely identifying the two effects he ascribed to them a common basis :

> " Electrical effects are exhibited by the same bodies, when acting as masses, which produce chemical phaenomena when acting by their particles ; it is not therefore improbable that the primary cause of both may be the same, and that the same arrangements of matter, or the same attracting powers, which place bodies in the relations of positive and negative, i.e., which render them attractive of each other electrically, and capable of communicating attractive powers to other matter, may likewise render their particles attractive and enable them to combine when they have full freedom of motion."[17]

Years later (in 1826) he denied ever having attached " much importance to this hypothesis ", but he appears never to have revoked or altered it[18].

In connexion with this modified form of his theory, Davy carried out an experiment which deeply interested him, and to which he referred on at least three different occasions[19]. This was the " electrization of mercury ". As first described, this involved placing globules of mercury in a vessel of common pump-water in which dip two wires from a Voltaic battery. When the circuit is completed, each globule becomes agitated, electrically polarized and distended into a pear shape, with the narrow (negative) end orientated towards the positive wire. Violent motion of these globules occurs and oxide of mercury appears at the wider end and rapidly flows across the surface to the narrow end. During this time, hydrogen is not liberated. Should either the " charge of the battery " or the concentration of mineral salts be increased, the motion of the globules ceases, they resume a spherical form and hydrogen appears at the negative pole.

[16] *Ibid.*, 1807, [**97**], 40.

[17] *Elements of Chemical Philosophy*, London, 1812, vol. i, p. 164.

[18] *Phil. Trans.*, 1826, [**116**], 390.

[19] (*a*) *Elements of Chemical Philosophy*, London, 1812, vol. i, p. 172 : (*b*) *ibid.*, p. 485 ; (*c*) *Phil. Trans.*, 1826, [**116**], 417.

Further details were given in 1826 of a revised form of this experiment[19c]. The solution was very dilute (1 : 2000) sulphate of potassa, and evidence was obtained of the existence of potassium amalgam which Herschel had shown[20] in 1824 to be strongly attracted to a negative surface. The peculiar effects were destroyed by the addition of a little muriatic acid[21].

Whatever may be the true nature of these remarkable effects, they had a great significance for Davy, who considered it a useful illustration of the way " chemical changes tend to restore the electrical equilibrium "[22]. He inferred that " the same power which gave motion to the mercury was neutralised by, or employed in, the evolution of hydrogene "[23]. That is, it confirmed for him the existence of a common cause for electrical and chemical phenomena ; the electrical attractions gave place to evolution of hydrogen, the negative energy of the added acid " destroying all the phenomena of attraction by the positive pole ".[24]

Davy attempted to put his theory on a semi-quantitative basis. He realized that his list giving relative powers of metals in a Voltaic cell also applied to their degrees of reactivity with the fluids concerned, " the electrical powers being in all cases apparently connected with the power of chemical combination "[25]. That is, the higher the attracting powers of metals, the more positive they would be. Since the order in which metals replace one another approximates to the order of affinities for oxygen, it follows that the replacement series was also the electrical series:

> " The order in which metals precipitate each other from solutions is nearly in the ratio of their attraction for oxygene ; and in all cases of neutral compounds, the precipitating metal takes the oxygene and acid of the metal thrown down. Iron readily precipitates copper ; zinc readily throws down tin, lead, tellurium, bismuth, etc., and in general the metallic substances . . . . . . attract oxygene, and precipitate each other in a ratio connected with their electrical relations ; those that are positive with respect to others having the highest attractive powers for oxygene and acids."[26]

[20] J. Herschel, *Phil. Trans.*, 1824, [**114**], 180.

[21] In connexion with this examination of Davy's work, some of these phenomena have been re-investigated experimentally. The polarization, shape and motion of the globules have been confirmed, but the effects of increased salt concentration have not been reproduced so far.

[22] *Elements of Chemical Philosophy*, London, 1812, vol. i, pp. 172–3.

[23] *Ibid.*, p. 173.

[24] *Phil. Trans.*, 1826, [**116**], 418.

[25] *Elements of Chemical Philosophy*, London, 1812, vol. i, p. 159.

[26] *Ibid.*, pp. 500–501.

In a similar way it would follow that since oxygen attracted metals, " the attractions of acids for salifiable bases would be inversely as the quantity of oxygen which they [the bases] contain "[27].

Hence analysis of the ratio acid/base in a neutral salt would give an index of the proportion of oxygen in the base ; in this way Davy inferred that baryta must contain the least and silex the most oxygen.

As a consequence of these speculations, Davy was able to suggest electrochemical methods for determining affinity :

> "Allowing combination to depend upon the balance of the natural electrical energies of bodies it is easy to conceive that a *measure* may be found of the artificial energies, as to intensity and quantity produced in the common electrical machine, or a Voltaic apparatus, capable of destroying this equilibrium ; and such a measure would enable us to make a scale of electrical powers corresponding to degrees of affinity."[28]

The apparatus Davy employed he described in a footnote :

> " In this experiment, the water was contained in a circular glass bason two inches deep, the communication was made by pieces of amianthus of about the eighth of an inch in breadth. The saline solution filled a half ounce measure, and the distance between the solution and the water, at both points of communication, was a quarter of an inch."

He added that amianthus (a form of asbestos) was found much better than " glass siphons ". Figure I is a diagrammatic attempt to convey what Davy may have had in mind. The distances marked " x " and " y " could be varied so that the solution could be moved around in the water, which was presumably impure, so as to be conducting. Davy assumed " the strength of opposite electricities diminish from the points of contact of the wires to the middle of the water, which is necessarily neutral ". He seems to have believed that the distances between the solution being decomposed and the two electrified wires would indicate the ease of decomposition of the material in the inner vessel. Later, in 1826, he affirmed that the use of the " multiplier " in this connexion " will be the first step towards fixing chemistry on the permanent foundation of the mathematical sciences ".[29] Unfortunately, he gave no further details of this, and one must presume these words represent more a hope than an accomplishment.

Although Davy was thus able to produce some kind of electrical theory of chemical reactions, he was certainly aware that *by itself* this explanation would not carry one far. In particular, he knew that relative concentrations had to be taken into account. As early as 1802 he had

[27] *Phil. Trans.*, 1810, [**100**], 62.
[28] *Ibid.*, 1807, [**97**], 41–2.
[29] *Ibid.*, 1826, [**116**], 419.

said[30] that forces between bodies " depend not only upon the difference between the primary affinities of the substances for each other, but likewise upon the difference of their masses, that is, upon the number of different particles brought into the sphere of activity ". To this idea, which Berthollet had also expressed in 1801[31], Davy added an electrochemical twist of his own : " The combined effect of many particles possessing a feeble electrical energy may be conceived equal or even superior to the effect of a few particles possessing a strong electrical energy "[32]. In favour of this view he quoted his own findings that

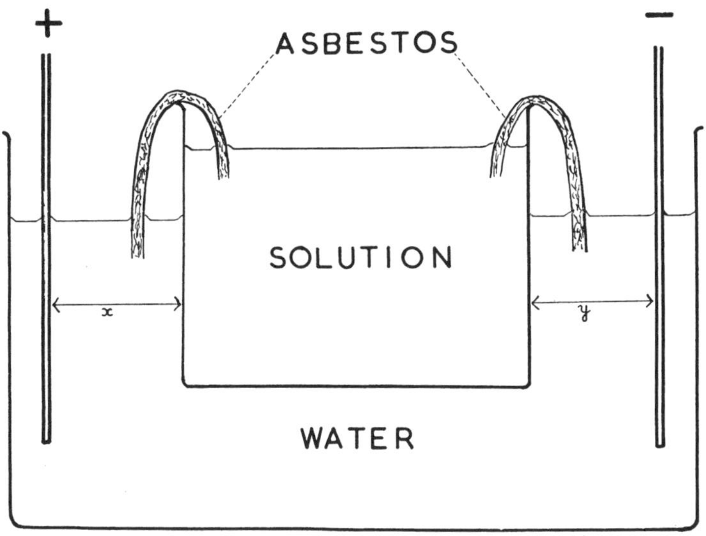

Fig. 1.

rate of transfer of sulphuric acid through alkalis depends on the dilution of the latter. He used the apparatus shown in Fig. 3 of Part I (p. 11), connected to a powerful array of Voltaic cells. Vessel A contained distilled water, while a solution of sulphate of potash was in B and a weak solution of ammonia in C. Under these conditions, litmus at the positive point was reddened within five minutes ; if the ammonia was replaced by strong potash or soda the effect took much longer, " but

[30] *Syllabus of a Course of Lectures at the Royal Institution*, London, 1802, p. 31.
[31] C. L. Berthollet, *Recherches sur les lois de l'affinité*, Paris, 1801.
[32] *Phil. Trans.*, 1807, [**97**], 41.

even with the most saturated alkaline lixivium, it always appeared in a certain period "[33].

One final aspect of Davy's theory on the relation between chemical and electrical changes remains for discussion. For a current to flow, the water in A must have been slightly contaminated, e.g., from the asbestos or the vessel (*cf.* Davy, *loc. cit.*, pp. 1–11).

The evolution of heat (and light) by both " intense chemical action " and electrical discharges he regarded as further evidence for a close connection between the two. " It is not a little in favour of this hypothesis, that heat, and sometimes light, result from the exertion of both electrical and chemical attractive powers "[34]. Similarly, a milder action of each type may produce heat only. He was further aware that a change of physical state, often involving a change of temperature, is often accompanied by electrification[35]. The heat appeared to have a twofold role in these situations.

In the first place, it serves " to exalt the electrical energies of bodies ". He was familiar with the well-known cases of pyro-electricity (glass, tourmalin, sulphur) where this is an obvious result of the rise in temperature. Further, he examined the course of the reaction between copper and sulphur and obtained the following results :

56° F. : little evidence of electrification ;

100° F. : intensity of electrification began to increase ;

melting-point (235° F.) : maximum electrification, heat evolved, reaction about to commence.

From this he concluded that " there is the same exaltation of power, in all cases of combustion ". But these effects will vary with the compounds formed : " In general, when the different energies are strong and in perfect equilibrium, the combination ought to be quick, the heat and light intense, and the new compound in a neutral state "[36]. Certain measurements of the electrical states of various salts did show an electrification in some cases, where presumably these energies were not equally strong :

neutral :	nitrate and sulphate of potash ; muriate of lime ; oxymuriate of potash.
slightly negative :	subcarbonate of soda ; borax.
slightly positive :	alum ; superphosphate of lime.

[33] *Ibid.*, 1807, [**97**], 24–5.
[34] *Elements of Chemical Philosophy*, London, 1812, vol. i, p. 164.
[35] *Ibid.*, pp. 138–139.
[36] *Phil. Trans.*, 1807, [**97**], 43.

In addition to this, Davy held that there was a second and more general function of heat in chemical changes, namely, to increase the motion of the particles. Implicit in these ideas were his views on the nature of heat. He was inclined to deny the material nature of heat on the basis of his early ice-block experiments, in which he attempted to melt ice by rubbing blocks of it together[37]; these experiments were very crude and of far less significance than once thought[38]. A more powerful reason presented itself to him in the form of the ability of platinum wires to glow indefinitely while conducting a current[39]. He concluded :

> " It seems possible to account for all the phaenomena of heat, if it be supposed that in solids the particles are in a constant state of vibratory motion, the particles of the hottest bodies moving with the greatest velocity and through the greatest space . . . . Temperature may be conceived to depend upon the velocities of the vibrations."[40]

He regarded heat as " the power of repulsion ", opposite in action to cohesion, the balance between forces determining whether the material concerned was solid, liquid or gaseous. Thus heat facilitates reaction by giving to the particles additional freedom of motion.

The precise relationship between this and electricity was not made at all clear, however, except that evolution of heat was regarded as a confirmation of his ascription of a common cause to both kinds of change. The nearest Davy got to a definite statement is that " most of the substances that act distinctly upon each other electrically, are likewise such as act chemically, when their particles have freedom of motion "[41], the latter being presumably imparted by the heat. Thus, if the degree of electrification were high enough and the particulate velocity sufficiently great, then chemical action would occur. In this way, it appears that he regarded chemical change as almost a *special case* of electrical change. If this is so, as he seemed to imply, how can it " exalt electrical energies "? On the assumption that the theory had been worked out in detail, it seems inevitable that a kind of cyclic process (or " chain reaction ") was envisaged. Thus in electrical ignition the effects are *caused* by electrification and " connected with the motions of the particles of the substances affected "[42]. But the electrification and ultimate reaction of copper and sulphur are caused by heat. The following is a schematized

[37] *Contributions to Physical and Medical Knowledge, principally from the West of England,* collected by Thomas Beddoes, M.D., Bristol, 1799, p. 13.

[38] See, e.g., E. N. da C. Andrade, *Nature,* 1935, **135,** 359.

[39] *Elements of Chemical Philosophy,* London, 1812, vol. i, p. 138.

[40] *Ibid.,* pp. 95–96.

[41] *Ibid.,* p. 158.

[42] *Ibid.,* p. 138.

summary of his ideas on the relations of heat, electricity and chemical change :

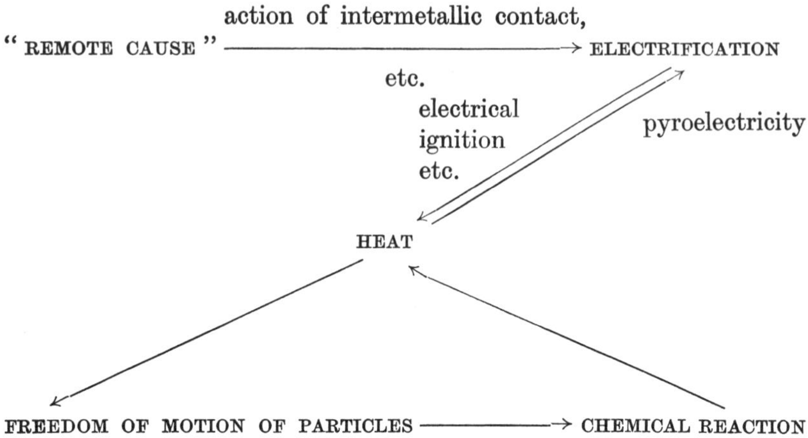

In these ways Davy was the first to link the well-known phenomena of chemistry with the new facts of Voltaic electricity. Electrical interpretations of chemical phenomena have now become a commonplace, but the pioneer in attempting a union of the two branches of science was Davy. One might thus expect that the early nineteenth century would see a gradual development of electrical theories of chemistry shaped essentially by the ideas of Davy and owing its inspiration to him. In fact it saw nothing of the kind. There *is* a dominating figure in chemical science, but it is that of Berzelius, not Davy. The electrochemical system of the Swede, not the Englishman, was to direct, and bedevil, chemistry for half a century. It seems, then, that the greatest achievement of Davy in *theoretical* chemistry was to give to Berzelius the initial impetus for the latter's system of electrochemistry.

The debt of Berzelius to Davy in this respect was realized by the latter[43] and seems to be acknowledged by the former[44]. It has been said that " Davy opened out the province in which Berzelius was to work with such effect in the decade following "[45]. Nearer the time, Whewell stated that " the most important attempt to appropriate and employ the generalization which these [Davy's] discoveries suggested, was that of Berzelius . . . . in which he was followed by a large body of the chemists of Germany "[46].

[43] *Phil. Trans.*, 1824, [**114**], 153.

[44] *J. de Phys.*, 1811, **73**, 256.

[45] E. von Meyer, *History of Chemistry*, trans. McGowan, London, 3rd edn., 1906, p. 206.

[46] W. Whewell, *History of the Inductive Sciences*, London, 2nd edn., 1847, vol. iii, pp. 194–5.

In other words, Davy's function in matters of electrochemical theory was to supply his continental rival with the nascent hypothesis, and to leave it to him to permeate the scientific world with its influence. And to say this is to imply no disrespect for Davy, whose achievements in the realm of experimental science have left him with an unshakable reputation. Had circumstances impelled him to pay closer attention to matters of theory, and had his own temperament given him the regard for detail that so characterized Berzelius, he might have " advanced chemistry by one hundred years "[47]. In the event, Davy's name is now remembered chiefly for other accomplishments, but for his place in the history of electrical theories of chemistry he should not be forgotten.

[47] Letter from Berzelius to Wöhler, 1831 ; published in *Berichte*, 1864, **15**, 3166.

IV

THE ELECTROCHEMICAL THEORY OF SIR HUMPHRY DAVY

PART III: THE EVIDENCE OF THE ROYAL INSTITUTION MANUSCRIPTS

Introduction

In the two previous papers of this series [1] some consideration was given to the electrochemical theory held by Sir Humphry Davy and, in particular, to his views on the voltaic pile, electrolysis and chemical change. The discussion has been based entirely upon printed sources, and attention has been paid only to the ideas that Davy chose to publish in the literature or to deliver as lectures at the Royal Institution and the Royal Society. These are all available in his collected *Works*.[2] It is, of course, true that the inner workings of a man's mind are not likely to be revealed very accurately by what he sees fit to print, and this is particularly true of Davy whose speculations tended to proliferate faster than he cared to admit publicly. Consequently, there may be a disparity between Davy's private conclusions and his more tentative public announcements. Nevertheless, a man's place in the development of any corpus of ideas is largely determined (*a*) by what he says in public (especially if this is well reported afterwards), and (*b*) by what he writes for publication. His influence is conveyed chiefly through these media, although this is not to deny the importance of personal contacts and correspondence.[3] It is for this reason that the present account has so far been confined to printed sources.[4]

[1] *Ann. Sci.*, 1959, **15**, 1–13, and pp. 15–25.

[2] *The Collected Works of Sir Humphry Davy, Bart.*, ed. by John Davy, London, 1840, 9 vols., including a *Life*.

[3] Davy's correspondence with Berzelius was of the greatest importance and has been referred to by the present author in *Ann. Sci.*, 1963, **19**, 121–122.

[4] A comment by L. Pearce Williams as a foot-note to an article on ' The physical sciences in the first half of the nineteenth century : problems and sources ' (*History of Science*, 1962, **1**, 15) has criticized this decision. Exclusive reliance on printed sources is ' inadequate ' and ' most certainly so when Davy is concerned. Injured by his flights of fancy when still a youth, Davy was careful ever after to conceal his speculations. These can be found only in the notebooks now at the Royal Institution '. The extent to which these sources add substantially to our understanding of Davy may be inferred from the present account. However, I am grateful to Professor Williams for his comment, for this has initiated a re-examination of the documents concerned in connexion with the present paper.

Annals of Science 19, 1963, 255–71. Reprinted by permission of Taylor and Francis Ltd (http://www.tandf.co.uk/journals).

In the case of Davy, however, there is also available the rich treasure of manuscript material in the Library of the Royal Institution, London. From this much can be gleaned of his private views and of the experiments that were to bring him immediate and permanent fame. It is, therefore, the purpose of this paper to examine the bearings of this large amount of source-material upon the assessment already made of his electrochemical theory. Some indication of the relevant manuscripts may be a useful preliminary point.

The most famous manuscript items are Davy's 'Laboratory Note-books', containing his much-quoted discoveries of potassium and sodium. These books are dated 1805–09, 1809–12, 1813–21, and 1821–59 (the last, of course, continuing long after Davy's time). Lamentably, no similar note-book covers the vital years before 1805. There is, however, a 'Physical Journal' covering a few months from November 1800 and including the first four weeks of Davy's work at the Royal Institution (from 11 March 1801). Probably Davy's intense activity over the next three or four years precluded any systematic recording of those experiments for which he was able to find time. The resumption of this practice in 1805 may not be unconnected with Davy's final abandonment at this time of the prospect of a medical career and with some relaxation of his responsibilities in the field of applied chemistry and geology.

Between the years 1800 and 1805 there is a series of books catalogued as 'Scientific Experiments, 1800–05'. These are not systematic practical note-books, however, but are a miscellany of unrelated jottings. They include plans for lectures at the Royal Institution (suitable for 'the uninstructed'), his own painful reactions to the task of lecturing to an audience with widely different abilities, outlines of a book on 'Cornwal', notes on the classical poets and much else. They contain almost nothing on electrochemistry.

Probably the most valuable source-material is to be found in Davy's 'Personal Note-books', which cover (though with some breaks) the period from 1795 to 1829. It is also the most difficult to unearth and assess. The books were usually started from both ends, entries are often undated and sometimes totally illegible. They are a bizarre mixture of poetic copyings and originals, admissions of failure and new resolves, geological sketches and personal caricatures, philosophical dogmatism and scientific uncertainty, precise observations and seemingly wild flights into the realm of hypothesis. Yet it is here that one seems to meet the philosopher as he really was, and he gains rather than loses in stature from the encounter. For in these note-books he may be seen more genuinely as a man of many more parts than his public may have guessed. The conflict between the boldness of his imagination and the caution of true scientific restraint is emphasized by the futile reproduction of hypotheses, only to be followed by elaborate plans for their testing, and often by their rejection. And to these note-books, rather than to any other source, we owe our knowledge of his thinking in the period from 1800 to 1805 when his electrochemical theory was germinating.

Other manuscript material has less bearing on this aspect of Davy's work. The 'Commonplace Book' (mainly of 1814) and the papers collected by the faithful Faraday (from 1815) are rather too late. However, Faraday's introduction to the latter throws an interesting light on the probable origin of some of the gaps in the earlier material :

> 'It was part of my duty to copy these papers after he had written them roughly. Sir H. Davy frequently wrote over these and then I had to copy again until a fair paper was produced which went to the Society or the printer. The originals used to be torn up or destroyed until I begged I might be allowed to keep them for myself and here they are.' [5]

There are also numerous manuscript lectures, Davy's annotated copy of his *Elements of Chemical Philosophy* [6] and a great number of letters. Here again there is much of interest, but relatively little of even indirect significance for Davy's electrochemistry. Some of this material has in any case been published in the collected *Works*.[2]

In public Davy persistently played down the importance of hypotheses, valuing them solely as a guide to further experiment.[7] Perusal of his unpublished writing reveals that this was indeed his philosophy, and he was therefore able to exchange one view for another as experience seemed to dictate. Berzelius could never have done this as easily. The following quotations, without comment, show his attitude to his own progress. They date from the Clifton period (about 1799), 1805 and 1821 respectively :

> 'Most of my errors have arisen from being placed in new situations, situations with which I am unacquainted . . . Chief use of physical science that it gives us definite ideas.' [8]

> 'On looking into the Book after at least three years and a quarter —now 1805, Feb^y., what an alteration do I find in my general sentiment and in power of composition. The Metaphysics were amongst my earlier speculations.' [9]

> 'It is now 11 years since I have written anything in this book.— I take it again Feb 17, 1821. I have gained much since that period and I have lost something, yet I am thankful to infinite wisdom for

[5] Faraday's introduction to the volume of Davy's collected papers, dated July 1832.
[6] London, 1812.
[7] *E.g.*, *Phil. Trans.*, 1826, [**116**], 390.
[8] Note-book no. 15j.
[9] Note-book no. 22c.

blessings and benefits and I bow with reverence before his chastisements which have always been in mercy. May every year make me better, more useful, less selfish and more devoted to the cause of humanity and science. . . .'[10]

Theory of the Voltaic Pile

Davy's interest in galvanism began as soon as he heard of Volta's work with his pile[11] and, while still at Bristol, he wrote that he had 'been pursuing Galvanism with labour and with some success'.[12] In those early months, speculations on the new phenomenon tended either to follow Volta in his 'contact theory',[13] or to be based on the views of Fabroni[14] and others that the final explanation is a chemical one. Davy, at first inclining to the 'chemical' theory, was by 1806 advocating the 'contact' theory, and finally emerged with a synthesis of the two.[15]

The manuscript evidence enables us to fill in several details of his early thinking. It becomes clear that, although at Bristol he had been inclined to attribute galvanic phenomena to 'metallic oxydation', a more general explanation was being sought. Thus within two days of his arrival at the Royal Institution his note-books contain the first reference to electrical experiments using 'sulphure of potash' and a metal (13 March 1801). Three days later he 'thought that other chemical processes might produce galvanism', but queried whether the effects arose 'from the combination of at^{c} oxygene with the sulphure'; sulphuration of the silver remained another possible cause.[16] These results were reported in June that year.[17]

There are four note-books at least, probably five or six, covering the period from 1800 to 1802, but none of them has any clear explanation to offer for Davy's conversion to the 'contact' theory of galvanism. There is a genuine problem here, because of his remark in 1806, that Volta's experiments 'immediately' disproved the 'chemical' theory.[18] He implies that he abandoned the latter as soon as he had read Volta's paper,[19] which was published in full in mid-1800. Yet his paper in November of that year[20] was still advocating Fabroni's theory of

[10] Note-book no. 15g.

[11] Letter to D. Gilbert, dated 8 March 1801, quoted by Anne Treneer in *The Mercurial Chemist*, London, 1963, p. 75.

[12] *Nic. J.*, 1800, **4**, 275.

[13] *Phil. Trans.*, 1800, **90**, 403.

[14] *Nic. J.*, 1799, **3**, 308.

[15] See C. A. Russell, *Ann. Sci.*, 1959, **15**, 1–10.

[16] Physical Journal.

[17] *Phil. Trans.*, 1801, [**91**], 397.

[18] *Ibid.*, 1807, [**97**], 49.

[19] *Ibid.*, 1800, **90**, 403.

[20] *Nic. J.*, 1800, **4**, 341.

chemical combination, and his note-books confirm that he did not abandon it until much later.

Thus, in November and December 1801, he describes several experiments on galvanism, and queries the effect of the atmosphere on the galvanic production of acids and alkalis. The nature of some of these experiments is not clear, but after one of them he remarks : ' does this not prove the dependence of galvanism upon oxydation ? '[21] In the same note-book are to be found the ' Outlines of a paper on Galvanism ' that he seems to have decided not to publish. It can be dated as late 1801 or early 1802.[22]

This extract is important because it contains a clear refutation of the ' contact theory ' on the grounds that Volta's contact experiments ignore any possible role for the atmosphere, that fluids in a pile must be more than mere conductors and are always ' chemical agents ', and that ' the amount of electricity is determined by the power of the substance to act ' (*i.e.*, the reactivity of the electrolyte). Indeed, he refers to an ' expt proving the dependence of electricity upon oxidation in which when a gold wire is made to compleat the circuit on the oxygene side very little hydrogene is given out ; but when silver wire very much '.[21] This is possibly an experiment that he had reported in February 1801.[23] He concludes : ' When water, acid or a solution of sulphuret acts upon a metal or upon charcoal, a change in Electrical equilibrium takes place, and the fluid becomes negatively electrified and the metal positively ; this occasions a current '.[21]

It is thus fairly certain that Davy did *not* reject the ' chemical ' theory quite as soon as his first Bakerian Lecture suggested. Unfortunately, the manuscripts at the Royal Institution give no further information, and all that can be safely said is that it must have been some time after late 1801 and before the autumn of 1806. Now in his Bakerian Lecture of that year [24] and in his *Elements of Chemical Philosophy* of 1812,[25] he gave several reasons for considering the ' chemical theory ' inadequate.[15] Were these the facts that first swayed his own judgment ? They all appear in publications of 1806 or later, but this does not preclude the possibility that Davy had performed the experiments years before he published them, and so in this case manuscript evidence would be important. As we have noticed, the records prior to 1805 do not tell

[21] Note-book 13c.

[22] This follows from the context in the book, and also from his identification of ' galvanism ' and ' electricity ', which he had not made earlier in 1801 (Note-book 22c).

[23] *Nic. J.*, 1801, **4**, 527.

[24] *Phil. Trans.*, 1807, [**97**], 49.

[25] *Elements of Chemical Philosophy*, London, 1812, vol. i, pp. 163–164.

us a great deal about either Davy's experiments or his speculations in electrochemistry. But a search is not entirely fruitless.

First, one might expect to discover some account of contact experiments in which chemical action was unobservable. Electrification in such circumstances would seriously discredit the 'chemical theory', as Davy pointed out.[24, 25] Yet all reference to his own contact experiments seems to have vanished, at least from the Royal Institution collection. The Laboratory Note-book for 1805–09 (the first) begins a record of electrochemical experiments only a few weeks before delivery of the 1806 Bakerian Lecture, and these do not include any contact experiments. There is, however, a reference in one of his personal note-books that seems to be from the period 1805–06. It runs as follows:

> 'Zinc and quicklime opposite after contact upon a large surface one positive the other negative electricity if the contact is not preserved for an amalgamation to take place; but if this chemical effect occurs the electricities are no longer manifest—It would seem that in these instances what before had been an attraction of the mass becomes an attraction of the particles and is no longer capable of—.'[26]

The MS. breaks off here; despite the lack of punctuation, it is fairly easy to see what he was trying to say. It is an explicit separation of contact and chemical phenomena, and an assertion that when one occurs the other does not. Since this was to be the basis of his theory of chemical combination, one may perhaps detect here a faint clue to Davy's *volte-face*. Could it be that in developing a theory of chemistry he came to clarify his thoughts on galvanism, with a consequent reappraisal of Volta's original view? The passage occurs shortly after a poem dated '1806, Aug.'. It is immediately preceded by a reference to galvanic influence on the germination of seeds, and followed by one to the 'expts of Guyton'. Both of these topics occur in the Bakerian Lecture of 1806, where also may be found the sense (though not the words) of the passage quoted. Hence the autumn of 1806 is not improbable as a date for this extract.

Davy's second argument for the 'contact theory' was based upon the electrical excitation of different gases. On this, the manuscript evidence is straightforward and clear. The experiments he quoted are recorded in the Royal Institution Laboratory Note-book for 9 September 1809. Hence they could have had no effect on the views reported in 1806 and were, in fact, part of a programme of research into the electrical behaviour of gases in general. It is of interest that he notes that Wollaston, the originator of the experiments concerned, was present at their repetition in 1809.[27]

[26] Note-book 13j.

[27] Not only had Wollaston concluded that oxidation was the 'primary cause of the electric phaenomena' on the basis of his experiments with the electrical machine (*Phil. Mag.*,

This is almost as far as the Royal Institution papers will allow us to go in the matter of Davy's views on galvanism. One must conclude that there are still records to be discovered relating to his early years at the Royal Institution, and we hope that they are not lost for ever.

Theory of Electrolysis

Davy's theory of electrolysis was chiefly important in that it cleared away the deadwood of speculation that had originated in the alleged production of various acids in the electrolysis of water. By a rigid exclusion of extraneous materials in solution, he showed that 'water, chemically pure, is decomposed into gaseous matter alone, into oxygene and hydrogene '.[28] No longer was it necessary to take into account a mass of experimental material now shown to be irrelevant and misleading. For this, perhaps, Davy is chiefly to be praised.

As a result of this work the problem of electrolysis became for the first time capable of enunciation in a simple form : why did the products appear only where the current entered and left the solution ? In reply to this question Davy has left us with a theory that was less involved and more static than his views on galvanism. The Royal Institution Note-books do not contain much unpublished matter that has a bearing on this, yet the dearth of material is significant. The Note-books between 1802 and 1806 reflect his preoccupation with other topics. The Managers had indicated leather-tanning, metal assaying and mineralogy as suitable subjects for their young protegé, who was thus diverted temporarily from electrochemistry. Scattered references show, however, that it was never entirely forgotten. By 1806 he resumed an active interest in the topic, possibly under the stimulus of other work being published in the field. He said later that he was unaware of Berzelius's work at this time.[29]

The only important manuscript antecedent of the celebrated Bakerian Lecture of November 1806 is the astonishing account of 108 experiments, nearly all on electrolytic transference, performed in less than five weeks during that autumn.[30] The cumulative effect of these must have been an overwhelming conviction of a series of decompositions and recompositions throughout the solutions, with only the 'extreme parts' of the conducting chain active. They illustrate the importance Davy attached to testing Ritter's theory[31] that decomposition took place

[28] *Phil. Trans.*, 1807, [**97**], 12.
[29] *Ibid.*, 1826, [**116**], 388 note.
[30] Royal Institution Laboratory Note-book, 1805–1809, pp. 22 ff.
[31] *Nic. J.*, 1801, **4**, 511.

1801, **11**, 206 ; *Phil. Trans.*, 1801, [**91**], 427) ; he also was the presumptive writer of a letter signed ' W. W.', describing electrification during evaporation (*Phil. Mag.*, 1802, **13**, 231), which gave added force to Davy's denial of the oxidation theory (*Phil. Trans.*, 1807, [**97**], 49).

only at the electrodes. Yet even the decisive evidence against Ritter afforded by these experiments did not altogether remove Davy's interest in the idea. Indeed Ritter's explanation that the hydrogen and oxygen evolved are compounds of (elementary) water with positive or negative electricity seems to have fascinated him. His Note-books frequently allude to this, and an undated reference between two notes on 'potassium'[32] shows that he was still toying with the notion after 1807. This is in accord with its appearance in several letters to Berzelius,[33] in a copy of a letter dated 12 March 1809, now in the Royal Institution,[34] and (most surprisingly) in a paper read to the Royal Society in the previous year.[35]

The division of the liquid into two zones of opposite charge with a neutral point between them is a notable feature of Davy's expositions of his theory from 1807 to 1826. That he had private doubts over this may be seen in a note dated 13 August and inserted on an added sheet in the Institution's Laboratory Note-book after the entries for 1812. An experiment is envisaged 'to ascertain whether there is according to the received belief a neutral point in the voltaic circle',[36] but is unfortunately not described in any detail. It seems that misgivings arose from some experiments using the electric spark, which satisfied him that here, at any rate, no neutral point existed. Again, the account is incomplete and we do not read how the matter was concluded.

This was the chief difference between the electrolytic theories of Davy and Grotthus,[37] although in other respects they were very similar. But again the Royal Institution Note-books neither illumine this relationship nor give any more precision to Davy's own schemes.

The laboratory notes for 1807, beginning in the autumn after a gap of some months, reveal a continuing interest in electrolysis. For a few days Davy was examining slurries of very sparingly soluble materials and their behaviour in an electrolytic cell; he noticed, for example, that, if magnesia was placed under water round the positive pole, a line of the gelatinous material moved towards the negative pole. He said little to suggest why this was being done, although he did suggest that the non-appearance of the line of solid in the electrolysis of magnesium sulphate might be due to the solvent action of the acid present. It seems

[32] Note-book 13j: 'potassium' first appears in the MSS. in the Laboratory Note-book for 30 June 1808, where it emends the phrase 'basis of potash' which is crossed out.

[33] *Jac. Berzelius Bref*, ed. by H. G. Söderbaum, Uppsala, 1912, vol. i, part 2, pp. 8, 10, 11, etc.

[34] Letter beginning 'My dear Sir', possibly intended for Berzelius?

[35] *Phil. Trans.*, 1808, [**98**], 369, note.

[36] Royal Institution Laboratory Note-book, 1809–1812, p. 684.

[37] It has been suggested that their theories differed in a more complex manner than merely in this respect (Dr G. V. Bykov, private communication).

that he was feeling his way towards some notion of the polarity of solids, possibly for further clarification of his understanding of the mechanism of electrolysis. These experiments[38] were followed by two days of intensive work on the insulating or conducting powers of a variety of solid bodies (acids, soap, salts, etc.), liquids and flames.[39] Bodies that discharged one kind only of electricity were said in those days to show ' unipolarity ', a phenomenon that is still imperfectly understood. Quite clearly, polarity was the key to these experiments also. Apart from two days of work on gases, there is no further record of experiments until 19 October.

On that day Davy returned to a substance that he had examined thirteen days before. Then he had found that pure potash, very dry, discharged negative, but insulated positive, electricity. Now, however, he repeated the experiment with slightly moist material. The resultant decomposition into elementary potassium established Davy's reputation for all time.[40] It is often represented that this experimental *tour de force* was the direct outcome of his electrochemical theory, but the manuscript evidence suggests that the relationship was rather more oblique than that. The employment on 6 October of ' very dry ' potash (a very inaccessible commodity) suggests that Davy had tried with unsatisfactory results the usual moist material. This would not give a satisfactory indication of ' unipolarity ', but may well have given a hint of possible decomposition. Even if this were not so, however, it is difficult to avoid the conclusion that uppermost in Davy's mind at the time was the thought that electrolysis in some way involved the polarity of solids, and that potash, being unipolar, was likely to break down under this treatment when moist. The ability of traces of water to overcome insulating power had been shown for oxalic acid, carbonate of ammonia and soda.[39]

Electrical Interpretations of Chemistry

An account of Davy's published views on this aspect of electrochemistry has already been given,[1] but the most important points may be conveniently repeated here. In 1806 he clearly identified ' the remote cause of the electrical energy ' with ' chemical affinity ',[41] and elaborated his ideas with particular reference to different types of chemical behaviour and to cohesion. Since, however, electrification produced by contact between substances disappeared when they combined, he became able no longer to identify the chemical and electrical effects, and instead

[38] Royal Institution Laboratory Note-book, 1805–1809, pp. 50–52.
[39] *Ibid.*, pp. 53–57.
[40] *Ibid.*, p. 61.
[41] *Phil. Trans.*, 1807, [97], 39. Lecture delivered 20 November, 1806.

ascribed a common basis to them both.[42] This was confirmed by the ability of an electric current either to give motion to globules of mercury (an electrical effect) or, under different circumstances, to cause hydrogen to be liberated at their surface (a chemical effect).[43] He even tried to put his theory on a semi-quantitative basis by using the metal-displacement series,[44] although he admitted also the influence of mass on chemical behaviour.[45] Finally, he conceived that heat could facilitate reaction by increasing both the degree of electrification[46] and the motion of the combining particles.[47]

All this may be found in the published literature. Nevertheless, there are many gaps in our knowledge which the Royal Institution manuscripts might be expected to fill at least partially. These include the route by which Davy arrived at his theory, the factors that led him to modify it, his own private assessments of the experiments that he quoted in public, and so on. In fact, however, the Davy Note-books are singularly uninformative on these matters. Perhaps others have been destroyed or exist elsewhere. But, assuming the Royal Institution collection of Davy's scientific work is reasonably complete, the absence of information is itself significant.

There is, for example, no clear reference to the experiment on 'electrization of mercury',[43] also referred to in his *Elements of Chemical Philosophy*.[48] In a comment on the latter, Berzelius expressed polite scepticism :

> 'You recount one particular experiment from which you draw a conclusion that you yourself call formidable. This observation is of great significance, but it is also, I confess, scarcely probable. And you will certainly pardon me if, having found so many little inaccuracies in your work, I remain amongst those who doubt the correctness of this observation until you have given a detailed account of the experiments from which you draw this conclusion.'[49]

In Davy's reply to this,[50] the requested account was not forthcoming, and for many years Davy did not refer to it again. It is likely that he was unable to obtain reproducible results or that he had failed to record his original experiment. The silence of his Note-books (before and after

[42] MS. of a course of lectures delivered in 1808, entitled 'Introductory to Electro-chemical Science', now at the Royal Institution ; printed in Davy's *Works*, London, 1840, vol. viii, p. 283.

[43] *E.g.*, *Phil. Trans.*, 1826, [**116**], 417.

[44] *Elements of Chemical Philosophy*, London, 1812, vol. i, p. 500.

[45] *Phil. Trans.*, 1807, [**97**], 41.

[46] *Ibid.*, p. 43.

[47] *Elements of Chemical Philosophy*, London, 1812, vol. i, p. 95.

[48] London, 1812, vol. i, pp. 172, 485.

[49] *Jac. Berzelius Bref*, ed. by H. G. Söderbaum, Uppsala, 1912, vol. i, part 2, p. 58.

[50] *Ibid.*, p. 59.

1812) rather suggests the latter. Only in 1826 did he mention it again, and then probably under the stimulus of Herschel's paper on a related topic published a few months before.[51] This work by Herschel is noted in Davy's own annotated copy of his *Elements* now at the Royal Institution. From this lack of information we infer that an experiment so inadequately documented is unlikely to have played a significant part in the development of his theory.

In his first Bakerian Lecture Davy identified contact electrification with chemical affinity.[41] The connexion between the two is so stressed there that one is inclined to suppose that he arrived at his theory of chemical change as a direct outcome of his contact experiments. Yet these experiments also are almost completely missing from the Royal Institution Note-books, and once again it seems doubtful that such a comprehensive theory could possibly have been based upon experiments that were never recorded. There is one clear reference to them, although not in a Laboratory Note-book. In a miscellaneous collection of poems, scientific speculations, queries and other matters appears the unfinished comment already quoted.[26] The date seems to be about 1806.

It is evident that Davy regarded electrification by contact as a confirmation of his theory, but unlikely that he was first led to it by these experiments. The origin of his views is more probably to be located in his preoccupation with electrolysis. This topic more than any other fills the early pages of his Laboratory Note-books, and an attempt to understand this led him to his work on polarity and hence, in all probability, to the isolation of potassium and sodium. At the same time it was natural for the idea to dawn that chemical combination was, in Armstrong's phrase, 'reversed electrolysis'.[52]

From 1806 to 1812, while he was at the height of his powers, many other items claimed his attention. The new alkali metals provided so powerful a tool for further explorations that for some time the electrochemical theory receded into the background. Moreover, Davy was still seriously troubled over the problem of the elements and was having private misgivings about the nature of water and of ammonia. Stimulated by the news of Berzelius's success with a mercury cathode[53], he returned to electrolysis in 1808 and obtained barium, strontium, calcium and magnesium.[54] He tried to initiate gas-phase reactions with electricity[55]

[51] *Phil. Trans.*, 1824, [**114**], 162.

[52] *Proc. Chem. Soc.*, 1885, **1**, 40.

[53] He seems to have learned of this by letter from Berzelius, the receipt of which he acknowledged on 10 July 1808 (Ref. 33, p. 8).

[54] *Phil. Trans.*, 1808, [**98**], 339.

[55] Royal Institution Laboratory Note-book, 1809–1812 : many experiments between pages 8 and 684.

and was able to displace both metals[56] and acids[57] by more reactive members of the same classes. But of his view that chemical reactions are electrical in character very little is said, and one is left with the firm impression that his theory was held lightly, was never allowed to impede his progress in experiment and remained at all times as a useful generalization to be kept strictly in the background.

One final point of Davy's theory gains clarity from a perusal of his manuscripts. At several points in his printed works a corpuscular interpretation of electrochemistry can be discerned. This is especially prominent in the distinction that he drew between forces that act upon particles and those that act upon masses. By this means he circumnavigated the difficulties of his early scheme in which electrical and chemical forces were identified. Thus he wrote :

> 'Electrical effects are exhibited by the same bodies, when acting as masses, which produce chemical phaenomena when acting by their particles ; it is not therefore improbable that the primary cause of both may be the same. . . .' [58]

He then went on to interpret the effect of heat on chemical reactions in corpuscular terms.[47] Before this, in 1808, he had asked : 'Are not chemical attractions of particles and electrical attractions of masses owing to one property and governed by one simple law ? '.[42] The same idea may be seen in the extract from a Note-book of about 1806 already quoted.[26]

The Note-books reveal that some kind of corpuscular philosophy had intrigued Davy well before this time. One of these ('before 1800') contains a 'Prospectus of a Work on the Laws of Corpuscular Motion, or the Philosophy of Chemistry '. The list of simple bodies to be discussed includes light (this was the 'phosoxygen' period), but not caloric. He will not accept this without proof 'for that theory which admits not of caloric is equally adequate to the explanation of the phenomena '.[59] This sounds remarkably like an echo of Nicholson who had suggested a corpuscular theory of heat in his *Dictionary*[60] that had been among Davy's earliest chemical reading.

Another Note-book contains a quotation and comment[61] on an article on 'Attraction' in the *Encyclopædia Britannica*.[62] This was

[56] *E.g., ibid.*, p. 349.

[57] *Ibid.*, p. 394.

[58] *Elements of Chemical Philosophy*, London, 1812, vol. i, p. 164.

[59] Note-book 13e.

[60] *A Dictionary of Chemistry, etc.*, London, 1795.

[61] Note-book 13c. The extracts follow a poem dated 1799 and precede a reference to Berthollet's 'Essay on Affinity', presumably his *Recherches sur les Lois de l'Affinité*, Paris, 1801.

[62] The reference is to an unsigned article in *Encyclopædia Britannica*, 3rd edn., Edinburgh, 1797, vol. ii, p. 670.

an account of the Newtonian view of affinity, in which affinity was virtually identified with gravitation, a view that had become popular both in Britain and France. Davy's comments included the remark that ' gravity is the Attraction between bodies so distant that the masses alone influence the result '. but ' adhesion supposes a distance too small for our senses '. Moreover cohesion (' only between particles of the same nature ') must, like adhesion, depend on ' the figure of the particles '. This last phrase has a Boscovichian flavour about it, but is more likely to have been derived from Guyton de Morveau, whose works were referred to by Davy a little later[63] and may well have been known to him at this time. Guyton had spoken of the influence of ' the varieties of figure and of density of the primitive molecules '.[64] He had also postulated fluidity as an essential condition for molecular interaction,[65] and Davy appears to echo this in a ' query ' immediately following the last extract : ' How far must the old-established law of chemistry be admitted that no two bodies act on each other except one of them be fluid ? '[61]

With this corpuscular element in his early thinking, it is not surprising that Davy was able to draw upon these ideas when he came to publish an electrochemical theory. His separation of ' mass ' and ' particle ' effects was a shrewd attempt to have the best of both worlds by uniting in one theory the gravitational and corpuscularian schools of thought. Between 1806 and 1812 he was sufficiently confident to allow his view to be published.[42, 47] It is unlikely that at this time he owed anything to either Dalton or Boscovich, for his Note-books contain no hint of this. His aversion to Dalton is well known,[66] but it has been suggested that Boscovich's theory furnished Davy with the framework of a programme of research.[67] If this is so, the Royal Institution manuscripts suggest that it must have been at a later period than this. Undoubtedly the *Elements of Chemical Philosophy* utilizes molecular form as a partial explanation of chemical specificity.[68] But in the years immediately prior to 1812 Davy's Note-books reveal that he had grave misgivings on one important characteristic of the Boscovich approach, namely, the dynamic theory of heat. Although his early views [59] were in accord with this theory, his Note-books of later date contain frequent speculations that the basis of all things, heat included, was electricity. Indeed, he attempted at least once to decompose heat into the two electricities of

[63] Note-book 13j (about 1806).

[64] *Élémens de Chymie*, Dijon, 1777, vol. i, p. 540.

[65] The arrangement of his *Élémens* (64) is based upon this assumption.

[66] See, *e.g.*, *Ann. Sci.* 1963, **19**, 122.

[67] L. Pearce Williams, ' Boscovich and the British Chemists ' in *Roger Joseph Boscovich*, ed. by L. L. Whyte, London, 1961, pp. 153–167.

[68] London, 1812, vol. i, pp. 146, 587, etc.

which he supposed it was made.[69] A manuscript note of about 1806 [70] refers to heat as 'destruction of repulsion', which is soundly Boscovichian, but his *Elements* six years later speaks of it in almost Daltonian tones as 'the power of repulsion'.[71] It is impossible to avoid the conclusion that during those momentous six years he did not have one theory of heat, but many. In this respect at least no Boscovichian frame-work can be discerned.

Nevertheless, chemical reactivity as a function of electrical change and particulate motion remained the corner-stone of Davy's theory as expounded in his *Elements of Chemical Philosophy*. Like some of his earlier views, however, this work did not gain universal applause, and his electrochemistry was roughly handled by Berzelius. Of the key-stone of Davy's argument he wrote: 'What is the difference between the action of the particles and the action of the mass? When a force acts on each particle of a body, has it at the same time no action on the mass, *i.e.*, on the whole body?' [72]

This may go some way towards explaining why thereafter Davy rarely spoke in corpuscularian terms publicly, and why years later he denied that he had ever attached much importance to the theory.[73] (One must except his paper on the diamond, where there is a very general statement on 'corpuscular arrangements' in matter giving rise to different crystalline forms.)[74] A few interesting passages in the Royal Institution Note-books, however, show not only that Davy retained the idea of non-Daltonian atoms, but also that after 1812 these had become the point-centres of force postulated by Boscovich.

In the Commonplace-Book a specific reference to Boscovich occurs, the date being the latter half of 1814. After commenting on his points of attraction and repulsion, Davy went on to suppose that 'chemical combination will depend upon particles meeting so as to be in polar relations—so that their spheres of attraction may coincide. . . . Heat may assist chemical action by enlarging the sphere of action, *i.e.*, by expanding bodies'.[75] Another allusion to Boscovich occurs in a late *Dialogue*, not presumably intended for publication in his lifetime but included by John Davy in the collected *Works* of his brother. Davy's mouthpiece, 'The Unknown', denies that 'I suppose the elements to be physical molecules endowed with the properties of bodies not believed to be decomposable. . . . I consider them with Boscovich merely as points

[69] Note-book 13j; note dated 'Sept. 22' (probably 1808).

[70] Note-book 15h.

[71] *Elements of Chemical Philosophy*, London, 1812, vol. i, p. 80.

[72] *Jac. Berzelius Bref*, ed. by H. G. Söderbaum, Uppsala, 1912, vol. i, part 2, p. 43.

[73] *Phil. Trans.*, 1826, [**116**], 390.

[74] *Ibid.*, 1814, [**104**], 557.

[75] Commonplace-book.

possessing weight and attractive and repulsive powers '.[76] In another *Dialogue* he speaks of the inability of gravitation to destroy 'the attraction of cohesion or chemistry ', reflecting how far he had moved from the Newtonian position ; consistent with this is the view that 'chemical attraction itself seems to be but a peculiar form of electrical attraction '.[77]

Even Davy's late Note-books do not, however, indicate a complete conversion to Boscovich. Following a famous indictment of Dalton, published in his *Works*,[78] there occurs a passage in this *Dialogue* which John Davy did not print. Davy's spokesman, Philochemus, speaks of his assumption that atoms must be near in order to combine. Now, however, he is not so sure for '1000 parts of one substance having a weak affinity for another will not separate a portion of that substance from one part of another substance having a strong affinity for it '.[79] Not for the first time in the nineteenth century the problem of chemical specificity was to remain unsolved in the face of increasing evidence of mass-action.

If the ideas of Boscovich played any part in Davy's electrochemical scheme, and there can be little doubt that they did, it is of interest to inquire how Davy came into contact with them. Certainly they were not popular with English chemists at the turn of the century, and it has been suggested that Davy was the first to put them to serious use in chemistry.[67] A possible influence in this matter was that of Thomson. The first two editions of his *A System of Chemistry* included an exposition of Boscovich's theory, but the third edition of 1807 favoured Daltonian atomism, though with some reservations.[80] Now a Note-book of Davy's contains a few remarks showing that Thomson's book was familiar to him.[81] Unfortunately, the passage is undated ; its context implies a date of *about* 1807 ; and the remarks are not specific enough to enable the edition to be identified. Hence this cannot be taken as unequivocal evidence that Davy met Boscovich in the pages of Thomson. Another possibility is Priestley, who mentions Boscovich's views in two of his works.[82] Davy praised him publicly, perhaps finding an unconscious reflection of himself in one aspect of his character : 'his hypotheses were rapidly formed, and relinquished with an ardour almost peurile '.[83]

[76] *Works*, ed. by John Davy, London, 1840, vol. ix, p. 388.

[77] *Ibid.*, p. 370.

[78] *Ibid.*, vol. v, pp. 328–330, note.

[79] Davy Letters and Papers—9 : *Archeus—or Summer Days*, 3rd Dialogue.

[80] Edinburgh, 1st edn., 1802; 2nd edn., 1804; 3rd edn., 1807 (later edns. also).

[81] Note-book 13g. Thomson's book "contains too much history."

[82] *Heads of Lectures on a Course of Experimental Philosophy, particularly including Chemistry*, London, 1791 ; *History of Optics*, London, 1792.

[83] Lecture delivered 1810 ; in *Works*, London, 1840, vol. vii, p. 116.

His private notes reveal that his admiration was genuine ; they observe with evident satisfaction that Priestley had come to chemistry from electricity.[84] And a copy of at least one cordial letter from the older man to Davy has been preserved in the Royal Institution.[85] Priestley's rearguard action in defence of phlogiston no doubt struck a sympathetic chord in Davy.[86] But all this must be speculation and one can merely record the possibility that Priestley mediated the Slav philosopher's atomism to Humphry Davy.

Conclusion

The volume and nature of the Davy manuscripts at the Royal Institution have necessitated a considerable discussion, even although only one aspect of Davy's work, his electrochemical theory, was the prime object of interest. Some light is undoubtedly thrown on the development of this theory by a study of the Note-books, although other topics are illumined more strongly.[87] It is perhaps unnecessary to state that the manuscripts show the issues of debate to have been more complex than Davy's publications suggest, for this would be expected. Many conclusions on matters of detail must be tentative, and it is often impossible to go beyond mere speculation. Yet one impression stands out from all the mass of detail : Davy's renowned contempt for detailed theoretical schemes. This allowed him to hold lightly to hypotheses, and he saw no incongruity in recording on consecutive pages two or even three different hypotheses on the same subject. He was thus inhibited from making known his theories, not only to avoid rebuffs but also on account of the simple fragility of those theories.

His own words can best conclude these remarks. Characteristically, these were crossed out, but still remain legible : 'Were a description indeed to be given of all the experiments I have made, of all the difficulties I have encountered, of the doubts that have occurred, the hypotheses formed—'.[88] Here the manuscript breaks off abruptly. The candid confession was never finished. But we may be grateful for its reminder of the intensity of thought and feeling that lay behind the glittering success and lasting achievements of Humphry Davy.

I wish to offer my sincere thanks to the Managers of the Royal Institution, London, for permission to quote from the Davy papers in

[84] Note-book 22a.

[85] Letter dated 31 October 1801, in 'Davy Letters and Papers—9'.

[86] Priestley's *Doctrine of Phlogiston Established*, Northumberland, U.S.A., 1800, was presented by Davy to the Royal Institution on 18 May 1801 and is still in the Library.

[87] *E.g.*, his views on the nature of the elements.

[88] Royal Institution Laboratory Note-book, 1805–1809, p. 185 ; entry dated 15 February 1809.

their possession. It is a pleasure also to acknowledge much assistance kindly given by the Librarian of the Royal Institution, Mr K. D. C. Vernon, F.L.A.

ERRATUM in Part II of this series :

p. 22 : the sentence ' One final aspect . . . . discussion ' on lines 3 and 4 should have appeared at the beginning of the *following* paragraph.

V

THE ELECTROCHEMICAL THEORY OF BERZELIUS

Part I: ORIGINS OF THE THEORY

There is little doubt that the electrochemical theory of Berzelius has had a lasting influence in chemistry, and that considerable traces of this are still discernable. The purpose of the present paper is to discuss the origins of the theory, using this phrase in two senses. First, an attempt will be made to analyse the influences operating upon Berzelius in the early years of the last century and supplying him with concepts that he was to embody in his own scheme; then the external events that drew forth his main expositions of it will be briefly outlined. Later papers will describe details of the theory and offer an assessment of its significance for chemistry as a whole.

Like many others of his day, Berzelius came under the spell of the new anti-phlogistic philosophy which gave a new significance to the chemical balance and, as a result, brought into clearer focus the concept of an element. This view of science, which he first learned from Girtanner's *Elements of Antiphlogistic Chemistry*,[1] affected all his thinking. Its effects may be seen in his electrochemical classification of the elements, as well as in his preoccupation with quantitative chemistry, of which more will be said later. To Lavoiser's influence may be ascribed the centrality of oxygen in Berzelius's electrochemical scheme, and also the fundamental dualism implicit in the word ' radical '.

There seems to be good evidence for supposing that the phlogistic controversies left their mark on Berzelius in a way that is perhaps less obvious. Since the days of alchemy chemistry had been preoccupied with the phenomena of fire and combustion. Now that a new explanation for these had been given this emphasis continued unabated. Berzelius wrote:

> 'The theory of combustion (and accompanying phenomena) has always been the basis of chemical theory, and it is probable it will never cease to be. At every period in the development of chemistry the inadequacy of our ideas on this subject has been recognised; in trying to make these agree with the facts, ever increasing in number, efforts have been made to give them the highest possible degree of probability. But never as now has one felt so much how incomplete is our knowledge

[1] C. Girtanner, *Anfangsgründe der antiphlogistischen Chemie*, Berlin; possibly the 2nd edition, 1795.

Annals of Science 19, 1963, 117–26. Reprinted by permission of Taylor and Francis Ltd (http://www.tandf.co.uk/journals).

> in this respect ; and never has it been more necessary to discover the extent to which the principles now accepted may be inexact and contrary to experience.' [2]

Obviously combustion was not regarded as the only kind of chemical reaction. But to a mind accustomed to thinking along these lines the similarity with 'electrical fire' would suggest that electrical neutralization might have an importance for chemistry as a whole.

A further influence clearly operating upon Berzelius was that of Berthollet, whose work extended and modified the ideas on affinity held by Berzelius's compatriot Bergman.[3] The most important aspect of this influence was the generalization that mass plays a determinative part in 'all combinations and decompositions produced by elective affinity'.[4] This view, amplified in Berthollet's *Essai de Statique Chimique*,[5] was upheld by Berzelius, particularly in his *Essai sur la Théorie des Proportions Chimiques*.[6] The electrochemical view of affinity here expounded owes much to Berthollet for its quantitative conception as well as for its detailed treatment of cohesion, dissolution and other matters. Of the influence of external conditions upon chemical reactions Berzelius wrote :

> 'These circumstances, the knowledge of which we owe almost uniquely to the researches on affinity that Berthollet has carried out with such wisdom, are as follows :
>
> (i) The influence of temperature on degrees of affinity ;
> (ii) The degree of volatility of the bodies ;
> (iii) Their different degrees of solubility ;
> (iv) The different combinations that can give rise to bodies.' [7]

It appears that Berzelius was conscious of the debt to Berthollet, to whom he referred as one of Lavoisier's 'most celebrated co-workers', expounding affinities 'in a truly philosophical manner', and having 'decisively proved that the intensity of chemical action of bodies on each other is not uniquely determined by the degree of their affinity, but depends also on the quantity of body which acts, that is, on the mass'.[8] He further expressed his regard in the dedication of his *Essai sur la Théorie des Proportions Chimiques* 'à l'Auteur de "L'Essai de Statique Chimique", comme un témoinage de ma profonde estime et de mon inviolable attachement'.[9]

[2] *Traité de Chimie*, Paris, 1831, vol. iv, pp. 554–5.
[3] *Opuscula physica et chemica*, Uppsala, 1783, vol. iii.
[4] C.-L. Berthollet, *Recherches sur les lois de l'affinité*, Paris, 1801.
[5] *Essai de Statique Chimique*, Paris, 1803.
[6] *Essai sur la Théorie des Proportions Chimiques et sur l'Influence Chimique de l'Électricité*, Paris, 1819.
[7] *Traité de Chimie*, Paris, 1831, vol. iv, p. 516.
[8] *Ibid.*, pp. 527–8.
[9] Paris, 1819, p. [v].

If it was Berthollet whose influence directed Berzelius's thinking on the reactions of bodies, his ideas on their composition were derived in the first instance from John Dalton. The manner in which this happened may best be stated in his own words :

> 'While occupied with publication of an elementary textbook of chemistry, I was perusing (among other works which one does not generally read) Richter's *Memoirs* . . . and was impressed by the light I found there on the composition of salts and the precipitation of metals by one another . . . It follows from Richter's researches that by means of good analyses of several salts, the compositions of all the others could be precisely calculated . . . At the same time I formed the project of analysing a series of salts so as to make it unnecessary to examine the others . . . While this project was being carried out, the composition of alkalis was discovered by Mr. Davy. Like other chemists, I have found that ammonia leaves at the negative pole of an electric pile a body with metallic properties, from which I conclude that this alkali should also be regarded as an oxide ; the amount of its oxygen, although it may be impossible to prove it by direct experiment, should be calculable from the phenomena of metallic precipitation . . . The study of these phenomena had thus to be part of my experiments ; and when I heard of Dalton's ideas on multiple proportions, I found in a number of analytical results which I already had such a confirmation of this theory that I could not prevent myself from examining the said phenomena . . .' [10]

The reference to Davy dates this work at about 1808, and in fact Berzelius's first acquaintance with the Daltonian theory came from Wollaston's paper of that year.[11] He received from the author a copy of his *New System of Chemical Philosophy*[12] in 1812, and immersed himself in its ideas. From a re-interpretation of his own gravimetric data he passed to an involved electrochemical scheme at the root of which lay the Daltonian atoms. Often he was to praise the atomic theory (though differing from Dalton on several details); for him it was 'without exaggeration one of the greatest steps chemistry has ever made towards perfection '.[13]

Berzelius seems to have been in no small measure responsible for the popularization of Dalton's theory. His depth of insight is shown clearly in his reconciliation of this with Berthollet's work on the influence of mass on the course of chemical reactions. To most chemists, these stood in

[10] *Traité de Chimie*, Paris, 1831, vol. iv, pp. 532–3.

[11] *Phil. Trans.*, 1808, [98], 96. Berzelius was much impressed with Wollaston after their encounter in England in 1812, especially with his 'atomic way of looking at the composition of bodies ', (Letter to Davy, August or September, 1812, reproduced in *Jac. Berzelius Bref*, ed. by H. G. Söderbaum, Uppsala, 1912, vol. i, pt. 2, p. 32). He had 'gained more profit from one hour's conversation with him than often by reading lengthy printed books ' (Letter to Berthollet, October, 1812, *ibid.*, vol. i, pt. 1, p. 41).

[12] Manchester, vol. i, pt. 1, 1808 ; pt. 2, 1810.

[13] *Traité de Chimie*, Paris, 1831, vol. iv, p. 530.

complete antithesis, largely owing to Berthollet's extension of his theory to the doctrine of continuously varying composition. But Berzelius welded the two ideas together to form a substantial basis for an electrical theory of chemical reaction. Significantly, it was in a discussion of his electrochemical theory that he wrote the following :

> 'Although the results of M. Berthollet's experiments have seemed at first so opposed to the adoption of a general system of chemical proportions, we now find that they follow as necessary consequences from the views of the corpuscular theory ; and we owe some recognition to the work of a scientist who, although envisaging matter from another theoretical point of view, has nevertheless traced out a way opposite to that of recent chemical speculations . . . Thus it is seen that M. Berthollet's ingenious researches in his *Statique Chimique* conflict with neither the theory of definite proportions nor the views of the corpuscular theory.'[14]

The new views on combustion, affinity and atomism were of course widely discussed during the first decade of the last century, and influenced many besides Berzelius. Amongst the more conspicuous figures at that time was Sir Humphry Davy, and he also produced an electrochemical theory.[15] His reactions to contemporary influences, and therefore his own theoretical schemes, were, however, rather different from those of Berzelius. Some sympathy with a phlogistic approach can be detected (possibly with the intention of reducing the number of elements then postulated) ;[16] his acceptance [17] of Berthollet's 'influence of quantity' was not followed by as detailed a development as that of Berzelius ; and he always retained a sceptical mistrust for Dalton's 'speculations', as opposed to his experimental work.[18] In all of these differences one may perhaps detect Davy's dislike of clear-cut conceptual models, sometimes expressed as a simple reluctance to commit himself to a theory. In any case it is certain that Davy's electrochemical scheme was worked out in far less detail than that of his Swedish contemporary, and the question of their mutual indebtedness must be examined.

In constructing his own electrochemical theory Berzelius was aware of, and used, Davy's discoveries in electrolysis. It seems possible to particularize Davy's influence to the years 1806 to 1811. Berzelius once implied that in 1806 he was 'unaware of the reducing power of Davy's electrical piles and troughs'.[19] However, in 1807 the publication of

[14] *Ibid.*, pp. 527–8.

[15] This has been discussed by the present author, *Ann. Sci.*, 1959, **15**, 1–13 and 15–25.

[16] *Cf.* R. Siegfried, 'The phlogistic conjectures of Humphry Davy', *Chymia*, 1964, **9**, 117–24.

[17] *E.g.*, *Elements of Chemical Philosophy*, London, 1812, vol. i, p. 122.

[18] *E.g.*, *Phil. Trans.*, 1811, [**101**], 17.

[19] *Ann. Phil.*, 1816, **8**, 263.

Davy's Bakerian Lecture of 1806,[20] for which he was awarded the Institut de France prize of 3000 francs, caused a widespread stir in Europe, and Berzelius soon became familiar with it. A correspondence between the two men, dating from that year, shows a considerable degree of mutual respect.[21]

As a result of Davy's Bakerian Lecture[20], Berzelius modified his original views [22] on the action of the Voltaic pile, and started new experiments of his own.[23] His paper of 1811 contains several flattering references to his English colleague and reveals a thorough acquaintance with his work on contact electrification and electrolysis.[24]

From about 1811, however, Davy's influence on Berzelius waned perceptibly as the gap between their differing views grew wider. First came Davy's announcement that he could find no oxygen in oxymuriatic acid (chlorine),[25] from which others were quicker than he to infer that it was an element, and which led eventually to the recognition of muriatic acid as hydrogen chloride,—an acid without oxygen. This struck at the roots of the dualism that Berzelius was then developing, and their correspondence from now on became more polemic and slightly less polite. A letter of 1811 betrays how strongly dualism was colouring Berzelius's whole attitude to chemistry, even at this early stage ; writing to Davy about muriatic acid he asserted that ' saline combinations of this acid prove that it contains oxygen '.[26] To Davy this priority of theory over experiment was anathema, and his opponent was unable to keep the disagreement as a private matter.[27] Umbrage was taken on both sides, and Berzelius wrote to Young, Secretary of the Royal Society :

> ' Count de la Gardil, who spent a few months last winter in London, came to tell me that Davy was displeased with me on account of my memoirs on chlorine ; he even goes so far as to complain of my omission of some measure of the respect due to his distinguished talents.' [28]

Feelings were further exacerbated by Davy's rejection of atomism—as uncompromising as his empiricism on chemical elements. Although he wrote to Berzelius that he had ' embraced the doctrine of definite proportions ' in his Bakerian Lectures of 1809 and 1810, he meant simply

[20] *Phil. Trans.*, 1807, [**97**], 1.

[21] *Jac. Berzelius Bref*, ed. by H. G. Söderbaum, Uppsala, 1912, vol. i, pt. 2.

[22] *Afhandl. Fys.*, 1807, **2**, 14 ; *J. Chem. Phys.* (Gehlen), 1807, **3**, 177.

[23] *Kongl. Vet. Handl.*, 1812, **33**, 180 ; *J. Chem. Phys.* (Schweigger), 1812, **6**, 136 ; *Nic. J.*, 1813, **34**, 161 ; *Ann. Chim.*, 1813, **86**, 165.

[24] *J. de Phys.*, 1811, **73**, 253.

[25] *Phil. Trans.*, 1811, [**101**], 1.

[26] Letter dated 10 June 1811, in *Jac. Berzelius Bref*, ed. by H. G. Söderbaum, Uppsala, 1912, vol. i, pt. 2, p. 28.

[27] *Ann. Phil.*, 1813, **2**, 254.

[28] Letter in *Jac. Berzelius Bref*, ed. by H. G. Söderbaum, Uppsala, 1920, vol. iii, pt. 2, p. 294.

that he had adopted the notion of equivalents, and went on to castigate Dalton's views as ' more ingenious than correct '.[29] This ' ingenuity ' had been damned with faint praise in the second of those Lectures :

> ' It is impossible not to admire the ingenuity and talent with which Mr. Dalton has arranged, combined, weighed, measured, and figured his atoms ; but it is not, I conceive, on any speculations upon the ultimate particles of matter, that the true theory of chemical proportions must ultimately rest.' [30]

In Davy's own highly personal account of chemistry, his *Elements of Chemical Philosophy*, he sedulously avoided any atomic interpretation, and stated :

> ' In the doctrine of proportions derived from facts it is not necessary to consider the combining bodies either as composed of indivisible particles, or even as always united.' [31]

At one time he had promised to send to Berzelius a copy of Dalton's second volume (his ' mechanico-chemical theory '),[32] but it was never received.[33] The omission did not deter Berzelius from becoming one of atomism's most ardent promoters. His later electrochemical work made extensive use of a corpuscular model, conceived from a standpoint of extreme dualism.

The rift between the two men became wider still in 1812. In the summer of that year Berzelius came to London, but arrived only a few days before Davy's departure for Scotland.[34] Nevertheless, Davy took the opportunity of sending him a copy of his *Elements of Chemical Philosophy*, and invited his comments for a second edition, which was never published, but for which the preliminary work was done on an annotated copy of the first edition, now in the Library of the Royal Institution, London.[35] As Sir Harold Hartley has observed,[36] ' Berzelius, with his encyclopaedic knowledge and perhaps a little nettled by Davy's cavalier treatment of him, took the invitation seriously '. His reply, accompanied by a disarmingly frank note, dealt with over a hundred topics from the *Elements* on which Berzelius disagreed.[37] In particular, there was a detailed criticism of Davy's conception of an element, a subject

[29] Letter dated 24 March 1811, *ibid.*, 1912, vol. i, pt. 2, p. 23.

[30] *Phil. Trans.*, 1811, [**101**], 17n.

[31] London, 1812, vol. i, p. 114.

[32] Letter dated 18 October 1809, in *Jac. Berzelius Bref*, ed. by H. G. Söderbaum, Uppsala, 1912, vol. i, pt. 2, p. 18.

[33] Letter to Berthollet, *ca.* 1810, *ibid.*, vol. i, pt. 1, p. 16.

[34] Letter to Berthollet, October 1812, *ibid.*, vol. i, pt. i, p. 141.

[35] Letter from Davy to Berzelius, 25 July 1812, *ibid.*, vol. i, pt. 2, p. 30.

[36] *Proc. Roy. Soc.* (A), 1960, **255**, 168.

[37] Letter, late 1812 or early 1813, in *Jac. Berzelius Bref.*, ed. by H. G. Söderbaum, Uppsala, 1912, vol. i, pt. 2, p. 35.

on which the latter held very strong views. He sent back a stiff reply, and, apart from a few brief and rather formal notes, mainly at the end of Davy's life, their correspondence came to an unhappy stop.

From this time onwards the electrochemical ideas of Berzelius developed independently of Davy. Their different philosophies of science probably made such an independence inevitable, although the circumstances of the actual parting of the ways must always be seen as a matter of great regret. By this time, Davy's energies were being diverted away from electrochemistry both by his social activities and by his scientific commitments. He did not return to it for twelve years. Meanwhile, Berzelius proceeded to extend his own theory to virtually the frontiers of chemistry.

It will now be convenient to summarize the chief publications in which Berzelius expounded his theory. With the exception of an early account of galvanism,[38] which was mainly historical and descriptive and had little theoretical content, these fall into three groups, corresponding to his interest in electrolysis, reform of nomenclature, and chemical proportions.

In 1804 there appeared a paper by Berzelius and Hisinger from Stockholm, dealing with electrolysis of salt solutions.[39] Some fifteen different experiments were reported with a variety of solutions and metal conductors, in several kinds of apparatus including U- or V-tubes. As Sir Harold Hartley has pointed out, Berzelius, then in his early twenties, was fortunate at this time in having the friendship and help of Hisinger.[40] The work described was less spectacular than Davy's more famous researches in electrolysis, but in fact predated all of them. The amount of 'theory' in the paper was small, but some important empirical generalizations were made. No doubt the discoveries were a powerful stimulus to the formulation of a theoretical scheme, at least for electrolysis. As Berzelius was not then aware of Davy's work in this field, he had to devise his own techniques. Thus, with his assistant M. M. de Pontin, he developed the idea of a mercury cathode, today an indispensable feature in such diverse fields as the alkali industry and polarographic analysis. With this technique applied to a solution of ammonia they obtained what seemed to be a compound of mercury and the 'deoxigenated compound basis of ammonia', as it gave off hydrogen and absorbed oxygen. These results Berzelius communicated by letter to Davy in June 1808, and the English chemist at once took the matter up.

[38] *Afhandling om Galvanismem*, Stockholm, 1802.
[39] *Ann. Chim.*, 1804, **51**, 167.
[40] *K. Svenska Vet. Arsbok*, 1948, **34**.

He repeated the experiments, including those involving ' ammonium amalgam ', ' with perfect success '.[41]

With this, pioneering work in electrolysis became more and more the province of Davy, while Berzelius began to concentrate on other matters. Nevertheless, as we have seen, he followed with interest the work done by Davy (and others, no doubt) and continued to develop his own ideas.

The next development of his theory was prompted by a revision in chemical nomenclature. The Swedish Government had ordered a revision of the country's Pharmacopeia, and this had underlined for Berzelius the urgent need for a reform in nomenclature, which he and his fellow-chemists hoped to incorporate in the new Pharmacopeia. Accordingly, he communicated a paper to de la Metherie's *Journal de Physique* in 1811 as ' An essay in chemical nomenclature '. Governed by the principles that he would ' make no other changes in words than those which it could not do without ', he was particularly anxious to establish Latin ' which is indispensably necessary to chemists to give direction and determination to the nomenclature of each country, and to place one in a position to compare and to understand fully nomenclatures in different languages '.[42]

It is not necessary to go into detail on this system, but the classification of substances is important and is reproduced in a schematic summary in the following table :

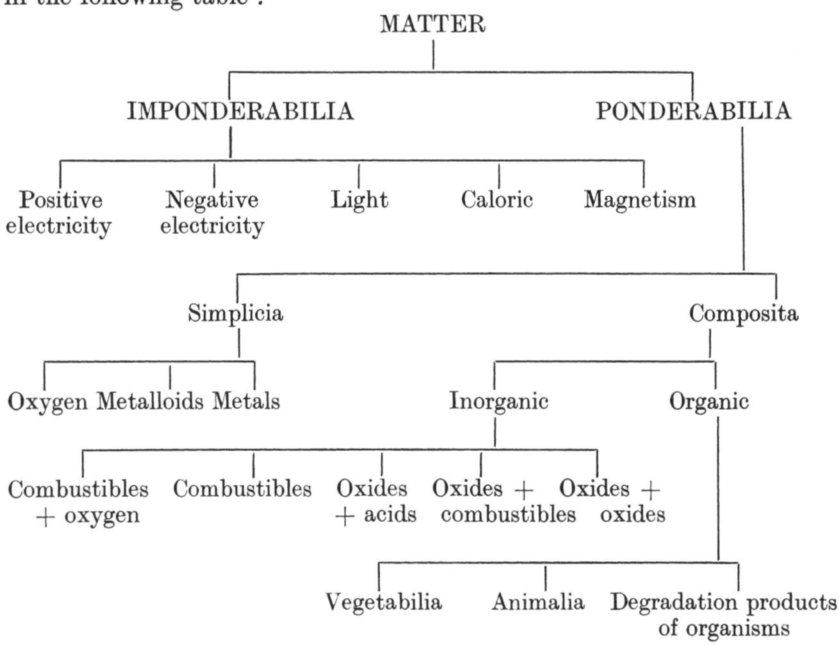

[41] *Phil. Trans*, 1808, [**98**], 339.
[42] *J. de Phys.*, 1811, **73**, 253.

This scheme did perhaps clarify the conception of an element. Lavoisier's earths were excluded (baryta and others having now been decomposed), and heat and light were relegated to 'Imponderabilia'. But its chief importance for electrochemistry was its classification of 'Ponderabilia' according to their behaviour in electrolysis.

Following the publication of this paper, Berzelius issued a number of statements explaining in further detail the electrochemical ideas he had incorporated in it. The first of these was a note published with the main article.[43] In this he defended his reversal of Davy's sign-convention for an electrochemical division of substances, and included an account of his theory of electrolysis. A further statement occurred in a lengthy memoir the next year.[44] It contained an electrochemical interpretation of combustion and chemical affinity in general, his characteristic views on affinity, aspects of his theory of electrolysis and of the Voltaic pile, and speculations on the nature of electricity and heat. It concluded with analytical data on four oxides of antimony. The final major work in this period was a treatise on mineralogy,[45] in which Berzelius's electrochemical classification of bodies is made the basis of 'A Scientific System of Mineralogy'. This contains a further short description of the electrochemical theory itself.

The third subject whose study stimulated development of this theory was chemical affinity, or chemical proportions. During the preparation of an elementary text-book[46] (1807) Berzelius was reading Richter's memoirs on stoichiometry[47] (as he remarked in his *Traité*)[10, 48]. Richter himself had almost reached the concept of equivalents, and the fertile mind of Berzelius saw in this half-formulated idea something that would integrate, check and make sense of a multitude of diverse analytical results. Accordingly he began to apply the touchstone of equivalence to existing analytical figures, but the result was confusion. The unreliability of contemporary results reflected basic defects in technique, and stimulated Berzelius to an examination of the whole question of analytical methods. While still in the early stages of this task, he first encountered Dalton's atomic theory,[10, 12] with the result that analysis was vested with a yet deeper significance.

Berzelius's many results were collected together in his famous *Essai*

[43] *Ibid.*, p. 277.

[44] *Kongl. Vet. Handl.*, 1812, **33**, 166, 233 ; *J. Chem. Phys.* (Schweigger), 1812, **6**, 119 ; *Nic. J.*, 1813, **34**, 142, 153, 240, 313 ; *Ann. Chim.*, 1813, **86**, 146.

[45] *Försök att genom . . . system för mineralogien*, Stockholm, 1814 ; English trans. by J. Black : *An attempt to establish a pure Scientific System of Mineralogy by the Application of the Electrochemical Theory and the Chemical Proportions*, London, 1814.

[46] *Lärbok i Kemien*, Stockholm, 1808, vol. i.

[47] *Anfangsgründe der Stöchyometrie oder Messkunst chymischer Elemente*, Breslau, 1792–4.

[48] *Traité de Chimie*, Paris, 1831, vol. iv, pp. 525, 533.

sur la Théorie des Proportions Chimiques.[49] As its sub-title shows, this included a synthesis of his views on electrochemistry with those on chemical proportions. It was the first detailed exposition of his electrochemical theory.

Much of this essay was bodily incorporated in the first French edition of his textbook on general chemistry.[50] Differences exist between parallel passages in the two publications and form an interesting commentary on the development of his ideas over the dozen years separating them. Apart from minor points, the chief alterations are a modified view on the thermal suppression of reactivity, and a considerably expanded treatment of organic chemistry (the latter transferred to volume v). Most of the differences will be discussed in later papers in the present series.

[49] *Essai sur la Théorie des Proportions Chimiques et sur l'Influence Chimique de l'Électricité*, Paris, 1819. The Swedish original of this was *Lärbok i Kemien*, Stockholm, 1818, vol. iii.

[50] *Traité de Chimie*, Paris, 1831, vol. iv ; a revised edition of the *Essai* was issued from Paris in 1835.

VI

THE ELECTROCHEMICAL THEORY OF BERZELIUS

PART II: AN ELECTROCHEMICAL VIEW OF MATTER

THE previous paper in this series[1] gave some account of the origins of the electrochemical theory of Berzelius, and enumerated the expositions of this in works on electrolysis, on reform of nomenclature and on stoichiometry. In the second group of these a tendency can be seen that was very strong in Berzelius, as in all systematizers of science, namely, a propensity to bring as many diverse data as possible within one comprehensive scheme of classification. This was the driving-force behind his reforms of chemical nomenclature and of mineralogy. As we have seen, the former of these involves a classification of matter into *Ponderabilia* (possessing weight) and *Imponderabilia* (weightless, and including electricity, light and magnetism). Of the first class Berzelius wrote:

> ' Concerning their relations with the electricities, we divide them into ' electropositive ' and ' electronegative ' substances. I call substances ' electropositive ' if, in the decompositions brought about by the electric pile, they gather round the positive pole, and ' electronegative ' if they gather round the negative pole '.[2]

At this early stage in the evolution of his theory Berzelius stumbled across a difficulty inherent in his subject, the matter of a sign-convention. Recognizing this, he appended a note as follows:

> ' Most of those who treat of these matters employ these denominations in the opposite sense. Mr. Davy has found that what I call electropositive substances, taken in contact with electronegative substances, both acquire an electricity opposite to that after which I have named them '.[3]

His reason seems to have been that there are two complementary views of electrolysis. It could be argued, for example, that positive electricity attracts oxygen and neutralizes its excess negative energy; or one could

[1] *Ann. Sci.*, 1963 (1965), **19**, 117–126.
[2] *J. de Phys.*, 1811, **73**, 257.
[3] *Ibid.*, p. 277.

Annals of Science 19, 1963, 127–45. Reprinted by permission of Taylor and Francis Ltd (http://www.tandf.co.uk/journals).

say that negative electricity liberates oxygen and then repels it. The choice of terms would then depend upon the assumption of repulsion or attraction.

Shortly afterwards Berzelius reverted to the more usual, and opposite, convention: ‘ My subsequent reflections upon this object, obliges [*sic*] me, at present, to change these denominations for each other ’.[4] No further reason was given. In his book on mineralogy he referred to his change of opinion which sprang ‘ from the consideration I have recently bestowed on this subject ’.[5] Doubtless there seemed to be no good reason for prejudicing the acceptability of his scheme by arbitrarily adopting a minority position. In the following discussion, the terms ‘ electropositive ’ and ‘ electronegative ’ will be employed in the later sense, the convention observed today. Consideration will be given first to *Ponderabilia* (elements, oxides, salts and organic compounds). The views of Berzelius on the nature of ‘ imponderable ’ electricity will form a brief conclusion.

1. *Elements*

Perhaps the most consistent feature throughout the development of this theory was that oxygen was laid down as the extreme of electronegativity (electropositivity in his old notation). Thus

> ‘ Oxygen is the most electronegative of all bodies. As it is never positive relative to any other, and as, according to all chemical phenomena known at the present time, there could probably exist no more electronegative element on our globe, we shall recognize that it has an absolute electronegativity ’.[6]

However, other less electronegative substances could be positive to some bodies although negative to most; it was a relative matter only. Thus sulphur was positive to oxygen but negative to metals. In view of this relativity a criterion was needed for division into the two main classes, and Berzelius found this in electrolysis: substances were obviously negative if they collected at the positive pole, and also ‘ if in combining with oxygen they gather round the positive pole during the operation of the pile ’.[7] Consequently, all known elements could be included, as those that were not liberated in electrolysis gave rise to oxides that were. Since acids collected at the positive surface, ‘ acidic oxide ’ became synonymous with ‘ oxide of an electronegative element ’.

[4] *Nic. J.*, 1813, **34**, 156–7.

[5] *An Attempt to establish a Pure Scientific System of Mineralogy by the Application of the Electrochemical Theory and the Chemical Proportions*, trans. from the Swedish by J. Black, London, 1814, p.12n.

[6] *Traité de Chimie*, Paris, 1831, vol. iv, p. 569.

[7] *J. de Phys.*, 1811, **73**, 257.

Further divisions could be convenient, and Berzelius proposed a tripartite classification in 1811.[8] First came oxygen, distinctively unique; next were the ' metalloids ', understood as combustibles readily forming acidic oxides; finally came the metals, including ' ammonium '. Hydrogen and nitrogen did not appear. As Davy had found already,[9] the constitution of ammonia was to prove a stumbling-block. Berzelius regarded this as ' composed of a metallic base and oxygen ',[10] the base being ' ammonium ' and giving rise to the *oxides* ' hydrogen ' and ' nitrogen ', so these were naturally omitted from the list of elements. The postulate of a hypothetical base was extended to ' sulphuricum ' and ' phosphoricum ' to distinguish them from the commercially available forms of sulphur and phosphorus.

In his paper of 1811 brief reference was made to a further fivefold division,[11] and this was elaborated, though without many examples, in the papers of 1813.[12] The third and fourth classes of 1811 were reversed two years later; the fourth group seems not to apply to elements:

(i) ' Absolutely electronegative: oxygen alone.'

(ii) ' Electronegative in general ': this consisted of all combustibles forming acid oxides, including the ' metalloids ' and ' the metals arsenic, molybdena and wolfram '.

(iii) ' Bodies of a variable electrochemical nature ': these formed oxides intermediate in character between electropositive and electronegative, or were alternatively capable of giving rise, in different states of oxidation, to basic and acidic oxides.

(iv) ' Indifferent ': such substances were neither acidic nor basic, and included certain oxides and salts.

(v) ' Electropositive: combustible bodies and their oxides, which, during the action of the pile, are never collected round the positive pole '. This, of course, included most metals.

These divisions were convenient, but did not obscure the existence of a gradation throughout the elements. For example, Berzelius held that, if oxygen were condensible to a solid, it would resemble sulphur, the most negative metalloid. Again, metals will combine with metalloids with a vigour determined by their electropositive character, and will vary between those forming exclusively basic oxides and those giving rise to acids.[11] There is no absolute extreme for electropositive elements, even in the radicals of the alkalis and alkaline earths. Neither can hydrogen be

[8] *Ibid.*, p. 259.
[9] E.g., *Phil. Trans.*, 1808, [**98**], 1.
[10] *J. de Phys.*, 1811, **73**, 282.
[11] *Ibid.*, pp. 258–9.
[12] *Nic. J.*, 1813, **34**, 142, 153, 240, 313 ; **35**, 38, 118, 159 ; **36**, 129.

allotted such a distinction, for it plays an electronegative rôle in its compound with potassium, and its oxide (water) gathers at the positive pole in the electrolysis of the alkaline earth ' hydrates '.[13]

The precise order in which Berzelius arranged his elements naturally varied as his knowledge increased. The details of the variations are instructive and form an interesting commentary on the developments in inorganic knowledge in the early nineteenth century as well as on his own grasp of them. For this reason the following summary is given of the five chief arrangements (1811,[14] 1814,[15] 1819,[16] 1831,[17] 1845[18]). Modern English names have been used except for the radicals.

1811 [14]	1814 [15]	1819 [16]	1831 [17]	1845 [18]
oxygen	oxygen	oxygen	oxygen	oxygen
sulphur	sulphur	sulphur	sulphur	sulphur
				selenium
phosphorus	nitrogen	nitrogen	nitrogen	nitrogen
muriatic rad.	muriatic rad.	muriatic rad.	fluorine	fluorine
	phosphorus		chlorine	chlorine
fluoric rad.	fluoric rad.	fluoric rad.	bromine	bromine
boron	boron		iodine	iodine
carbon	carbon	phosphorus	selenium	phosphorus
	hydrogen	selenium	phosphorus	
arsenic	arsenic	arsenic	arsenic	arsenic
	chromium		chromium	chromium
molybdenum		molybdenum		vanadium
chromium	molybdenum	chromium	molybdenum	molybdenum
tungsten	tungsten	tungsten	tungsten	tungsten
		boron	boron	boron
		carbon	carbon	carbon
tellurium	antimony	antimony	antimony	antimony
osmium	tellurium	tellurium	tellurium	tellurium
tantalum	silicon			
silicon	tantalum	tantalum	tantalum	tantalum
titanium	titanium	titanium	titanium	titanium
zirconium	zirconium	silicon	silicon	silicon
antimony	osmium	osmium		
bismuth	bismuth	hydrogen	hydrogen	hydrogen
tin		gold	gold	gold
			osmium	osmium
iridium	iridium	iridium	iridium	iridium
platinum	platinum	rhodium	platinum	platinum
gold	gold			

[13] *Traité de Chimie*, Paris, 1831, vol. iv, pp. 569–570.

[14] *J. de Phys.*, 1811, **73,** 259.

[15] See note 27 below.

[16] *Essai sur la Théorie des Proportions Chimiques et sur l'Influence Chimique de l'Électricité*, Paris, 1819, pp. 77–78.

[17] Ref. 6, vol. iv, pp. 570–571.

[18] *Traité de Chimie*, Paris, 2nd edn., 1845, vol. i, pp. 116–117.

rhodium	rhodium	platinum	rhodium	rhodium
palladium	palladium	palladium	palladium	palladium
mercury	mercury	mercury	mercury	mercury
silver	silver	silver	silver	silver
lead	lead	copper	copper	copper
nickel	tin	nickel		
	nickel	cobalt	uranium	
copper	copper	bismuth	bismuth	bismuth
		tin	tin	tin
		zirconium		
		lead	lead	lead
		cerium	cadmium	cadmium
uranium	uranium	uranium	cobalt	cobalt
zinc	zinc		nickel	nickel
iron	iron	iron	iron	iron
		cadmium		
		zinc	zinc	zinc
manganese	manganese	manganese	manganese	manganese
				uranium
cerium	cerium		cerium	cerium
			thorium	thorium
			zirconium	zirconium
		aluminium	aluminium	aluminium
				didymium*
				lanthanum
yttrium	yttrium	yttrium	yttrium	yttrium
beryllium	beryllium	beryllium	beryllium	beryllium
aluminium	aluminium			
magnesium	magnesium	magnesium	magnesium	magnesium
calcium	calcium	calcium	calcium	calcium
strontium	strontium	strontium	strontium	strontium
barium	barium	barium	barium	barium
sodium			lithium	lithium
potassium	sodium	sodium	sodium	sodium
ammonium	potassium	potassium	potassium	potassium

* ‘ didymium ’ : praseodymium + neodymium

The spacing between elements is arbitrarily arranged to facilitate comparison between the five lists. There are no ‘ gaps ’ in any of the original series.

Berzelius was insistent that these series were approximate only; but despite imperfections he regarded such a system as ‘ more correct than any other in giving an idea of chemistry ’.[19] Extensive details were not given for determining the order, but ‘ simple substances are arranged relative to their electrochemical properties and those of their strongest oxides ’.[20] Electrolysis was used for the basic division into electro-

[19] *Traité de Chimie*, Paris, 1831, vol. iv, p. 569.
[20] *Ibid.*, p. 570.

positive and electronegative elements, particularly the behaviour of oxides:

> 'I shall consider all substances as electropositive [later 'electronegative'] if in combining with oxygen they collect round the positive pole during the operation of the pile, and *vice-versa*. The demarcation between substances is such that the most electronegative [later 'electropositive'] substances even after being saturated with oxygen do not cease to be drawn by the negative* pole'.[21]

Later he asserted that 'although differing degrees of oxidation of several bodies present some exceptions, the electrical order of combustible bodies in general agrees with that of the oxides'.[22]

Davy's contact experiments were probably of use here; they were certainly spoken of in favourable terms.[23] It is also likely that the behaviour of metals in Voltaic piles gave further help. Hardly satisfactory as a general method would be an assessment of the vigour of reaction between metal and 'metalloid' as an index of the electropositive character of the former.[24] Acidic, basic, amphoteric or neutral characteristics of oxides provided another guide, though even acidic character was relative; an oxide could be acidic towards another but basic to a third.[25] Determination of relative acid strength was apparently from displacement of one 'acid' by another, but Berzelius was well aware of the hazards of such methods, arising from the effects of other factors like solubility and volatility.[26] A further yard-stick might have been provided by the stabilities of compound molecules, though application might be difficult: 'combinations cohere with a force which is proportionate to the degree of opposition in the electrochemical nature of the ingredients'.[27]

Berzelius does not seem to have used a displacement series of the metals as an index of their positions. He also rejected the affinity for oxygen as a criterion; it was known, for example, that this could vary with temperature and was in any case a composite term not to be equated with the true electrochemical character.[28] Later, however, he appears to have thought that complicating factors can be allowed for, and that 'a metal endowed with the greatest affinity for oxygen constantly becomes

* The actual reading in the paper of 1811 is 'positive', but the sense clearly necessitates 'negative' in view of the notation Berzelius was then using.

[21] *J. de Phys.*, 1811, **73**, 257–258.

[22] *Traité de chimie*, Paris, 1831, vol. iv, p. 569.

[23] *Ibid.*, p. 564 ; *J. de Phys.*, 1811, **73**, 256.

[24] *J. de Phys.*, 1811, **73**, 258.

[25] *Ibid.*, pp. 258–259.

[26] *Traité de Chimie*, Paris, 1831, vol. iv, pp. 521–522.

[27] *An Attempt to establish a pure Scientific System of Mineralogy by the Application of the Electrochemical Theory and the Chemical Proportions*, trans. from the Swedish by J. Black, London, 1814, p. 11.

[28] *Traité de Chimie*, Paris, 1831, vol. iv, p. 572.

positively electrified in losing negative electricity to the metal which has less '.[29]

2. *Oxides*

The properties of oxides thus proved to be a frequently used guide to the electrochemical character of the element combined with the oxygen. It is now desirable to explore more fully Berzelius's attitude to this relation between the electrical properties of elements and their oxides.

Berzelius was at pains to counter the accepted view that 'the electrical series of combustible bodies differed from that of their oxides', contending rather that 'the electrical order of combustible bodies in general agrees with that of their oxides'.[30] This view was repeated and expanded later in the same book, and at the same time the important point was made that oxygen was the principle of bases as well as acids, a fact previously demonstrated by Davy.[31] Berzelius wrote:

> 'The oxide is ordinarily electronegative relative to other oxides when its radical is negative with respect to their radicals, and similarly *vice-versa.* For example, sulphuric acid is electronegative with respect to all metallic oxides because sulphur is negative with respect to all metals. On the other hand the oxides of potassium and zinc are electropositive with respect to all those oxides towards whose radicals potassium and zinc are positive. This fact, the cause of which we cannot explain, puts right an incorrect idea on the principle of acidity which in antiphlogistic theory was thought to be oxygen. We now find that this resides in the radical of the acids, and that there the oxygen plays such an indifferent part that it enters equally into the strongest salifiable bases (that is, electropositive oxides) and into the strongest acids (or electronegative oxides). However, it sometimes happens that a positive oxide acquires by higher oxidation less electropositive properties which approach electronegative properties, as for example stannic oxide and manganese acids; but in the strongest bases, as potash and soda, addition of oxygen can well destroy the positive reaction, without producing a negative; it is thus that superoxides of the strong salifiable bases are formed.'[32]

That it was the *radical* that determined the chemical behaviour of an oxide had been recognized at an early stage. Following a passage similar to the above he wrote: 'It is the radical itself and not the oxigen which determines whether the oxide shall be an acid or a base'.[33] This dictum followed from a well-known fact. Suppose a radical can combine with a

[29] *Ibid.*, 2nd edn., Paris, 1845, vol. i, p. 79.

[30] *Ibid.*, 1st edn., Paris, 1831, vol. iv, p. 569.

[31] *Phil. Trans.*, 1808, [**98**], 1.

[32] *Traité de Chimie*, Paris, 1831, vol. iv, pp. 579–580; *Essai sur la Théorie des Proportions Chimiques et sur l'Influence Chimique de l'Électricité*, Paris, 1819, pp. 90–91.

[33] *Nic. J.*, 1813, **34**, 163–164.

base and at the same time with varying amounts of oxygen, giving rise to a series of compounds. Then, although the ratio base/oxygen will alter, the ratio base/radical will usually be constant. For example we have the sulphide, sulphite and sulphate of potassium. Since in each case the non-basic part *must* be acid, it is not likely that these parts depend chiefly upon oxygen for their acidic character.

Five classes of oxides of combustibles were recognized by Berzelius. They were:

(i) ' Suboxides ': with so little oxygen that they were neither basic nor acidic;

(ii) ' Oxides ': devoid of acid character, but combining with other oxides;

(iii) ' Acids ';

(iv) ' Superoxides ': containing excess oxygen and not combining as such with other oxides;

(v) Oxidation products of ' ammonium '.

His list shows that his class (ii) includes our ' amphoteric ' oxides.[34]

3. *Salts*

In 1811 Berzelius gave as the ' chief difference ' between inorganic and organic compounds that ' the former contain only binary combinations, alone or combined amongst themselves in several different ways, whereas the latter always contain combinations that are tertiary, quaternary etc.'[35] A note on this passage adds:

> ' Inorganic combinations which are composed of three, four or several simple bodies are in effect only combinations between two or several binaries; for example salts are combinations of oxides. The metallic sulphides, which are made of sulphur with two, three or several metals ought to be considered as composed of different sulphides. It is evident that in the same way the crystalline metallic alloys (which often contain several metals) can be considered as being made up of binary combinations.'[36]

In the statement that ' salts are combinations of oxides ' may be seen the germ of Berzelius's dualistic theory. This had profound and far-reaching results for chemistry as a whole, extending right up to the present time. For Berzelius in 1811, however, its immediate purpose was to provide a basis of classification. Perhaps this is why it was

[34] *J. de Phys.*, 1811, **73,** 262–264.
[35] *Ibid.*, p. 260.
[36] *Ibid.*, p. 283.

introduced almost as an aside. The classification scheme evolved divided salts (' combinations of oxides with acids ') as follows:

(i) neutral salts;
(ii) acid salts (supersaturated with acid);
(iii) basic salts (supersaturated with base);
(iv) double salts (such as the alums);
(v) combinations of acids with acids (the feebler acid acting as base to the stronger, as in borofluoric acid);
(vi) combinations of water with acids, oxides and salts (including oxyacids, alkalis and hydrated salts).[37]

This system was further developed in 1814 when it was made the basis of a ' pure scientific system of mineralogy '. Ideally, minerals were to be classified ' according to the order of their electrochemical properties, from the most electronegative oxygen to the most electropositive potassium; and to place every compound body according to its most electropositive ingredient '.[38] Every mineral is assumed to have the components ' base ' and ' acid ', and to be thus classifiable as a salt. This did not imply that the usual ' acid ' properties were evident, or that in a combination of two acids one could not act as a base towards the other. The essential point was that each compound consisted of a positive and a negative part, and its mineralogical family was determined by the positive part. Each family thus consisted of a body and all its combinations with more electronegative bodies, and was further divided into orders corresponding to the latter. It seems that one reason for the success of this scheme was the recent general recognition of the acidic character of silica.[39] The bulk of the book was concerned with the application of these ideas to minerals then known.

Berzelius thus regarded salts from the same dualistic point of view as that from which he had looked at oxides; they were divisible into positive and negative parts, each of which maintained its normal electrical relations within the salts.

' Decomposing salts ' were those whose constituents became united in a new ratio, the change depending ' on the specific electrical reactions of the particular elements which tend to more perfect neutralization '. On the other hand ' combining salts ' tended to form double salts by virtue of the electrical reaction of the entire compound atom. If the resultant electrochemical condition of a single salt is positive, it will

[37] *Ibid.*, pp. 266–269.

[38] *An Attempt to establish a pure Scientific System of Mineralogy by the Application of the Electrochemical Theory and the Chemical Proportions*, trans. from the Swedish by J. Black, London, 1814, p. 48.

[39] *Ibid.*, p. 14.

combine with a salt having resultant electronegativity. Finally there were 'indifferent salts'. In these 'such amounts of the substances have combined together that a perfect neutralization results and nothing else can enter the combination. Such a compound will add to no others but may undergo replacement of parts of it'.[40]

4. *Organic Compounds*

The views of Berzelius on the nature and composition of organic compounds had a more profound, and less helpful, influence on chemistry in general than his ideas on other substances.

His classification of 1811 had of course to embrace organic compounds:

> 'These are composed of two or sometimes several combustible bodies combined with a quantity of oxygen, which is common to them, and is usually sufficient only to oxidize one of these combustibles. Their constituent parts cannot be separated without combining together in several new ways to produce binary combinations and to share the oxygen between them according to their very complicated affinities. Elements which constitute organic bodies obey the same general law as that regulating the formation of inorganic combinations; but the innumerable variations which this law admits in organic nature are not yet sufficiently clearly defined.'
>
> 'The chief difference between these two classes is that the former contains only binary combinations alone or combined amongst themselves in several different ways, whereas the latter contains combinations that are tertiary, quaternary etc. Bodies composed according to the principle of organic formations can only very rarely be produced from purely inorganic substances by means of chemistry ... The chief condition for organic formation seems to be an electrochemical modification in the elements which differs from that which they normally have in inorganic nature. After they are removed from the organic body which has produced them they tend always to regain this original electrochemical modification; it is a result of this tendency that they ferment and decompose on contact with air and water and on exposure to a high temperature.'[41]

In 1814 Berzelius started a series of analyses[42] to determine whether the laws of combination familiar to inorganic chemistry applied to the 'innumerable variations' to which he referred in 1811. As a result he concluded that they did apply, and so in 1817 extended the dualistic system to them:

> 'The peculiar distinction exists that in inorganic nature all oxidized bodies have a simple radical, whereas on the other hand all organic

[40] *Essai sur la Théorie des Proportions Chimiques et sur l'Influence Chimique de l'Électricité*, Paris, 1819, pp. 80–81.

[41] *J. de Phys.*, 1811, **73**, 260–261.

[42] *Ann. Phil.*, 1814, **4**, 323, 401; 1815, **5**, 93, 174, 260.

> substances consist of an oxide with compound radicals. In the case of plant substances the radical generally consists of carbon and hydrogen, and in the case of animal substances of carbon, hydrogen and nitrogen.'[43]

It can be seen that this view was foreshadowed to a certain extent in 1811. It was repeated in the *Essai*; dualism ' is applicable to organic chemistry, and each organic product can be regarded as divisible into oxygen and a compound radical '.[44] But this bold assertion was immediately qualified: the multiplicity of particles in a compound atom and the probable non-existence of the positive radical outside its compound rendered these speculations valuable ' only as an idea '. Thus at this stage Berzelius was still the master rather than the servant of his theory. The time was yet to come when the existence of his radicals became the issue of bitter conflict.

In this same work a fuller statement appeared of the nature of this correspondence between the two branches of chemistry. At first a limited conception was presented: ' the only phenomenon analogous to these [inorganic] laws is that substances having entirely the same properties have also the same composition '.[45] However, Berzelius proceeded to note a closer resemblance in the reactions of organic compounds with inorganic: ' when organic atoms of the first order combine with inorganic atoms of the first order they follow (according to actual experiment) the same laws as compound inorganic atoms amongst themselves '.[46]

Nevertheless, striking differences remained in practice, and some explanation was patently necessary. At this stage in the progress of his conceptual scheme, Berzelius was inclined to answer in electrochemical terms. These did not greatly differ from his earlier statement:[41]

> 'The structure of organic atoms is quite different from that of inorganic atoms. Also, their electrochemical relations are essentially different. In inorganic nature, elements have a certain specific polarization which does not vary . . . but in organic nature the different elements are quite outside this relation . . . Organic nature has its own particular way of producing oxides of compound radicals, and of giving to their chief constituents an electrical polarity that is quite independent of the different polarity that they usually have in inorganic nature. This is maintained in most cases only under the organic influence, but sooner or later all give it up to resume that which is proper to them in inorganic nature. Thence in all organic products spring the phenomena of destruction that we call fermentation and corruption, by means of which the

[43] *Lärbok i Kemien*, 2nd edn., Stockholm, 1817, vol. i, p. 544. Quoted in *Annalen*, 1839, **31**, 5n.

[44] *Essai sur la Théorie des Proportions chimiques et sur l'Influence chimique de l'Électricité*, Paris, 1819, p. 98.

[45] *Ibid.*, pp. 40–41.

[46] *Ibid.*, p. 45.

> elements gradually take up again their ordinary electrochemical characters and organic nature ceaselessly restores to the other that which it has borrowed from it.'[47]

On the basis of this, oxygen must obviously be present in every organic substance. Certainly at this time Berzelius believed that 'compound atoms of the first order contain at least three elements (oxygen, hydrogen and carbon)'.[48] But he was aware that some substances (such as turpentine) seemed to contain the last two only; he wondered 'if a small amount of oxygen . . . might have escaped the attention of the experimenter'.[49]

When Berzelius was preparing the first edition of his *Traité* (1831) he incorporated large sections of the *Théorie des Proportions Chimiques* (1819) in the fourth volume. It is interesting, therefore, that the above extracts are completely missing from that volume. Probably he wished to give a fuller treatment to such matters in the volume on organic chemistry. In the fifth volume this is done, although the treatment of organic chemical proportions is completely rewritten. Indeed, Berzelius had by now moved away from his position of 1819, and had a rather altered point of view to express. The radical theory, however, was still maintained and, despite the anomalous existence of several volatile oils which seemed to be hydrocarbons, he considered that 'as oxygen is one of the essential elements in organic nature, organic products can be regarded as oxides of compound radicals'.[50]

By now the differences between the two branches of chemistry were more perceptible. The properties of inorganic compounds

> 'depend upon the primitive electrical state (electric polarity) of these bodies and we presume that this state is the cause of their reciprocal action and their affinities. On the other hand in organic nature compounds of the same elements show chemical properties so different that, looked at from the same point of view, it is impossible to regard them as formed from the same elements.'[51]

Regarding these electrochemical ideas, he wrote further:

> 'Since we believe that we have found that the electrical state of bodies and the electricities in general are the prime movers of all chemical activity, we can reasonably form this conjecture: organs which living bodies possess can determine in a special way the electric polarity of the bodies they produce; but unless we wish to be lost in vain speculation this is all we can say on the subject.'[52]

[47] *Ibid.*, pp. 95–96.
[48] *Ibid.*, p. 42.
[49] *Ibid.*, p. 43n.
[50] *Traité de Chimie*, Paris, 1831, vol. v, p. 5.
[51] *Ibid.*, p. 9.
[52] *Ibid.*, pp. 10–11.

It is to be noted that Berzelius did not consider that his electrochemical theory had foundered at this point. Instead, he assumed another factor behind it. In identifying this factor with a Vital Force he was utilizing a principle suggested by others before this time.[53] Yet his own account of this (the most extensive discussion of vitalism in organic chemistry so far) makes it clear that he was not altering his electrochemical model but rather implying a limit to its applicability.

The following passage, in a vein reminiscent of the philosopher-poet Davy, shows that the reason for Berzelius's adherence to vitalism lay in the facts of mortality:

> 'It is true that the elements of organic nature are indestructible, but the existence, strictly speaking, of organic bodies ceases irrevocably. The individual that dies and yields up its elements to inorganic nature never returns. From this it follows that the essence of a living body is based not upon its inorganic elements but upon some other principle which disposes the inorganic elements common to all living bodies to co-operate in producing a particular determinate result, different for each species.
>
> 'This principle, which we call Vital or Assimilating Force, is not inherent in the inorganic elements, and does not constitute one of their original properties such as density, impenetrability, electrical polarity etc., but we do not know what it consists of, how it was formed nor how it ends. Thus we can foresee that, if the terrestrial globe existed with its inorganic elements, without living nature but otherwise in the same circumstances, it would continue to remain without living beings. An incomprehensible force, unknown to dead nature, has introduced this principle into inorganic matter. And this has not happened as a result of chance, but, in admirable variety, through a profound wisdom and with the aim of producing a particular determinate result: an uninterrupted succession of mortal individuals, born of each other, the destruction of some serving to sustain the others. Everything pertaining to organic nature proclaims a wise objective and is distinctly the product of a superior understanding.'[54]

This teleological insistence, reminiscent of Paley, is further emphasized by a diatribe against that 'narrow philosophy' which 'pretended to be profound in asserting that all was the work of chance'.[55] Views like this were of course common in certain quarters in the early nineteenth century, but it is not very clear why Berzelius went out of his way to interweave Vitalism and teleology. Perhaps the matter may be put in this way: the alternative to a vitalistic explanation of nature lay then not in a 'materialistic' view applying the same laws to living and non-living matter; rather it could be found in this invocation of blind chance. Now Berzelius's electrochemical system was a comprehensive systematization

[53] E.g., Davy, *Elements of Chemical Philosophy*, London, 1812, vol. i, p. 175.

[54] *Traité de Chimie*, Paris, 1831, vol. v, pp. 2–3.

[55] *Ibid.*, p.3.

of many diverse facts, and the exact antithesis to any view that atoms combined on account of chance. It was therefore in complete harmony with the electrochemical theory for Berzelius to oppose 'random causality', just as it was in accord with the whole trend of his chemical discoveries for him to postulate a common factor behind all living phenomena, even if its nature was quite unknown:

> 'It will always be more honourable to admire the wisdom we cannot follow, than to raise ourselves by arrogant philosophy and feeble argument to a supposed knowledge of things which are probably always out of the reach of our understanding.'[56]

In an article on 'Vitalism and organic chemistry during the first half of the 19th century', J. Jacques has aptly spoken of the 'religious vitalism of Berzelius', as compared with the 'uncertain vitalism of Liebig' and the 'ordered vitalism of Gerhardt'.[57] It reflects an interesting aspect of the Swedish chemist's character not often discussed.

Nevertheless, the progress in organic syntheses during the fifteen or so years following the first French edition of his *Traité* caused Berzelius to withdraw from his position of dogmatism on the Vital Force. In the second French edition he was content to write as follows:

> 'In the phenomena of organic chemistry, if one means anything by the "vital force" other than a particular set of different circumstances setting in play the natural forces of organic nature, or if one intends by it a particular chemical force inherent in living nature, then one is making a mistake.'[58]

Of this retreat from mysticism Berthelot and others seem to have been unaware, as Jacques has pointed out.[57] He rightly places it in the context of successful efforts in synthetic organic chemistry antedating those of Berthelot himself. It would also seem valuable to recall that at this time the electrochemical theory itself was in such serious difficulties that its survival appeared to many to be impossible. For this reason the now hard-pressed Berzelius was faced with the necessity of modifying that theory almost to the point of unrecognizability in order to retain his position of leadership in chemical thought. These efforts were, of course, largely in vain. Yet in making them Berzelius may have found withdrawal to a merely nominal vitalism a small concession in comparison, and one calculated to favour his position generally. At all events, his early vitalistic ideas were unlikely to survive in an atmosphere so charged with the influence of Comte as was that of Paris at this time. And from Paris came some of his chief opposition.

[56] *Ibid.*, p. 4.
[57] *Rev. Hist. Sci.*, 1950, **3**, 32.
[58] *Traité de Chimie*, 2nd edn., Paris, 1849, vol. v, p. 5.

The controversy between Berzelius and the Parisian chemists (Laurent, Gerhardt and others) is too well known to need detailed recapitulation here. Under the strong pressures of new experimental findings and a growing rebellion against his long-established authority, Berzelius was forced to make major concessions. At least, that is how opponents and historians have tended to regard his reactions, although what was conserved was far greater than what was thrown away.

The oxide theory of organic chemistry was maintained at the end more firmly than at some other times. At an early date it had been enlarged to accommodate other molecules linked to the oxide within the compound. For example, glycerol was regarded as a hydrate of ' lipyl oxide ', while ' acrolein ' (acraldehyde) consisted of two ' lipyl oxide ' units joined together.[59] The theory depended of course on the existence of oxygen in all organic compounds, and this was being asserted (' with very few exceptions ') as late as 1845.[60] Was this really tenable at this time? The only known exceptions were simple hydrocarbons (methane, ethylene, acetylene, etc.) and more complex compounds of the same two elements present in various essential oils. Berzelius regarded the first group as inorganic ' because they are decomposition products of organic bodies ',[61] and thus had no need to accommodate them within his organic scheme. His early doubts regarding the others[49] appear to have been resolved and he was content to regard them as unexplained anomalies.

Yet it is doubtful if the oxide theory could have survived much longer. For one thing, the series of researches initiated by Kolbe and Frankland to isolate the Berzelian radicals in fact produced a succession of methods whereby hydrocarbon *compounds* were formed. Admittedly these were not always recognized as such at first, but some were immediately designated as compounds. Thus Frankland's discovery of the reaction between zinc and ethyl iodide[62] opened up the way not only to organo-metallic chemistry (and thence to valency), but also to a general method for preparing hydrocarbons of the paraffin series; in the following decade many of these were made. Had Berzelius lived beyond 1848, it is doubtful if even he could have maintained that these were all inorganic. Moreover, in the year after his death the primary amines were discovered,[63] and shortly afterwards Hofmann revealed the existence of secondary and tertiary amines.[64] These facts, together with an increasing use of alkyl halides and cyanides which resulted in their widespread recognition as

[59] *Traité de Chimie*, 2nd edn., Paris, 1850, vol. vi, pp. 307, 441, 452.
[60] *Ibid.*, 1845, vol. i, p. 124.
[61] *Ibid.*, p. 279.
[62] *J. Chem. Soc.*, 1850, **2**, 263 and 297.
[63] C. A. Wurtz, *Compt. Rend.*, 1849, **28**, 223.
[64] *Phil. Trans.*, 1851, [**141**], 357.

organic, spelt inevitable eclipse for any view founded on the universal occurrence of oxygen in organic compounds.

Other aspects of Berzelius's theory had also come in for sharp criticism, and he himself had replied with several modifications. The discovery of benzoyl (written as $C^{14}H^{10}O^{2}$) in 1832[65], though welcomed by Berzelius,[66] in fact presented him with the difficulty of an oxygen-containing radical. In the following year Berzelius revised his opinion and formulated benzoyl as an oxide of $C^{14}H^{10}$, benzoic acid being an oxide with more oxygen.[67] He also withdrew an earlier opinion on alcohol and ether[68] (that they were hydrates of ' etherin '), and now proposed them as oxides of two different hydrocarbon radicals.[67] This last change of view was prompted by the recognition of the non-alkaline nature of ethylene (Berzelius's ' etherin '); additive compounds of etherin would have been expected to show a resemblance to additive compounds of ammonia, *i.e.*, the ammonium salts. Therefore the marked difference between ethylene and ammonia soon brought Berzelius to abandon that hypothesis and return to the older point of view that they were oxides.

This fragmentation of radicals into smaller units was deeply significant. Since each of these simple radicals was subject to the electrochemical ' rules ', the latter were seen to have an importance not restricted to the two major divisions of a molecule. Also, the analytical view of substances implicit in this approach was to lead ultimately to the concept of valency. But for the moment its use in the last skirmishes of Berzelius's final rearguard action brought it generally into disrepute.

The details of this celebrated conflict have often been rehearsed. In a replacement reaction, an atom could be replaced only by another of the same kind of polarity—according to the electrochemical theory. But Dumas had shown that hydrogen (electropositive)was replaceable by chlorine (electronegative), as in the chlorination of acetic acid.[69] Now Berzelius had written the latter as an oxide of a hydrocarbon radical:[70]

$$C^{4}H^{6}+3O+H^{2}O.$$

He was not prepared to formulate the chlorinated product as

$$C^{4}Cl^{6}+3O+H^{2}O,$$

so suggested instead:[71]

$$H^{2}O+C^{2}O^{3}+C^{2}Cl^{6},$$

[65] J. Liebig and F. Wöhler, *Annalen*, 1832, **3**, 249.
[66] *Ibid.*, p. 282.
[67] *Jahresberichte*, 1833, **13**, 190.
[68] *Annalen*, 1832, **3**, 282.
[69] *Compt. Rend.*, 1838, **7**, 474.
[70] *Jahresberichte*, 1832, **11**, 210.
[71] *Ibid.*, 1840, **19**, 370.

in this way avoiding direct replacement of H by Cl. Unfortunately for Berzelius the discovery that trichloracetic acid could be easily converted back to acetic acid[72] underlined the essential similarity between the two compounds, and rendered unlikely so great a disparity in their formulae. Even in those days formulae were regarded as in some sense expressing chemical behaviour.

These suggestions were regarded as a face-saving retreat and received much bitter criticism.[73] Yet, in rearranging the atoms within a molecule in accordance with the dictates of his theory, Berzelius was anticipating our own preoccupation with atoms rather than radicals as the ultimate chemical units. And as Japp pointed out in his Kekulé Memorial Lecture,[74] his practice was ' perfectly legitimate ' since the ' copulae ' were not ionizable and the electrochemical theory strictly applies only when ionization takes place. Hence Berzelius was merely withdrawing from a position of excessive zeal for his own theory.

5. *Electricity*

Speculation on the nature of electricity had been a favourite relaxation of eighteenth-century men of science. A disciplined rejection of this indulgence was one of Priestley's less obvious accomplishments,[75] and in this respect he was followed by Davy whose views are likely to have influenced those of Berzelius. At first Davy had regarded the ' one fluid theory ' of Franklin, Aepinus and Cavendish as ' adequate to the solution of all phenomena '.[76] Later, however, he recognized the utility of the rival ' two fluid theory ' of du Fay and others.[77] But he became much concerned to emphasize the limitations of each view and to stress the incompleteness of both models. Thus he dismissed as inconclusive those experiments with sparks claimed to have a bearing on the controversy;[78] he deplored as ' hasty generalisations ' those assertions that even one fluid ' exists ', and regretted the use of ' positive ' and ' negative ', with their implications of something in excess or deficient.[79]

The same healthy scepticism may be seen in Berzelius. In 1811 he

[72] Melsens, *Compt. Rend.*, 1842, **14**, 114.

[73] E.g., A. Laurent, *Chemical Method*, trans. by W. Odling, London, 1855, p. 204.

[74] *J. Chem. Soc.*, 1898, **73**, 109.

[75] E.g., *The History and Present State of Electricity, with Original Experiments*, 3rd edn., London, 1775.

[76] *A Syllabus of a Course of Lectures delivered at the Royal Institution of Great Britain*, London, 1802, p. 61.

[77] *Phil. Trans.*, 1826, [**116**], 415.

[78] *Elements of Chemical Philosophy*, London, 1812, vol. i, p. 177.

[79] Lecture ' Introductory to Electrochemical Science ', delivered at the Royal Institution, London, on March 12th, 1808 ; included in Davy's *Works*, ed. John Davy, London, 1840, vol. viii, p. 290. The MS. is now in the Library of the Royal Institution, London.

classified electricity, together with magnetism, heat and light, into a group of substances termed ' Imponderabilia '. Of the nature of these he wrote:

' I shall not touch the question as to whether these should be called ' matter ', ' force ' or ' phenomena '. These researches have occupied and still occupy our philosophers, whose most ingenious ideas on these objects it is almost as hard to prove, as it is to refute in an evident and constant manner those which seem to be the most absurd.'[80]

Whether electricity and light and ' caloric ' were ' so many different substances ', and what might be their relationship, he simply professed not to know.[81]

A year or two later he seemed a little more sure of the interconversion of heat and electricity, for ' caloric . . . is capable, in certain circumstances, of disappearing as caloric, and reappearing in the electric state '.[82] But we find him still speculating on the fundamental nature of each. He remained perfectly definite on the imponderability of heat, and continued:

' From the relation that exists between caloric and the electricities, it is clear that what may be true with regard to the materiality of one of them, must also be true with regard to that of the other. There are, however, a quantity of phenomena produced by electricity, which do not admit of explanation, without admitting at the same time that electricity is matter. Electricity, for instance, very often detaches everything which covers the surface of those bodies which conduct it. It . . . penetrates non-conductors . . . and makes a perforation . . . We may therefore, at least, with some probability, imagine caloric and the electricities to be matter destitute of gravitation but possessing affinity to gravitating bodies.'[83]

Like most other contemporary workers, Berzelius could realize that such speculations have little value, and in 1819 remained ' silent on all hypotheses to which it has given rise; . . . none of these hypotheses throws light on its nature and all have their defective sides; it has been possible to see that this was not the right way to imagine so remarkable an agent.'[84]

The passage containing these words was embodied in the *Traité* of 1831[85], but with two small differences illustrating the progression of thought after 1819. In the later work Berzelius referred to magnetism as bearing to electricity the same relation that light has to heat. More significant was his comparison of electricity ' to less known bodies with

[80] *J. de Phys.*, 1811, **73**, 256.

[81] *Ibid.*, p. 257.

[82] *Nic. J.*, 1813, **34**, 160–161.

[83] *Ibid.*, p. 165.

[84] *Essai sur la Théorie des Proportions chimiques et sur l'Influence chimique d'Électricité*, Paris, 1819, p. 92.

[85] *Traité de Chimie*, Paris, 1831, vol. iv, pp. 580–581.

which it has moreover no relation; ' this phrase replaced the earlier words ' but this fluid has no relationship with any other ', and in the change we can see a reluctance to imply a ' fluid theory ' of electricity. However, in the same edition, but in the first volume, he definitely inclined to the view that electricity was matter on account of its passage through vacua (which Davy had shown was easier than through air).[86] His ultimate reflections we find in the second French edition:

> ' The nature of electricity is as little known to us as that of light and heat. We know electricity only by its effects . . . something unknown, which we call electric fluid, ether or any other name.'[87]

Yet, despite these statements Berzelius appeared to find an explanation approaching the ' two fluid theory ' most useful in formulating definite views on current electricity.[88]

[86] *Ibid.*, 1829, vol. i, pp. 114–115.
[87] *Ibid.*, 2nd edn., Paris, 1845, vol. i, pp. 69–71.
[88] E.g., *ibid.*, pp. 89*f*.

VII

Berzelius and the Development of the Atomic Theory

Introduction

It is a curious fact of history that the veneration of Dalton today is due in no small measure to the chemical expertise and influence of a rival from whose views Dalton detached himself with force and candour. The ideas of the Manchester philosopher were immortalized largely by the Swedish chemist Jöns Jacob Berzelius through whose efforts atomism became incorporated in a scheme of chemistry more comprehensive and far-reaching than any yet produced. His differences from Dalton were not so much reflections on the inadequacies of the latter's concepts; they were rather extensions of them made possible by Berzelius' exceptional powers of generalization and profound chemical insight.

Berzelius was, of course, the dominant figure in early nineteenth-century chemistry, and for several decades wrote with unequalled, and often unquestioned, authority. Of his place in the history of chemistry as a whole Sir Harold Hartley has well written:

> Chemistry was indeed fortunate that in the years when she was becoming an independent science and the stream of knowledge was growing so rapidly she had the encyclopaedic mind, the judgment, the craftmanship and the watchful eye of Berzelius to guide her career. It was that rare combination of qualities that made possible his massive contribution that is unique in the history of chemistry. His systematic mind saw the need for a structure in which chemistry could grow

with the precision and the articulation of a living organism. The basic principle of his design was atomic composition.[1]

Berzelius first encountered atomism in a *chemical* context; his introduction owed little to physics or meteorology. A short extract from his own account is as follows:

While occupied with the publication of an elementary textbook of chemistry, I was perusing (among other works which one does not generally read) Richter's *Memoirs* . . . . and was impressed by the light which I found there on the composition of salts and the precipitation of metals by each other. . . . It follows from Richter's researches that by means of good analyses of several salts, one could calculate precisely the compositions of all the others. . . . At the same time I formed the project of analysing a series of salts so as to make it unnecessary to examine the others. . . . When I heard of Dalton's ideas on multiple proportions, I found in a number of analytical results which I already had such a confirmation of this theory that I could not prevent myself from examining the said phenomena [of metallic precipitation]; and so it came about that, initially dealing with very limited chemical proportions, the plan of my work grew more and more to include proportions in their whole extent.[2]

Access to Dalton's ideas was far from easy, and for several years Berzelius suffered in a form acute even in those days from the difficulties of scientific communication. His correspondence with chemists outside Sweden contains many a *cri de cœur* for information, books and periodicals. We know[3] that he was aware of Dalton's 1807 paper on mixed gases,[4] and he observes[5] that he met English atomism first in Wollaston's ingenious essay on superacid and subacid salts the following year.[6] From then on every effort was made to acquaint himself further with the new doctrines. Thus to Berthollet he wrote in about 1810:

I should like to have knowledge of the ideas of Mr Dalton, who, according to all appearances, is one of the most ingenious physicians of our age. Mr Davy, whom I asked if he knew something more of his system, replied that Dalton had come to London to publish it, and that he would obtain his work for me as soon as it was published. But it is over two years since I have had communication with Mr Davy.[7]

In June 1811, a letter to Davy himself proceeded:

Pardon me, sir, if I trouble you again with a request that I have already made once, i.e., to acquaint me with some ideas on the system of Mr Dalton. Despite all my researches I have still been able to obtain other information on his doctrines only from a supplement to Murray's *System of chemistry* that Mr Brandel brought me on his return from London. The edition of Mr Thomson's *Chemistry* where Mr Dalton's ideas ought to be mentioned, translated into French by M. Riffault, which I have requested from Count Berthollet, has not yet reached me. There is every appearance that the ideas of Mr Dalton, of whom I have a very high opinion, will help to correct my own, in addition to which he must have done many experiments which will save me the trouble of doing them for myself.[8]

Murray's book[9] would have commended itself by its stand against the anti-Berzelian views on the nature of chlorine, and Berzelius confessed to Gahn[10] that he preferred it to Thomson's work. The supplement mentioned is in fact a Note[11] at the end of Volume ii dealing (in a disapproving way) with Dalton's theory of mixed gases. But it was only a snack not likely to satisfy for long Berzelius' voracious appetite for atomism. Eventually, in 1812, a copy of Dalton's *New system* reached him from the author a few days before his departure for England. His reactions were expressed to Gaultier de Claubry in these words:

Never has a present given me such pleasure as this one did at first. I have been able only to skim through the book in haste, but I will not conceal that I was surprised to see how the author has disappointed my hopes. Incorrect even in the mathematical part (e.g., in determining the maximum density of water), in the purely chemical part he allows himself lapses from the truth at which we have the right to be astonished.[12]

He goes on to lament Dalton's bondage to his preconceptions with all the unreliability (for Berzelius) that this involves. But if this was an anti-climax to his years of searching and grouping it also marked a redefinition of his own views over against those that he now discovered to be Dalton's. His original adoption of an atomic philosophy two or three years previously had not been followed by much success in convincing others: 'I have been publishing memoirs on the subject on the continent for more than three years, without having hitherto, as far as I know, made a single proselyte.'[13] Now, however, some headway was to be made as the divergence of his ideas from those of Dalton gave rise to a series of articles in Thomson's *Annals of Philosophy* under the title 'Essay on the cause of Chemical Proportions, and on some Circumstances relating to them: together with a short and easy Method of expressing them'.[14] After three instalments of this essay had appeared, Dalton published some 'Remarks on the Essay of Dr Berzelius on the Cause of Chemical Proportions'[15] which drew forth from Berzelius a characteristically forthright rejoinder. This was in the form of a letter[16] to Thomson which the latter published[17] after toning it down and amending it in a few other ways. No further reply seems to have been made by Dalton, and the whole episode now seems rather futile and pointless. Yet the polemics did draw attention to the real issues involved and gave Berzelius his first major opportunity for a systematic, public exposition of his own brand of atomism.

The atomic philosophy which he adopted underwent some change over the years, but most of its important features emerged in the debate of 1813–14. We shall now review the main points of departure from Daltonian atomism because they provide the clues to much of later chemical history.

Dalton v. *Berzelius: Emergent Atomism*

1 *'Figure' of the atoms*

In 1813 Berzelius supposed 'that atoms are all spherical, and that they have all the same size',[18] for only in this way could regular figures be obtained. Dalton, on the other hand, refused to be committed to any relation between sizes and weights of atoms, though inclined to the opinion that atoms were not all of the same size. Moreover he could not see why sphericity was probable, and imagined that an oxygen atom might be a regular tetrahedron, the nitrogen atom a cylinder and so on. But he did concede that all would be enveloped in roughly spherical atmospheres of heat.[19]

Many years later Berzelius expounded the same doctrine of equal-sized spheres as an explanation of isomorphism (discovered by his own assistant, Mitscherlich, in 1819[20]): identical numbers of these atoms arranged in the same way always produce the same solid figure, i.e. crystalline form.[21] Further, it should be possible for the same number of such spheres to be differently arranged giving products having different chemical properties. To him, therefore, we owe the idea of isomerism—the first successful prediction from the atomic theory alone.[22]

Now it is not of course necessary to assume that all atoms are spheres of equal size in order to provide an atomistic explanation of isomorphism or isomerism. But the first does demand a similar geometrical form in the atoms being compared, and the second requires an interchangeability of atoms. Both of these conditions were fulfilled in Berzelius' spheres.

Perhaps the chief significance of Berzelius' difference from Dalton on this point was methodological. Berzelius regarded Dalton's tetrahedral atoms of nitrogen, etc., as a flight of fancy beneath his notice,[23] even though they had been advanced with far greater caution than his own confident assertions. Essentially the supposition of Dalton was unnecessary and metaphysical since no way existed of distinguishing between the atom *per se* and an atom in its spherical heat envelope.

2 *Atomic symbolism*

In the third part of his English 'Essay'[24] Berzelius devoted himself to the first full account of a new chemical symbolism, intended to replace the circular signs of John Dalton. This was, of course, the alphabetical notation from which modern symbolism is derived.

Much has been written on this already.[25] We know that Berzelius was not the first to employ an alphabetical method, that it underwent a number of changes at his own hands, that its initial popularity was small, especially in England, and that he himself made little use of it in his published writings.[26] However, there are a few points that are worth re-examination here in view of the great importance of this symbolism for the spread of the atomic concept.

Thus it is desirable to stress that Berzelius was not the first to conceive formulae as expressing quantities. Certainly he wrote, 'The object of the new signs is not that, like the old ones, they should be employed to label vessels in the laboratory; they are destined solely to facilitate the expression of chemical proportions'.[27] But Dalton's signs had the same intention. The obvious advantage of the new symbols was ease of reproduction by hand or in print, with a consequent saving of time and cost. For this reason the service rendered to atomism by his new notation is precisely that given to the theory of valency by the notation of Frankland and Crum Brown. In both cases improved techniques of communication were followed by rapid assimilation of the ideas symbolized.[28]

The formulae of Berzelius had another, less obvious advantage over those of Dalton, and this was because of what they did *not* do. In the symbolism of *A new system of chemical philosophy* there lurked a quasi-structural significance that had no right to be there. Thus Dalton wrote 'carbonic acid' as

 rather than 

Berzelius did not commit himself to any arrangement of atoms in a molecule, and his symbols appeared to Dalton as 'horrifying':

> A young student of chemistry might as soon learn Hebrew as make himself acquainted with them. They appear like a chaos of atoms. Why not put them together in some sort of order? Is not the allocation a subject of investigation as well as the weight? If one order is found more consistent than another, why not adopt it till a better is found?[29]

Dalton seems to have been guided by an aesthetic feeling for the symmetry of the system in writing down certain preferred patterns. He often spoke of the beauty and simplicity of the atomic theory, and in this respect the Berzelian notation is lacking. But Berzelius saw the dangers of submerged assumptions of this kind, and in the absence of other evidence preferred his own non-committal formulae.

To suppose that aesthetic considerations were Dalton's sole guide, however, would be manifestly unjust. Of his own formulae he wrote:

> When three or more particles of elastic fluids are combined together in one, it is to be supposed that the particles of the same kind repel each other, and therefore take their stations accordingly.[30]

This was a remarkable anticipation of our own understanding of an important rule of molecular geometry. It might have found a sympathetic response in Berzelius whose devotion to the dualistic doctrine that 'like repels like' was almost unequalled. But for some reason it did not. One can only conclude that it was too speculative, and fundamentally, of course, Berzelius was right. One cannot have a theory of structure without a theory of valency.

3 *Rules of chemical combination*

Berzelius was the first of the great pioneers in the accurate determination of atomic weights. In this again he built upon the foundation laid by Dalton, and, like the English chemist, he proceeded from a set of axioms, termed in his case the 'corpuscular theory'. His early views may be generalized in the formula of a binary compound as AB_n, where A and B are elements and n is an integer up to 12. In other words, one element (A) is restricted to one atom in each molecule, so combinations of the type 2A + 3B, 2A + 2B and so on were excluded. These would be 'contrary to sound logic' because 'no obstacle, either mechanical or chemical' existed 'to prevent such an atom from being divided', and because 'such a composition would almost totally destroy chemical proportions'.[31] Dalton's sceptical rejoinder to this[32] brought forth the assertion that the rule was founded upon analyses of inorganic compounds ('and I have analysed a great number of them') and the challenge to offer 'one proof that this opinion is evidently incorrect'.[33]

Now the fact that Berzelius was wrong about this matters less than the reason behind his mistake. His conclusions, because they were based upon analysis, however limited, were sounder than the more arbitrary rules of Dalton. And subsequent experimentation caused him to change his mind, so that by 1831 he was conceding that 2 atoms of one element could combine with 3, 5 or 7 of another, and wondering whether combinations with 2, 4 or 6 atoms of the other might be found.[34] His theory, based as it was upon analysis, became the starting-point for a whole series of other analyses and gradually a self-consistent picture of atomic weights began to emerge. The superb quality of this experimental work received the homage of the author of the Stas Memorial Lecture who wrote:

> The high and highly deserved reputation of Berzelius for accuracy, never before surpassed in quantitative determinations, demanded extraordinary care on the part of any one who should call in question the numbers accepted on his

authority. . . . With apparatus in many respects inferior to that of the present day, and with scarcely any aid from chemical manufacturers in preparing pure materials and reagents, but with unsurpassed manipulative skill, and the most honest criticism of his own work and endeavour to improve its accuracy, he produced the first fairly trustworthy list of numbers representing the proportions by weight in which the elements combine.[35]

It is well known that Berzelius supplemented the rules mentioned above with the law of isomorphism and that of atomic heats. As a result, he obtained atomic weight values for some thirty elements which, corrected from his standard of O = 100 to our own, come very near the modern values. His most important error sprang from the false assumption that all strongly basic oxides were of the form M.O, so that he obtained for the alkali metals and silver double the correct values. This had a serious effect later on analyses of organic acids whose equivalents were generally obtained from ignition of their silver salts. Dalton's frank confession could have been also made by Berzelius:

After having the atomic principles in contemplation for ten years, I find myself still at a loss, occasionally, to discriminate between the combinations which contain two atoms of a given body from those which contain only one atom.[36]

This was a difficulty that remained until Avogadro was allowed to provide the solution.

Evaluation of atomic weights was only one fruitful result of the atomic interpretation of combining quantities. The very success of these investigations prompted Berzelius to wonder whether atomism was relevant to the vast and little understood realm of organic nature.

At one time he had doubted it. An analysis of oxalic acid implied far more than twelve atoms attached to a single atom of hydrogen—a result quite at variance with the expected composition of inorganic compounds. It would appear that this was a great stumbling-block to Berzelius who felt that the atomic theory could not be applied to organic compounds.[37] Dalton disagreed with the analysis figures, and produced his own which are in fact correct.[38] It is probable that he diagnosed Berzelius' fault correctly, suggesting that the lead oxalate used had not been dried at a high enough temperature. Since the presence of a small amount of hydrogen vitiated the first result, inadequate drying was probably responsible for the error. With Dalton's figures no difficulty arises with an atomic interpretation.

It was after this experience that Berzelius resolved to examine how far organic compounds could be viewed from a corpuscular standpoint. With an improved method of analysis he obtained data for thirteen compounds within a few months, and concluded that although the law of definite proportions did not seem applicable, formulae could be given in accordance with Dalton's atomic theory.[39] From this time onwards atomism became applied to every kind of ponderable matter, and this decisive step forwards was taken by Berzelius in 1814–15.

One other rule of chemical combination presents some interesting features. This concerned the reaction between two oxides, which according to Berzelius always took place in such a way that the ratio of oxygen in each was always a whole number. Permitted combinations could include, for example, (A + O) with (B + 2O) but not (A +3O) with (B + 2O). Dalton's atomic theory could not explain this restriction. Conversely, the 'oxide rule' allowed combinations that were absurd in Daltonian terms as (A + 3O) with (B + $1\frac{1}{2}$O).[40]

In fact the rule is not strictly correct, but does hold for the majority of combinations (partly because most basic oxides contain only one oxygen atom). Its chief interest is the construction that Berzelius placed upon it. Dalton was quick to point out that the difficulty involving the oxide (B + $1\frac{1}{2}$O) may be resolved by regarding it rather as (2B + 3O) by halving the atomic weight of B. On the question of the atomic theory permitting (A + 3O) + (B + 2O) he observed that it was no more the business of the theory to explain the non-occurrence of this combination 'than to show why all the metallic oxides do not mutually combine with each other'.[41] Here, surely, he was on solid ground. It was quite unreasonable for Berzelius to demand that any hypothetical scheme should *by itself* cover all phenomena, and it is now obvious that a complete explanation could not be forthcoming without the concept of valency. He returned to the matter in his reply to Dalton, and in a passage not printed by Thomson remarked that:

> when a theory explains only half of what should follow from a theory of these phenomena, that theory is incomplete. Where the theory is deficient a difficulty exists that cannot be shrugged off by the assertion of an individual pretending that he expects from it a satisfactory explanation of only one part of the phenomena presented.[42]

It is clear that Berzelius had no criteria to offer by which the valid limits of a hypothesis could be delineated, and this methodological blind-spot is an interesting indication of one of his chief limitations. Yet as he wrote to Thomson, 'When a corpuscular theory of chemistry is involved, I have believed that this theory ought to be the fundamental theory of the science, and that it ought . . . . to embrace all'.[43]

4 *Theory of volumes*

A brief reference must be made to another topic in which Berzelius differed from Dalton, the theory of volumes. Gay-Lussac's law of combining volumes was an important generalization on the chemistry of gases. It was seen by Berzelius as an experimental basis for his corpuscular theory which was thereby placed in a stronger position than Dalton's hypothetical atomism:

> What in the one theory is called an *atom* is in the other theory a *volume*. In the present state of our knowledge the theory of volumes has the advantage of being founded upon a well-constituted fact, while the other has only a supposition for its foundation.[43]

The inapplicability of volume theory to organic compounds was at that time deemed an advantage for it was not then clear that atomism was relevant to them. In 1814 Berzelius said that he regarded the theory 'as a leading-string to keep us in the way of truth'.[44] He used it to show that the ratio of oxygen to hydrogen atoms in water was 1 : 2, for example. Unfortunately he was unable to apply it to compound gases for this would have committed him to either a demi-atom of oxygen (for instance) or to a molecule of an element containing more than one atom. He deplored either possibility, polyatomic molecules of one element being a violation
268 of dualism. He concluded that several compound gases

'diminish in volume at the moment of chemical combination, since the repulsive force of one or all of the elements is diminished by the juxtaposition of an atom of another element',[45] so the theory of volumes does not apply to them. That was as near as he got to Avogadro's position. Had he been able to take the final step he would not only have given atomism the weight of his immense authority, but would also have advanced its cause by half a century.

5 *Dualistic atomism: the electrochemical theory*

In his Essay on Chemical Proportions of 1813 Berzelius had lightly touched on the supposition of spherical atoms in contact with each other and held together by neutralization of opposite electrical charges. This was not the first reference to the idea of an electrochemical union. The view that chemical affinity and electricity were essentially the same had been put forward in 1811,[46] and the idea had been expanded the following year.[47] In 1813 Nicholson had published a lengthy Note by Berzelius in his *Journal* as a 'beautiful generalization of facts'. It included the passage:

> As at the moment of union [of two reacting substances] there is a production of heat, which may vary from a very slight elevation of temperature to that of the most intense fire, we think we may conclude, that at the moment of the chemical combination there is a discharge of the opposite electric state of the bodies which, here, as in the pile, produces the phenomenon of fire, at the instant when the electricities disappear.[48]

Dalton's objection was to the imposition of these electrochemical speculations upon his own atomism. He protested that they make 'no necessary part of the atomic theory such as I maintain it'.[15] It is clear that this phrase in particular greatly offended Berzelius, for he quotes or parodies it several times in his letter to Thomson, not all of which came to be published in *Annals of Philosophy.* He contrasts Dalton's attitude with his own endeavours 'to strain towards the first principle of science' by broadening his theory as far as possible. It is true that he categorized his views as 'conjectures', but added that 'Ideas on the relation of atoms to their electrochemical properties ought in my opinion to constitute an essential part of the corpuscular theory as I have envisaged it'.[49]

From now on Berzelius committed himself to a view of matter in which atomism and electrochemical dualism were so intimately associated that they had to stand or fall together for the rest of his life. This partly explains his single-minded, dualistic philosophy even in the face of great opposition. The very success of atomism encouraged him in his devotion to its electrochemical interpretation. It also accounts for the intensity of feeling behind his long reply to Dalton's mild comments on his Essay.

I have given an account elsewhere of Berzelius' electrochemical theory,[50] and it will be sufficient to note here the main aspects of its later relation to atomism.

The first detailed exposition of it came in 1818–19, in the influential *Essai sur la théorie des proportions chimiques et sur l'influence chimique de l'électricité.*[51] From this we see that at the heart of his theory lay the concept of atomic polarization: 'we can represent each atom as having an electrical polarity'.[52] This meant that electricities were concentrated in opposite parts of the atom, rather like magnetic poles. To explain why some atoms tended to be positive and others negative, Berzelius supposed one pole might predominate—and this phenomenon he called unipolarity. Later on, he added the idea of several electrical axes in an atom in order to account for multiple combinations.[53] In several ingenious ways these ideas were developed to cover as many phenomena as possible. Thus he assumed that the intensity of polarity varied with different substances, that it depended also upon temperature, and so on. To account for the effects of environment and physical state he assumed that spatial effects operated, that reaction was facilitated when atomic poles were near each other and that certain relative orientations of the atoms were more favourable than others.

In all of these respects his scheme was irrevocably committed to a vivid, pictorial atomism. It covered all types of chemical reaction, mechanical cohesion, dissolution, electrolysis and so on. It ranged from simple binary inorganic compounds to complex organic substances. To use modern terminology, the extension of a fundamentally ionic concept to covalent carbon compounds had unfortunate consequences, though even here Berzelius' view of the unity of chemistry had profound justification.

These things need to be said because it is so easy to deride Berzelius, as he derided Dalton, for harmful extensions of his hypothesis. It cannot be denied that dualism held up the progress of the atomic theory in blinding its adherents to the rightness of Avogadro's hypothesis, but its other effects assisted that progress greatly. Yet so profound was the impact of such dualistic philosophy that as late as 1861 Kekulé needed to observe that, though atomic theories were compatible with dualism, they were not inherently dualistic themselves.[54]

Conclusion

In the foregoing account, much has been said of the differences between Berzelius and Dalton. This was inevitable in view of the importance of these divergences for the spread and growth of atomism. Yet it would not be fair to the memory of either man to leave it quite like that. Nor would it be right to categorize one or the other as the more truly scientific. Fortunately, we are not here to award prizes.

Both Berzelius and Dalton had their fair share of obstinacy and tenacity, but both had their vision of what science ought to be. It would be naïve in the extreme to imagine that either of them deliberately put forward one theoretical scheme in preference to an alternative necessitated by experimental facts. Their difference lay in matters of personal choice: how far they chose to consider certain factors as opposed to others, how widely-embracing they chose to make their theoretical schemes, even how far aesthetic considerations determined their ultimate decisions.

To Berzelius Marcet once wrote:

> How you love chemical theory! Shall I tell you of the fault I find in your writings? It is that they embrace too much at a time. Theoretical discussions come too often to interrupt the facts. You have the whole of chemistry before your eyes like a tableau; each little fact reminds you of a theory; and each theoretical idea reminds you of a crowd of facts. There results a richness in your writings which dazzles the philosophers of ordinary calibre.[55]

Let the last word come from Berzelius himself:

When I endeavoured to draw the attention of chemists to the difficulties in the atomic theory it was not my intention to refute the hypothesis. I wanted to lay open all the difficulties of that hypothesis that nothing might escape our attention calculated to throw light on the subject. I wished the experiments to verify the theory, and I should have considered it as absurd if I had taken the other road.[56]

These two comments, illustrating as they do both the vision and the philosophy of Berzelius, serve briefly to remind us of the debt owed to him by John Dalton and by us all.

Notes

Responsibility for translation into English of citations from other languages is accepted by the author.

1 *Kgl. Svenska Vetenskapsakad. Årsbok* (1948) 49.
2 J. J. Berzelius, *Traité de chimie* (Paris, 1831) iv, 532–3.
3 Letter to Gahn, 28 September 1807, in *Jac. Berzelius Bref*, ed. H. G. Söderbaum (Uppsala, 1922) iv, Part 2, 23.
4 *Nicholson's Journal*, xvi (1807) 4.
5 *Philosophical Magazine*, xli (1813) 4.
6 *Nicholson's Journal*, xxi (1808) 164.
7 *Jac. Berzelius Bref* (1912) i, Part 1, 16.
8 Ibid., Part 2, 29.
9 T. Thomson, *A system of chemistry* (Edinburgh, 1806–7) 4 vols.
10 *Jac. Berzelius Bref* (1922) iv, Part 2, 69.
11 Op. cit., (1806) ii, 3–26 of 'Notes' at end of volume.
12 *Jac. Berzelius Bref* (1920) iii, Part 2, 105.
13 *Annals of Philosophy*, ii (1813) 315.
14 Ibid., ii (1813) 443; iii (1814) 51, 93, 244, 353.
15 Ibid., iii (1814) 174.
16 *Jac. Berzelius Bref* (1918) iii, Part 1, 27.
17 *Annals of Philosophy*, v (1815) 122.
18 Ibid., ii (1813) 446.
19 Ibid., iii (1814) 175.
20 *Abhandlung Akad. Berlin* (1819) 427.
21 Berzelius, *Traité de chimie* (Paris, 1831) iv, 548.
22 Ibid., 549; *Jahresbericht*, xi (1832) 44.
23 *Annals of Philosophy*, v (1815) 123.
24 Ibid., iii (1814) 51.
25 E.g. M. P. Crosland, *Historical studies in the language of chemistry* (London, 1962) 270–81.
26 J. R. Partington, *A history of chemistry* (London, 1964) iv, 158–60. In an article on 'Dalton and Berzelius' (*Monit. Sci.*, i (1921) 4), M. Delacre has decried Berzelius' symbols on the grounds that they add nothing new, 'absolutely nothing', to those of Dalton and has condemned the assertion that Berzelius invented atomic symbolism as the 'grossest error in the history of chemistry'. This unusual view takes no account of both the main points in the following discussion.
27 *Annals of Philosophy*, iii (1814) 51.
28 C. A. Russell, 'The Influence of Frankland on the Rise of the

Theory of Valency', *Actes of the Xth International Congress of the History of Science* (Ithaca, 1962) 883.

29 Letter to Graham, 1837, in W. C. Henry, *Memoirs of the life of John Dalton* (London, 1854) 124.

30 Dalton, *A new system of chemical philosophy* (Manchester, 1808) i, Part 1, 216.

31 *Annals of Philosophy*, ii (1813) 447.

32 Ibid., iii (1814) 176.

33 Ibid., v (1815) 125.

34 Berzelius, *Traité de chimie* (Paris, 1831) iv, 542–3.

35 J. W. Mallet, in *Chemical Society Memorial Lectures* (London, 1901) Stas Lecture, 3, 12–13.

36 *Annals of Philosophy*, iii (1814) 178. The result given (H + 27C + 18O), seems to be a misprint for (H + 18C + 27O), as the C/O ratio must be 2/3.

37 Ibid., ii (1813) 450. There should be no hydrogen in the 'acid' (= anhydride) and therefore no need to include such large numbers of carbon and oxygen atoms.

38 Ibid., iii (1814) 179.

39 Ibid., v (1815) 260.

40 Ibid , ii (1813) 447; iv (1814) 324.

41 Ibid., iii (1814) 177.

42 *Jac. Berzelius Bref* (1918) iii, Part 1, 31.

43 *Annals of Philosophy*, ii (1813) 450.

44 *An attempt to establish a pure scientific system of mineralogy by the application of the electro-chemical theory and the chemical proportions* (London, 1814) 114.

45 Berzelius, *Traité de chimie* (Paris, 1831) iv, 553.

46 *Journal de Physique*, lxxiii (1811) 253.

47 *Kongl. Vet. Handl.*, xxxiii (1812) 166.

48 *Nicholson's Journal*, xxxiv (1813) 145.

49 *Jac. Berzelius Bref* (1918) i, Part 1, 28.

50 London University M.Sc. Dissertation, 1958. Also in *Annals of Science*, xix (1963) 117 et seq.

51 Paris, 1819.

52 Op. cit., 55.

53 Berzelius, *Traité de chimie* (2nd edn, Paris, 1845) i, 106.

54 *Lehrbuch der organischen Chemie* (Erlangen, 1861) i, 60.

55 *Jac. Berzelius Bref* (1914), i, Part 3, 43.

56 *Annals of Philosophy*, v (1815) 127.

VIII

Introduction to the Reprint Edition of *Essai sur la Théorie des Proportions Chimiques et sur l'Influence Chimique de l'Électricité*

1. CHEMISTRY IN 1819

In 1819, when Berzelius' *Essai sur la Théorie des Proportions Chimiques et sur l'Influence Chimique de l'Électricité* was published in Paris, Europe was still recovering from the Napoleonic wars. The science of chemistry had undergone an experience no less traumatic in its own way than the political upheavals of the times, only here the revolution in ideas was gathering rather than losing momentum. To summarize adequately the state of affairs in a few pages is therefore clearly impossible. But at this distance of time, certain features stand out so sharply against the mass of other detail that something may be gained by a glance at each of these aspects of the scene.

With the passing of the eighteenth century had come about the virtual extinction of a principle which had dominated the chemical scene for fifty years and more, and had been the first great unifying principle of the science. Owing largely to the discoveries of Lavoisier, combustion was now regarded as a chemical combination with oxygen, and not as liberation of *phlogiston*. The antiphlogistic philosophy was now widely accepted, though traces of the older mode of thought still lingered on in places.[1] With the overthrow of phlogiston there came into being a new conception of the chemical elements, and this clarification alone marks a major watershed. Yet one legacy of phlogistic chemistry lingered on: the preoccupation with combustion and fire. For Berzelius these phenomena were to have a singular importance.

[1] Davy is an important example of this phlogistic tendency; see, e.g., R. Siegfried, "The Phlogistic Conjectures of Humphry Davy," *Chymia*, *9* (1963), 117.

Meanwhile a further development had been taking place in the quest for an understanding of chemical affinity. This idea had emerged when it became clear that not all chemical combinations had the same stabilities, and one component of a compound could be displaced by a substance that could exert a stronger hold upon the rest. Chemical union was thus an exhibition of the affinities of the constituent parts. The subject had a long and respectable history, represented by the names of Geoffroy, Buffon, Macquer, Baumé, de Morveau, Fourcroy, and others—to mention only those of the eighteenth century, Tables representing "elective affinities" were drawn up in 1783 by Berzelius' compatriot Bergman.

Work of this kind did much to codify existing data but added little that was new and lacked any real quantitative basis. One of its most fortunate effects, however, was to lead to the stoichiometric laws. An attempt by J. B. Richter to produce affinity tables led him to create what was effectively a table of equivalents, with the implicit recognition of the law of reciprocal proportions. Richter's memoirs fell into the hands of Berzelius and gave to him the direct inspiration for his work on chemical proportions. More important still, the law of reciprocal proportions, like that of definite proportions formulated by Proust in 1799, could be derived from the atomic theory of John Dalton which emerged in the early years of the nineteenth century. The validity of these laws was thus a powerful argument for atomism, although the historical relationship between the stoichiometric laws and Dalton's atomism is a complex one.

The replacement of classical corpuscular theory by Daltonian atomism was far from complete by 1819. Yet one of the most insistent voices advocating a new atomic theory (though not under this name) was that of Berzelius. Differing though he did in some respects from the English Quaker,[2] he built a vast system of chemistry upon atoms that owed more to Dalton than to anyone else.

Side by side with the emergence of atomism came the last echoes of a controversy that also had its origins in the study of affinities. In 1798 the French chemist Berthollet accompanied Napoleon on his expedition to Egypt and was surprised to notice large deposits of

[2] For the relationship between Dalton and Berzelius see C. A. Russell, "Berzelius and the Development of the Atomic Theory" in *John Dalton, 1766–1844*, ed. D. S. L. Cardwell (Manchester, 1966).

soda on the shores of the Egyptian salt lakes. He concluded that they arose from a reaction between salt and limestone; this would be a reversal of the familiar laboratory process:

calcium chloride + soda = calcium carbonate + salt

or, $$CaCl_2 + Na_2CO_3 = CaCO_3 + 2NaCl.$$

The abnormally high concentration of sodium chloride must have been responsible in this case. He then developed this idea into the generalization (implicit in earlier writers) that *mass* could play a determinative role in all reactions produced by "elective affinity." It was therefore possible that "an excess of quantity of the body whose affinity is the weaker compensates for the weakness of the affinity."[3] Ultimately the recognition of the part played by mass in chemical reactions was to culminate in the law of mass action. Unfortunately, however, it led Berthollet to a false conclusion on the matter of chemical composition. Since relative masses influenced the proportions in which substances reacted together, could not the products have an indefinite composition? The continuous variations observed with solutions and alloys could not be lightly dismissed, for the distinction between mixtures and compounds was still far from clear. This incorrect extension of his own theory led Berthollet into much trouble, including his celebrated controversy with Proust whose law of definite proportions was formulated in 1799. But by about 1808 the Proust-Berthollet argument was over, most chemists deciding that Proust had established his case and that Berthollet's views were incompatible with the new atomism.[4] It fell to Berzelius in his *Essai* and elsewhere to provide an effective challenge to this last opinion.

Chemical thinking in the early nineteenth century owed much to these novel doctrines: an antiphlogistic view of combustion, a *chemical* atomism (as opposed to the corpuscular philosophy of eighteenth-century physics) and the investiture of mass with a new significance. Yet none of these had the spectacular effects provided by the unexpected arrival of a new science at the turn of the century. The new subject of electrochemistry came into being with the

[3] *Researches into the Laws of Affinity*, trans. M. Farrell (London, 1804), p. 45.
[4] See S. C. Kapoor, *Chymia*, *10* (1965), 53–110.

discoveries of the voltaic battery[5] and electrolytic conduction.[6] As Davy said, the first of these "was an alarm-bell to experimenters all over Europe,"[7] and he himself responded to the call with the utmost vigor. Not merely did electrochemistry provide him with a potent means for chemical decomposition and the isolation of the alkali and alkaline earth metals. It also led him to propose an electrical theory of chemical action which inspired Berzelius to develop his own far more detailed electrochemical scheme.

Fairly recently a fellow-countryman of Berzelius has hailed Davy's isolation of the alkali metals as "the foundation of electrochemistry."[8] Certainly from this point onward wide public interest was centered on the interactions between electricity and chemistry. Speculations abounded and many other workers sought to extend the range of experimental knowledge. But it was Davy who first produced a coherent electrochemical theory, compassing the phenomena of voltaic electricity, electrolysis, and chemical reactivity. This theory has been described elsewhere,[9] but some aspects of it are so necessary for an understanding of the position adopted by Berzelius that a brief recapitulation is desirable.

Davy's theory of the voltaic pile embraced, eventually, a synthesis of the two prevailing explanations in terms of chemical action or contact phenomena.[10] His theory of electrolysis was less important than his experimental clarification of a number of confused issues. However, he envisaged a chain of combinations and recombinations taking place throughout the solution which was divided into two zones of opposite charge with a neutral point between them.[11] But it was in his view of chemical change that he made the greatest impression. At an early stage he supposed that chemical affinity and electrical energy had a common "remote cause."[12] If substances

[5] A. Volta, *Phil. Trans. 90* (1800), 403.
[6] W. Nicholson and A. Carlisle, *Nicholson's J.*, *4* (1800), 179.
[7] *Works* (London, 1840), vol. viii, p. 271 (from a lecture in 1810).
[8] A. Tiselius, *Purity and Purification of Chemical Substances*, Royal Institute of Chemistry Lecture (London, 1958), p. 1.
[9] C. A. Russell, "The Electrochemical Theory of Sir Humphry Davy," *Ann. Sci.*, *15* (1959), 1, 15; *19* (1963), 255.
[10] E.g., *Phil. Trans.*, *97* (1807), 1.
[11] E.g., *ibid.*, *116* (1826), 416.
[12] *Ibid.*, *97* (1807), 39.

acquire opposite electrifications on contact, they lose their charges when they react chemically, so Davy could not equate electrical and chemical attraction. Instead he ascribed a common basis to them both.[13] Like Berzelius, he recognized the importance of mass as demonstrated by Berthollet,[14] and he tried to use the metal displacement series as a semiquantitative basis for his theory.[15] He supposed the effect of heat on a chemical reaction increased both the motion of the combining particles[16] and their degree of electrification.[17] But his corpuscular philosophy owed more to the eighteenth than to the nineteenth century (see footnote 9).

Davy is obviously an important figure in his own right.[18] The question at issue here, however, is the extent to which his ideas influenced those of Berzelius. The relationship has much of interest to offer the psychologist as well as the chemist or the historian. It is a sorry tale of blundering inflexibility, genuine difficulties in communication, and the most incredible tactlessness. Thomas Young, having gained access to Berzelius' private comments on Davy's *Elements of Chemical Philosophy*, was foolish enough to pass these on to Davy, who took instant umbrage. This incident completed the estrangement that had been largely caused by their differences on the nature of chlorine. After this, their slight correspondence was virtually confined to an exchange of formalities. Nevertheless, the years between 1806 and 1811 saw some fruitful dialogue between the two men. It was during this period that Berzelius' own ideas were maturing. It is probably, and unhappily, the case that the finest flowering of Davy's thoughts came in the work of two men for whom at times he showed scant respect: Faraday and Berzelius.

Thus the background to the appearance of this epochal *Essai* was one in which an emerging atomism, lately victorious over doctrines of continuous composition, bade fair to vanquish also the related concepts of mass action. But the potential simplifications

[13] E.g., *ibid.*, *116* (1826), 417.
[14] *Ibid.*, *97* (1807), 41.
[15] *Elements of Chemical Philosophy* (London, 1812), p. 500.
[16] *Ibid.*, p. 95.
[17] *Phil. Trans.*, *97* (1807), 43.
[18] On Davy generally, see Sir Harold Hartley, *Humphry Davy* (London, 1966, reprinted 1971).

implicit in the new attitude toward chemical elements were likely to be lost in the complexity of new phenomena revealed by the further study of electricity. Perhaps the greatest contribution of the *Essai* was that it clarified chemistry just at the time when clarification was most sorely needed. Without fully endorsing Thomson's opinion that at about this time "little is doing at present in chemistry, either in this country or on the continent of Europe,"[19] we may readily see that the science was in need of some direction. There is evidence that this was supplied by the *Essai*.

2. BIOGRAPHICAL SKETCH

Jöns Jacob Berzelius was born at Väversunda, Sweden, about 120 miles southwest of Stockholm, 20 August 1779.[20] His father was a clergyman-schoolmaster, but since both his parents died while he was quite young his early childhood was spent with relatives. In 1796 he went to Uppsala to study medicine, and graduated as an M.D. six years later. While holding several medical appointments he pursued the study of chemistry, and in 1803 (with Hisinger) he electrolyzed solutions of salts, anticipating some of Davy's work. In 1807 he became professor of medicine and pharmacy at the Stockholm School of Surgery. A year later he became a member of the Swedish Academy of Science, and (with Pontin) devised the mercury cathode for electrolysis, discovering "ammonium amalgam." About this time he began his immense task of studying the whole field of chemical proportions. Meanwhile, his electrochemical

[19] Letter dated 2 November 1820, to Silliman, in *Silliman's J.*, *3* (1821), 396.

[20] Since the material in this section is readily accessible and well known, no detailed references will be given. For biographical information on Berzelius generally, see:

(a) H. G. Söderbaum. *Jac. Berzelius Levnadsteckning* (Stockholm, vols. i and ii, 1929; vol. iii, 1931).

(b) J. E. Jorpes, *Jac. Berzelius, his Life and Work*, trans. B. Steele (Stockholm, 1966).

(c) *Autobiographical Notes*, trans. O. Larsell (Baltimore, 1934).

(d) J. R. Partington, *A History of Chemistry*, (London, 1964), vol. iv, pp. 142, *et seq*.

A biography by Sir Harold Hartley is also in course of preparation.

theory was developing under the successive stimuli afforded by the need to reform chemical nomenclature (1811), the constant pressure of his stoichiometric work, and his new classification of minerals (1814).

Possessing great experimental skill, he discovered three new elements: cerium (1803), selenium (1817), and thorium (1829). Assistants who worked in his laboratories added several more to this list.

Berzelius showed during this period (1808–1818) that organic substances could be brought within the scope of the atomic theory. During the rest of his life he examined many organic substances, including the products in blood, bile, and other natural materials. He discovered pyruvic acid, invented the term "protein," and introduced into chemistry the concepts of catalysis and isomerism. Extension of his electrochemical theory to organic substances brought about the controversies that marred his declining years. He held to certain viewpoints so tenaciously that he found it difficult to abandon them long after they had outlived their usefulness and had indeed been shown to be incompatible with new discoveries. His views on the nature of chlorine and nitrogen and his approach to the volume relationships of combining gases must be numbered among the less successful aspects of his work.

In 1818–1819 he made the epochal visit to Paris of which more will be said later. While he was away he was appointed Secretary to the Swedish Academy of Sciences, and in 1832 he resigned his Chair in order to concentrate on the Academy's affairs. In addition to several monographs (on the blowpipe, mineralogy, etc.), he produced the *Textbook* which, in six languages, ran into many editions and exerted an enormous influence upon European chemistry. This was further extended by the *Jahresberichte*, or annual reports, that began in 1822. The magnitude of the respect accorded to him is indicated by his membership in 94 learned societies and receipt of numerous decorations and honors.

In 1835 he married Elisabeth Poppius, the daughter of an old friend, and on that occasion he was created a baron. The marriage was happy though childless. He died on 7 August 1848, and was buried in Stockholm.

3. EVALUATION OF THE *ESSAI*

One of the outcomes of Berzelius' visit to Paris was this *Essai sur la Théorie des Proportions Chimiques, et sur l'Influence Chimique de l'Électricité*. The circumstances of its origin will be discussed later, as will its literary connections and its eventual impact upon contemporary science. Meanwhile, it is convenient now to inquire into its significance as a historical document. As will be seen, it had individual innovatory features of its own, but also broke new ground in including between the same covers a number of important novelties which, though not strictly unprecedented, had never appeared together before.

First may be mentioned Berzelius' own brand of atomism, termed rather oddly his *corpuscular theory*. This was in several ways an advance on the ideas put forth by Dalton (see footnote 2). Berzelius declined to accept nonspherical atoms, proposed to lay down certain rules of chemical combination, and for gaseous elements identified an atom with a volume. Certainly in the last two respects he has been shown to be wrong, but this matters less than the general impetus his dicta gave to the atomic theory itself. Moreover, his extension of this to organic compounds was a step forward of the utmost significance.

Not all of these features appeared for the first time in the *Essai*. Some of them had been discussed in a series of articles entitled "Essay on the cause of chemical proportions, and on some circumstances relating to them: together with a short and easy method of expressing them."[21] Other references were made in his textbook on mineralogy (see below). But there they were placed in a setting that displayed them to better advantage, being related to other aspects of atomism for the first time.

Directly derived from the corpuscular theory was the table of atomic weights and formulas that formed the climax to the book. This astonishing list, compassing about 2000 different compounds, included also the first fairly complete table of atomic weights ever to appear. It represents the distilled essence of a vast quantity of

[21] *Ann. Phil.*, *2* (1813), 443; *3* (1814), 51, 93, 244, 353.

experimental work by Berzelius and others.[22] An interesting comparison of some of the values with modern figures has been recently published,[23] revealing the considerable accuracy attained by Berzelius with equipment of the utmost simplicity. The tables are a monumental record of a decade of ceaseless activity devoted to one end: the unraveling of the mystery of chemical proportions. As Sir Harold Hartley has pointed out, they "are one of the milestones of the History of Chemistry."[24]

Impressive as this achievement is, however, it can be regarded as less important in the long run than the suggestions made here for a new systematic nomenclature and symbolism. These have been frequently described[25] and did not originate in this *Essai*. An important early exposition of the nomenclature came in 1811,[26] and the new symbolism was described in 1814 (see footnote 21). Nevertheless they were brought together for the first time here. There is evidence (to which we shall return) that this presentation of them was particularly influential.

A noteworthy feature of the *Essai* was its attempted reconciliation of the mass action concept of Berthollet with the doctrine of definite proportions. Berzelius was not the first to see this possibility clearly; Davy had written of it in his *Elements of Chemical Philosophy* (see footnote 14). How far this affected contemporary thinking is debatable, but it remains an innovation worthy of comment and applause.

Finally, the *Essai* contains the classic statement of early electrochemical dualism. Thus von Meyer speaks of "this memorable work for the history of chemistry" in which "he developed his conception of the atomic doctrine, and his ideas upon the relations between chemical affinity and electric polarity."[27] Recalling the

[22] W. M. MacNevin, "Berzelius—Pioneer atomic weight chemist," *J. Chem. Educ.*, *31* (1954), 207.

[23] See footnote 20 ref. (d), pp. 166, 226, 231, 232.

[24] *K. Svenska Vetenskapsakademiens Årsbok* (1948), p. 41.

[25] E.g., see footnote 20, ref (d), p. 158; M. P. Crosland, *Historical Studies in the Language of Chemistry* (London, 1962), p. 270.

[26] *J. de Phys.*, *73* (1811), 253.

[27] E. von Meyer, *A History of Chemistry from the Earliest Times to the Present Day*, trans. G. M'Gowan (London, 1891), p. 201.

later difficulties of the electrochemical theory, it is easy for one to regard it as an unnecessary excrescence upon Daltonian atomism. Yet nothing could be further from the truth so far as Berzelius was concerned, and this book reveals the extent to which the ideas had become fused together in his synthesis. Not only did he from now on regard electrical union atomistically; he also envisaged his atoms electrically.

Thus Berzelius was now in possession of a viewpoint that embraced the classical though vague notion of affinity, the new Daltonian atomism (suitably modified), the accumulating evidence for some kind of mass action, the widely received Lavoisierian view of combustion and the elements, and the recent discoveries of voltaic electricity and electrolysis. From this position of vantage "he succeeded in penetrating and dominating . . . the whole domain of chemistry."[28] The significance of the *Essai* (and its Swedish predecessor) is that for the first time this comprehensive synthesis was made known to the chemical world in general.

4. LITERARY RELATIONSHIPS

Berzelius was a chronic sufferer from difficulties in communication. A large factor here was undoubtedly the sheer problem of language, of which he complained bitterly.[29] This, together with the relative isolation of Sweden in post-Napoleonic Europe and a certain blundering tactlessness on the part of both Berzelius and his opponents, is probably sufficient explanation for the slow recognition of the Swedish chemist. The surprising thing, in fact, is the extent to which he did eventually command attention and respect. This can be seen now as a result of, first, the spread of his ideas by devoted disciples who had spent some years working in his laboratory, and, second, the prodigious literary output of Berzelius and his translators. Perhaps his textbooks were his most influential advocates, and it was from one of these that the present *Essai* was derived.

Berzelius' *Lärbok i Kemien* (Stockholm) began to appear in 1808. Details are as follows:

[28] *Ibid.*, p. 217.
[29] *Ann. Phil.*, *6* (1815), 49.

Vol. I: 1808 (2nd ed., 1817).
Vol. II: 1812 (2nd ed., 1822).
Vol. III: 1818.
Vol. IV: 1827.
Vol. V: 1828.
Vol. VI: 1830 (based on *Föreläsningar i Djurkemien*, 1806–1808).
The first three are on inorganic chemistry, the last three on the organic branch. German and French translations of early volumes began to appear before the later Swedish volumes, and the relationships between editions even in the same language are too complex to describe here. Translations into Spanish, Dutch, and Italian were also made, but not into English. At this time British chemistry was manifesting an extreme insularity, and Berzelius suffered more than most in this respect.

Thus behind the multiplicity of editions and re-editions in half a dozen languages lay the great six-volume *Lärbok i Kemien*, which with some justification can be described as one of the first modern textbooks of chemistry. From Volume III the present *Essai* is derived. The date of the Swedish original is therefore 1818, and not 1814 as is often stated.[30] This mistake may have arisen from a confusion between the *Essai* and the textbook on mineralogy[31] which also contains an exposition of Berzelius' electrochemical theory. The two works are, however, quite distinct.

The circumstances of the translation will be described in the next section, but it is convenient to note here that the *Essai* is *not* a complete translation of the relevant parts of the *Lärbok*. Passages have been omitted or rewritten and new material has been added. References to the Swedish *Afhandlingar i Fysik, Kemi och Mineralogi* are omitted and there are other minor concessions to the wider audience for whom the French version was intended. The most

[30] As, e.g., by von Meyer (cited in footnote 27), p. 201. The mistake is repeated by Partington in his *Short History of Chemistry*, 2nd ed. (London, 1951), p. 198 (but not in his *History of Chemistry* [London, 1964], vol. iv), and other modern authors as H. Sharlin, *The Convergent Century* (London, 1967), p. 65. W. G. Palmer, who quotes from the 1835 2nd ed., gives the date for the Swedish original as 1819 (*A History of the Concept of Valency to 1930* [Cambridge, 1965], p. 20n).

[31] *Försök att genom användandet af den elektrokemiska theorien och de kemiska proportionerna grundlägga ett rent vettenskapligt system för mineralogien* (Stockholm, 1814).

important variations are indicated in the Commentary and Notes below. It is also true that the standard of typographical presentation and layout was rather better in the *Essai* than in its Swedish original; headings, though sometimes still confusing, are a good deal clearer than they were.

The pages of the *Lärbok* (1–132) from which the bulk of the *Essai* is derived do not contain the table that forms the great climax of the latter. This also had a Swedish precursor, however, in a separate publication of the same year, *Bihang till tredje delen för Lärbok i Kemien* (Stockholm). This table was the joint work of Berzelius and his pupil Per Lagerhjelm,[32] to whom fell the arduous and exacting task of performing the calculations. Lagerhjelm had worked in Berzelius' laboratory on analyses of bismuth compounds, including the oxide and nitrate,[33] and was qualified for his present task by both an extensive mathematical knowledge and a tireless capacity for hard work. Born on 13 February 1787, he had first studied law at Uppsala before joining Berzelius in 1809. He became an authority on hydraulics, and from 1817 to 1831 was assessor at the Bergkollegium (College of Mines). He died on 18 July 1856.

It remains to point out the main connections between this *Essai* and later works by Berzelius. It was translated into German in the following year and also incorporated in Wöhler's translation of the *Lärbok*, appearing as *Lehrbuch der Chemie*, Vol. iii, Part 1 (Dresden, 1827). A French edition of the whole textbook appeared shortly afterward, and Vol. iv of this *Traité de Chimie* (Paris, 1831) contained a revised version of the *Essai*, the table appearing in the fifth volume (Paris, 1831) together with Berzelius' revised account of organic constitution. These translations, by Me. Esslinger, were based upon manuscript material by the author and the last German edition. A comparison between the texts of the *Essai* and *Traité* shows large areas of complete identity, and occasional variant readings which offer an interesting commentary on the development of Berzelius' ideas over the intervening years.

[32] (a) T. Dahl (ed.), *Svenska Män och Kvinnor: Biografisk Uppslagsbok* (Stockholm, 1948), vol. iv, p. 425 (portrait).

(b) J. C. Poggendorf, *Biographisch-literarisches Handwörterbuch zur Geschichte der exacten Naturwissenschaft* (Leipzig, 1858), vol. i, p. 1342.

[33] *K. Vet. Akad. Handl.*, 1813, p. 224.

Finally, a second French edition of the *Essai* was brought out in 1835: *Théorie des Proportions Chimiques et Table synoptique de poids atomiques* (Paris). There is no evidence to suggest that this had anything like the epochal significance of the first edition, and by now the main innovatory features of the Berzelian system were widely known.

5. CIRCUMSTANCES OF THE 1819 PUBLICATION

Berzelius completed his *Essai* in 1819 at the end of a visit to Paris. For about ten years he had worked intensively at the study of chemical proportions, and during that period he had issued the first two volumes of his *Lärbok* (1808 and 1812). Publication of further volumes, and revision of the existing ones, had been precluded by the great uncertainty in chemistry at that time, due in considerable measure to the discovery by Davy of the alkali and alkaline earth metals. With Berzelius' own work incomplete he had felt it necessary to withhold further issues of the book until about 1817, when the new pattern of chemistry began to emerge. In that year he issued a second edition of Volume i, and in 1818 published Volume iii, with its revolutionary account of chemical proportions and the electrochemical theory. But the efforts of the past decade had begun to take their toll upon his health, and by the spring of 1818 he was decidedly unwell. A chronic victim of migraine since a young man, he now experienced an unwonted lassitude and depression, showed other disturbing symptoms, and eventually underwent a worsening of his condition upon accidental inhalation of hydrogen selenide.[34]

A most fortunate circumstance intervened at this moment. Count Löwenhjelm had recently been appointed Swedish Minister in Paris, and Berzelius was invited to accompany him, at the Count's own expense, to the French capital. Indeed on his arrival at Paris, Berzelius discovered that he was to stay as the guest of Löwenhjelm. Moreover the king had added his efforts to facilitate Berzelius' convalescence by a grant of 2000 riksalders, ostensibly to enable him

[34] See footnote 20, ref. (c), pp. 95–97, 181.

"to obtain information concerning the newest foreign contributions to improvements in the manufacture of powder."[35] Consequently he was at least free from financial anxieties during his stay abroad.

The party left Sweden in July 1818 and traveled via London. Here Berzelius renewed acquaintance with Banks, Herschel, Roget, Wollaston, and Arfwedson, and was introduced to William Prout. After arrival at Paris on August 24 Berzelius was received "with extraordinarily great friendship" by Berthollet who introduced him to many of his colleagues including Ampère, Argo, Biot, Brongniart, Chaptal, Chevreul, Cuvier, Dulong, Gay-Lussac, Humboldt, Laplace, Laugier, Thenard, and Vauquelin. He attended lectures by some of these and was greatly impressed by their excellence. He also spent much time with the Abbé Hauy ("who declared himself in entire agreement with me in the principle of classifying minerals") and at Arceuil where he worked with Berthollet, Dulong, and others.[36]

Despite all this activity, the most lasting memorials to Berzelius' stay in Paris were the French translations of his books on mineralogy and on chemical proportions. Although the former (see footnote 31) had received an adequate translation into German by Gehlen,[37] the English version[38] had been disastrous. The deficiencies of the translator did not wholly justify the bad press accorded to the book in Britain ("a chaos of inverted reasoning and gratuitous assumption, such a labyrinth of analogies and generalisations"[39]) but they were bad enough. Consequently a translation of *this* version into French[40] could hardly hope to succeed, and it did not. Appalled by both the translation and its reception, Berzelius resolved to issue a new edition while he was in Paris, and this he did with the assistance of Brochant de Villiers as translator.[41] The revised edition contained four pages

[35] See footnote 20, ref. (c), p. 98–99.

[36] *Ibid.*, p. 100 *et seq.*
See also M. P. Crosland, *The Society of Arceuil* (London, 1967), pp. 411–4.

[37] *J. Chem. Phys.* (Schweigger), *12* (1814), 17.

[38] *An attempt to establish a pure scientific System of Mineralogy by the application of the Electro-chemical Theory and the Chemical Proportions*, trans. by John Black (London, 1814).

[39] *Quart. J. Sci.*, *1* (1816), 226 (and much else in this vein).

[40] *J. de Phys.*, *1* (1818), 86.

[41] *Nouveau Système de minéralogie, traduit de Suédois sous les yeux de l'auteur, et publié par lui-même* (Paris, 1819).

of corrections and a reference to the unfortunate Black as "someone without any idea of mineralogy."

The mineralogy treatise does not directly concern us here, though it laid the first chemical basis for mineralogical classification. But its history is interwoven with that of the *Essai*, and is a good illustration of the difficulties of communication to which Berzelius was subject. Why he undertook his second literary task is stated thus in his own words:

> The law of chemical proportions had at this time, as yet, won so little general recognition in Paris that in the lectures of the professors it was barely mentioned, much less expounded. The electrochemical theory had indeed not yet reached as far as Paris. This led me, while in Paris, to publish my work on this subject.[42]

The translations of both works are described as having been performed "under the eyes of the author"; Berzelius had obviously learned by bitter experience! In this case the translator was a colorful person whom history seems to have neglected: Constantin Mouradgea d'Ohsson, at that time Swedish Minister at the Hague. Of Armenian ancestry, d'Ohsson was born at Constantinople on 26 November 1779. Like his father, he became a distinguished orientalist and later produced an authoritative *Histoire des Mongols*.[43] Entering the Diplomatic Corps after a period at Uppsala, he became attaché in Paris in 1801, and held appointments at Berlin (1807), Madrid (1808), and Paris once more (1810). He saw war service in Norway, and, after adopting the Lutheran faith, in 1816 received his appointment at the Hague. Later he was created a baron (1828) and eventually became Swedish Minister in Berlin (1834). He died on 26 December 1851.[44]

Like several of his compatriots who distinguished themselves

[42] See footnote 20, ref. (c), p. 101.

[43] *Histoire des Mongols, depuis Tchinguiz-Khan jusqu' à Timeer-Lanc* (Paris, 1824). He also wrote *Des Peuples du Caucase et des Pays au Nord de la Mer Noire et de la Mer Caspienne dans le 10e Siècle, ou Voyage d'Abou-el-Cassim* (Paris, 1827)—a work of historical fiction, and other works.

[44] (a) J. C. Poggendorf, *Biographisch-literarisches Handwörterbuch zur Geschichte der exacten Naturwissenschaft* (Leipzig, 1863), vol. ii, p. 318.

(b) Nils Bohman (ed.), *Svenska Män och Kvinnor: Biografisk Uppslagsbok* (Stockholm, 1944), vol. ii, p. 269 (portrait).

(c) J. F. Michaud, *Biographie Universelle,* 2nd ed. (Paris, n.d.), vol. xxix, p. 473.

in nonscientific fields, d'Ohsson acquired a high regard for Berzelius and relished the opportunity of working with him. In 1817 he spent some time "doing serious work"[45] with Berzelius in Stockholm on the hypophosphorous acid discovered by Dulong.[46] Then, hearing of Berzelius' arrival in Paris, he hurried to meet him and was soon caught up in the whirl of scientific activity there. With Arwedson he accompanied Berzelius to the mineralogical sessions and proclaimed himself willing to translate the *Essai*.[47] A little later we hear how this incorrigible experimenter still "sometimes put on the apothecary's apron" when diplomatic duties permitted, and in 1821 had been back in Stockholm working on mineral analysis.[48] During the next two winters he sometimes joined the Crown Prince of Sweden in his twice-weekly lectures from Berzelius.[49] In 1828 at the Hague he entertained Berzelius, Mitscherlich, and H. Rose on their way to a scientific meeting in Berlin,[50] and seven years later was assisting in the despatch of Berzelius' *Traité*.[51] But it is doubtful if any of his numerous services can compare with that rendered in his translation of the *Essai sur la Théorie des Proportions Chimiques*.

Throughout the winter of 1818–1819 preparation of the two new books was actively in hand. Some delays were experienced and Berzelius stayed on to see them published. On 1 June 1819, he wrote to Marcet to tell him that both had been printed and were to appear "the day after tomorrow." One task, however, remained undone. In a letter to Berthollet, enclosing a copy of each book, he wrote:

> I could not resist the desire to profit from the occasion by publicly showing how I admire your talents and the happy outcome of your work. I beg you to forgive me that, without informing you beforehand, I have set aside a page in one of these brochures for the expression of the sentiments for you which I share with all those who are able to appreciate your true talent. I had always been afraid that if I had informed you in advance, your modesty—the inseparable companion

[45] See footnote 20, ref. (b), p. 67.

[46] *Ann. Chim.* 2 (1816), 141.

[47] Letter to Gahn, *Jac. Berzelius Bref.*, ed. H. G. Söderbaum (Uppsala, 1922), vol. iv, p. 182.

[48] Ref. cited in footnote 45, pp. 67–68.

[49] See footnote 20, ref. (c), p. 109.

[50] *Ibid.*, p. 130.

[51] *Jac. Berzelius Bref.*, ed. H. G. Söderbaum (Uppsala, 1916), vol. ii, p. 18 (Letter from Mulder, 20th April 1835).

of true virtue—would have forbidden me to profit by an occasion that has given me much pleasure and satisfaction.[52]

A cordial letter from Berthollet, dated 20 December 1819, reciprocates compliments, inviting him to return to Arceuil.[53]

This task accomplished, Berzelius took leave of Paris in June 1819. *En route* for home he spent a week at Dresden where Carl August Blöde, a privy finance-counselor, had been translating the *Essai* and its Swedish original into German. Having been dissatisfied with Blöde's efforts in turning the first volume of his *Lärbok* into that language, he now went through all that had been translated so far of the *Essai* and assured himself of Blöde's competence to complete it satisfactorily.[54]

Despite all his activity his health was now greatly improved, only the periodic headaches remaining. During his absence he had been elected Secretary to the Swedish Academy of Science, thus being assured an adequate income for the rest of his working days. His stay in Paris had been rewarding in many ways professionally, and he now had many new friends in the chemical world. Small wonder, then, that he wrote "With my heart filled with thankfulness to the highest Providence, I felt that I now was one of the most fortunate of men."[55]

Certainly his trip had been a turning-point in his personal fortunes. We shall now examine the suggestion that it was similarly pivotal for the fortunes of chemistry as a whole.

6. THE RECEPTION OF THE *ESSAI*

In assessing the influence of this book, it is necessary to recall that Berzelius had other means of communication, and that a general reception of his ideas is not necessary proof that the *Essai* was responsible. It will be convenient to consider the response of several countries in turn.

[52] *Ibid.*, i (1912), pp. 69–70.
[53] *Ibid.*, p. 70.
[54] See footnote 20, ref. (c), p. 106.
[55] *Ibid.*, p. 108.

(i) France

A measure of the importance attached to this work by Cuvier is the emphasis that he placed upon it in his Report to the French Académie in 1819. Of its 23 pages, 19 are devoted to a detailed analysis of the *Essai*, and several more to the treatise on mineralogy. Their importance lay in the new directions which they had given to chemistry. Work by Berthollet, Gay-Lussac, Lavoisier, and Erman was also discussed, and to all of these Berzelius was particularly in debt.[56]

When Berzelius came to present his *Essai* before the Académie, he was complimented by the President, Vauquelin:

> We consider it as only a proper reparation to say to you, that if your and Herr Hisinger's work on the chemical effects of the electric pile had been known to us when Davy received the great prize from us, we should have divided it between you and him.[57]

On specific matters, it is noticeable that the reform of nomenclature was delayed for some time in France. Possibly reforms initiated by Frenchmen would have been more welcome.[58] Berzelius was aware of this reluctance several years later.[59] Nevertheless, by 1824 his friend and admirer Thenard had introduced Berzelian symbols into his textbook *Traité de Chimie*.[60]

The theoretical aspects of the *Essai* were assimilated more readily. Berzelius himself wrote:

> This [enterprise] turned out however to be . . . remunerative because this publication, as I believe, was well received by French scholars. De Laplace, in the first words he spoke with me said: "You have attempted to explain chemical relations by electricity, but you cannot explain cohesion with it and accordingly you will see that this explanation accomplishes nothing." During a promenade in Berthollet's garden shortly before my departure he said concerning the translation made by d'Ohsson: "Votre mémoire, Monsieur, m'a presque fait prosélyte." [61]

[56] *Mém. de l'Acad.*, *4* (1824); report for 1819, pp. lxxxvix–ciii).
[57] See footnote 20, ref. (c), p. 40, note.
[58] M. P. Crosland (cited in footnote 25), p. 221.
[59] Letter to Jourdain in 1828, *Jac. Berzelius Bref.*, ed. H. G. Söderbaum (Upsala, 1920), vol. iii, part 2, p. 177.
[60] *Traité de Chimie*, 4th ed. (Paris, 1824), vol. i, p. 34.
[61] See footnote 20, ref. (c), p. 182.

The conversion of Laplace on the matter of cohesion is particularly interesting because of his well-formulated views that this phenomenon, like chemical binding generally, was gravitational in nature.[62]

It has been observed that Berzelius' dualistic theory became the only one to receive the sanction of the French Académie.[63] Skeptics remained, of course, and J. F. Persoz in a large textbook[64] rejected it, particularly with reference to its explanation of heat in chemical reactions. This may well have influenced Gerhardt,[65] who was to become an implacable opponent of dualism in organic chemistry.

(ii) Britain

It is difficult to avoid the impression that at this time British chemistry was exhibiting one of its worst manifestations of insularity. The rift between Davy and Berzelius was accompanied by a general antipathy to the Swedish chemist which certain Englishmen seemed to regard as their patriotic duty to maintain. To be more precise, one observes the attacks on Berzelius chiefly in those areas in which he opposed or surpassed Davy. Thus Children thought it worth his while to translate the treatise on the blowpipe, but also to omit the perplexing symbols[66]—which earned the congratulations of a reviewer,[67] who denounced them as "abominable" and went on to extend his disapproval to the *Essai*. He wrote:

> May we beg to dissent *in toto* from the assertion that these or any other signs are at all necessary. . . . We might almost fancy we had taken up a volume of Bonnycastle's Algebra.[68]

Perhaps the choicest example of this kind of polemics may be found in the following passage, which, if intended to be taken seriously, displays a parochialism of outlook that would be hard to equal:

> While we entertain the highest respect towards the Swedish chemist for his unwearied and valuable researches, we cannot help thinking

[62] *Exposition du Système du Monde* (Paris, 1796), vol. ii, p. 27.

[63] C. de Milt, *Chymia*, *4* (1953), 94.

[64] *Introduction à l'Étude de la Chimie Moléculaire* (Paris, Strasbourg, 1839), p. 805.

[65] See footnote 20, ref. (d), p. 429.

[66] *The Use of the Blowpipe in Chemistry and Mineralogy*, trans. by J. G. Children (London, 1822), p. viii.

[67] *Quart. J. Sci.*, *13* (1822), 319.

[68] *Ibid.*, p. 322.

> his manner of viewing chemical proportions to be unaccountably perplexed and defective. We can more easily understand why his partial mode of notation should be preferred by many writers in France to the universal system, as taught by Mr. Dalton and so clearly exhibited on Dr. Wollaston's scale, for the same reason that vaccination was so ill received in that country,—because it was discovered by an Englishman.[69]

In Scotland, there seems to have been less hostility to Berzelius. Perhaps the influence of Thomson may be discerned in this. The sixth edition of his *System of Chemistry* gives the electrochemical system favorable notice, though calls for a much closer examination before it can be accepted. He remarks that it "has been embraced, either entirely or with some modifications, by several of the most eminent chemists of the present day."[70] The ever-vigilant *Quarterly Journal of Science* could not let this pass, and declaimed in characteristic vein:

> We should like to know, what eminent chemists of the present day have embraced these shapeless chimeras of Berzelius. Wherever his notions of chemical affinity are intelligible, they are borrowed from the Bakerian Lecture.[71]

The luckless Thomson was also assailed for a tactless but fairly accurate remark that "Sir Humphry Davy took up the subject [electrochemistry] where Berzelius and Hisinger laid it down."[72] For a truer perspective the reviewer (anonymous as ever) suggested:

> Hisinger and Berzelius may be ranked among these scattered stars which preceded the day, but had no influence on the rising of our "philosophic sun," in whose beams they were absorbed.[73]

In 1821, *A Dictionary of Chemistry on the Basis of Mr. Nicholson's*[74] was produced by Andrew Ure, professor at Anderson's Institution, Glasgow, from 1804 to 1830. This used the symbolism advocated in the *Essai*. Aspects of Berzelius' theory appeared also in an *Introduction to the Study of the Laws of Chemical Combination and the Atomic Theory*, by Edward Turner, at that time a lecturer in chemistry at

[69] *Ibid.*, *12* (1822), p. 322.
[70] *A System of Chemistry*, 6th ed. (London, 1820), vol. i, p. 161.
[71] *Quart. J. Sci.*, *11* (1821), 135.
[72] Ref. cited in footnote 70, p. 171.
[73] *Quart. J. Sci. 11* (1821), 136–37.
[74] Ure, *A Dictionary of Chemistry* (London, 1821).

Edinburgh.[75] How far Berzelius was aware of the services of these men may be open to question, but he handled some of their ideas very roughly.[76]

These considerations lead us to the conclusion that, apart from a few authors, mainly in Scotland, most British chemists took little note of the ideas embodied in the *Essai*. Fortunately, this general impression can be made more definite by further inquiry, and we find evidence that its impact upon Britain was very small indeed. Firstly, its appearance was treated with a studied neglect in the scientific press; very few reviews (good or bad) may be found. Secondly, it has not been possible to locate *one* copy (or a record of one) in any university library that was in existence in 1819.[77] Thirdly, the same pattern appears in the learned societies of Britain,[78] with the exception mentioned below.

A copy of the book at the Royal Institution bears Faraday's bookplate.[79] This is interesting in view of Faraday's own attitude to Berzelius' electrochemical theory. In 1833 he spoke approvingly of the view that electrical neutralization was responsible for the evolution of heat and light in chemical reactions;[80] shortly afterwards, however, he categorized certain of these speculations as "doubtful knowledge."[81] But these reactions came long after the

[75] Turner, *Introduction to the Study of the Laws of Chemical Combination* (Edinburgh, 1825).

[76] For Berzelius' own assessment of British chemistry at the time see footnote 20, ref. (b), p. 99.

[77] This information is based upon replies to my enquiries addressed to the present librarians, to whom I wish to express my sincere thanks. No copies were reported in the universities of Oxford, Cambridge, Glasgow, Edinburgh, Aberdeen, or St. Andrews, or in any of the Oxford or Cambridge College libraries, or in those of the Royal Army Medical College, London Hospital or St. Bartholomew's Hospital. At Aberdeen, however, there was a copy of the 2nd ed. (1835), as also formerly at St. John's College, Cambridge (bequest, 1899).

[78] I am grateful also to the Librarians of the following institutions, all of whom have reported the absence of the *Essai* from their collections: the Royal Society, the Royal Society of Edinburgh, The Royal College of Physicians, the Royal College of Physicians (Edinburgh), the Royal Society of Arts, Chetham's Library (Manchester), Dr. Shepherd's Library (Preston), etc.

[79] Information kindly supplied by the then Librarian of the Royal Institution, Mr. O. G. W. Stallybrass. This copy probably came to the Institution in Faraday's bequest (1867), but there is a record of an earlier copy, now lost.

[80] *Experimental Researches in Electricity* (London, 1839), Art. 870 (31 Dec. 1833).

[81] *Ibid.*, Art. 959 (31 March 1834).

Essai had been published, and by the 1830's Berzelius' reputation in Britain had recovered from the unfriendliness of the previous decade. Just as his new (electrochemical) classification of minerals became in 1833 the basis for cataloguing the collection at the British Museum and won for him the Royal Society's Copley medal for 1836,[82] so his work on chemical proportions was rewarded in 1834 by the Fullerian Medal of the Royal Institution, in whose journal he had been so mercilessly pilloried a dozen years before.[83] Although aspects of his electrochemical theory were still criticized (as by Joule, who opposed his views on the part played by heat and light in combustion),[84] his studies of chemical proportions were at last seen in retrospect as of great value by the fellow-countrymen of John Dalton.

(iii) Germany

Of the reception in Germany of Berzelius' ideas Whewell made the comment:

> Berzelius . . . adopted at once the view of the identity, or at least universal connexion, of electrical relations with chemical affinity. He considered that in all chemical combinations the elements may be considered as electro-positive and electro-negative, and made this opposition the basis of his chemical doctrines; in which he was followed by a large body of the chemists of Germany.[85]

Several factors were responsible for this situation, but chief among them were undoubtedly the good offices of five men: Mitscherlich, Heinrich Rose, Gustav Rose, Magnus, and Wöhler. These five Germans had all worked with Berzelius in Stockholm before returning to their homeland. On this Söderbaum has commented:

> As a matter of fact, of all Berzelius' foreign pupils, it was Wöhler who came in person nearest to himself. Also, before long the acquaintanceship with Wöhler turned into a new and powerful bond of fellow-feeling between himself and the chemists working in Germany. However, it was, in fact, only the younger chemical school who enjoyed this sympathy; the representatives of the older traditions

[82] See footnote 20, ref. (b), p. 81.
[83] See footnote 20, ref. (c), p. 91 note.
[84] *Phil. Mag.*, *18* (1841), 308.
[85] W. Whewell, *History of the Inductive Sciences* (London, 1847), vol. iii, pp. 194–95.

were on the contrary every now and then subjected to his critical disapproval.[86]

Long before these emissaries arrived in Germany, a review of part of the *Essai* appeared in *Schweigger's Journal*. After a discussion by Meinecke of the new nomenclature, a transcript of the table appeared that was probably the first German table of atomic weights.[87] A few months later, Blöde's translation of the *Essai* appeared in Dresden, and successive issues of the *Lehrbuch* included substantial portions of it.

Some of the proposals met with a stiff resistance; those relating to symbolism were not generally adopted until the 1830's, for example.[88] But the fact of greatest significance is that Germany, more than any other country, became the home for Berzelius' underlying philosophy of dualism and the cradle of the theory of radicals. As Kekulé pointed out,[89] German chemical tradition favored the concept of radicals—long after French chemistry had come under the sway of the unitary hypothesis, one may add. From the *Essai* (or rather the *Versuch*) some of this influence must have come in the beginning. The achievements of Liebig, Wöhler, Bunsen, and Kolbe stand directly in this tradition. Later on, Blomstrand wrote his *Die Chemie der Jetztzeit vom Standpunkte der elektrochemischen Auffassung, aus Berzelius Lehre entwickelt*[90] from precisely the standpoint designated in the title. Even in the twentieth century, Stark, who came very near to the concept of covalent electron-sharing,[91] confessed his debt to Berzelius.[92] And it has been suggested[93] that Arthur Schopenhauer (1788–1860), the "philosopher of pessimism," may have derived his "two absolute primordial substances" from Berzelius. So deeply has Berzelian dualism been in *rapport* with the ethos and outlook of nineteenth-century Germany.

[86] See footnote 20, ref. (a), vol. ii, pp. 312–13.

[87] *Schweigger's J.*, *27* (1819), 113; table, pp. 122–84.

[88] M. P. Crosland (cited in footnote 25), p. 277.

[89] *Lehrbuch der organischen Chemie* (Erlangen, 1861), vol. i, p. 64.

[90] Heidelberg, 1869.

[91] *Phys. Zeitsch.*, *9* (1908), 85.

[92] G. V. Bykov, "Historical sketch of the electronic theories of organic chemistry," *Chymia*, *10* (1965), 199.

[93] E. Farber, "The theory of the elements—nucleosynthesis in the nineteenth century," *Chymia*, *9* (1964), 185.

(iv) Russia

Recent work has shown that the ideas of Berzelius were readily received in Russia.[94] In 1820 he was elected an honorary member of the St. Petersburg Academy,[95] and four years later the first Russian article appeared in which Berzelian symbolism was used.[96] A Russian textbook[97] appeared in 1827 which acknowledged a debt to Berzelius and other European authors and reprinted 120 pages of "Tables of Berzelius" taken from the *Essai.*[98]

N.B.: the 'Commentary and Notes' section of this introduction (xxviii–xlix) has been omitted for the purposes of this volume.

[94] Y. I. Solov'ev, "New materials for the scientific biography of J. J. Berzelius —the scientific relations of Berzelius with Russian scholars from unpublished letters," *Chymia*, 7 (1961), 109.

[95] *Ibid.*, p. 110.

[96] H. M. Leicester, *Chymia*, 5 (1959), 144.

[97] A. A. Iovskiĭ, *Chemical Equations*.

[98] Ref. cited in footnote 94, pp. 110–11.

IX

Chemistry on the edge of Europe: growth and decline in Sweden

The development of a Swedish chemical tradition

One of the most northerly countries in Europe, and once by far the largest with territory round much of the Baltic Sea, Sweden has experienced a turbulent political history that profoundly affected its contribution to science. By 1700 it included Lapland, Finland and the East Baltic countries of Ingria, Estonia and Livonia. After a disastrous War of the North, in 1721 the East Baltic states were ceded to Russia. It was the end of Sweden's role as a great European power. Strangely this turned out to be beneficial to science. In the nineteenth century Sweden soon lost Finland to Russia (1809), though in 1814–5 it was allowed to annex Norway by the Treaty of Vienna; they remained as two countries under one monarch until 1905.

The background of Swedish chemistry is as rich and colourful as its politics.[1] Swedish science has had a long history[2] but in the eighteenth century it was experiencing an unprecedented revival.[3] There are well-understood cultural reasons for this, not least the remarkable progress in two famous Swedish institutions. One was the University of Uppsala,[4] established in 1477, but not emerging until the eighteenth century from a long period of scholastic philosophy and ways of thinking. From then on medicine, astronomy and physics flourished mightily within its walls. The other was the Royal Swedish Academy of Sciences,[5] founded in 1739 and from the very beginning dedicated to progress in both pure and applied sciences. This reflected a new spirit of utilitarian advance,

[1] H. Oslo, *Kemiens historia i Sverige intill År 1800* [*History of Chemistry in Sweden to 1800*] (Uppsala: Almqvist & Wiksell, 1971).

[2] S. Lindroth, ed., *Swedish men of science 1650–1950* (Stockholm: Almqvist & Wiksell, 1952).

[3] C. A. Russell, Science on the fringe of Europe: Eighteenth-century Sweden, in D. C. Goodman and C. A. Russell, eds., *The Rise of Scientific Europe 1500–1800* (Sevenoaks: Hodder & Stoughton, 1991), pp. 305–32.

[4] S. Lindroth, *A History of Uppsala University, 1477–1977* (Stockholm: Almqvist & Wiksell, 1976).

[5] T. Frängsmyr, ed., *Science in Sweden: The Royal Swedish Academy of Sciences 1739–1989* (Canton MA: Science History Publications, USA, 1989).

itself stimulated by the new democratic Age of Freedom (from 1718) where mercantile, rather than military, expansion was the order of the day.

Chemistry was not the only beneficiary, of course. This was the age of Celsius and Linnaeus, to name but two of the giants of Swedish science in this period. Yet in the few attempts to offer a comprehensive account of science in Sweden, chemistry is rightly hailed as one of the most prominent examples of spectacular growth at this time. It is curious that Swedish chemistry as a whole has yet to receive a definitive history.

By the end of the century Sweden was at the very front of chemical research and discovery. The new chemistry of Lavoisier was accepted, though not uncritically, by A. G. Ekeberg (1767–1813), and particularly guided his reform of chemical nomenclature and use of chemical instrumentation.[6] In the chemistry of ores and minerals, in developing analytical techniques and in the work on chemical affinity by T. O. Bergman, the national output was unsurpassed for its time. An obscure Stockholm pharmacist named Carl Wilhelm Scheele, who later moved to Uppsala, brought to a triumphant conclusion a century or more of European research on pneumatic chemistry. He discovered (and breathed) chlorine, silicon tetrafluoride, arsine, nitrogen dioxide and hydrogen sulphide. Not surprisingly he died at an early age (43). Above all he is remembered as discovering oxygen, a year before its isolation by Priestley. Indeed it is the discovery of elements that so distinguishes the chemistry of Sweden for over a century. The dates in Table 1 are often quoted for discoveries of elements in that country. A few may be argued about (e.g. Scheele and oxygen, and the precise identity and purity of some of the rarest earths) but Table 1 gives a good general impression of the enormous contributions of Sweden to the discovery of elements.

It is possible to identify three distinctive characteristics of Swedish chemistry at the period of its greatest achievements. They may be put as follows:

Swedish chemistry stemmed largely from mineralogy

Apart from alchemy, the known ancestors of modern chemical science are medicine and mineralogy. The former included pharmacy and the latter included metallurgy. Whereas the University of Uppsala had a long tradition in medicine it is remarkable that most Swedish chemistry originated elsewhere (apart, that is, from the exceptional work of Scheele). The reasons are not hard to find: in Sweden's rich mineral deposits. Geologically complex and geographically isolated, Sweden had long depended for economic growth on the rich stores of minerals within

[6] A. Lundgren, The chemical revolution from a distance: Anders Gustaf Ekeberg, the antiphlogistic chemistry, and the Swedish scene, in B. Bensaude-Vincent and F. Abbri, eds., *Lavoisier in European context: negotiating a new language for chemistry* (Canton MA: Science History Publications, USA, 1995) pp. 19–41.

Table 1. *Swedish discoveries of the elements.*

Element	Discoverer	Date
cobalt	Brandt	1730
nickel	Cronstedt	1751
oxygen	Scheele	1771
chlorine	Scheele	1773
manganese	Gahn	1774
molybdenum	Hjelm	1781
tantalum	Ekeberg	1797
selenium	Berzelius	1817
lithium	Arfwedson	1818
zirconium	Berzelius	1825
thorium	Berzelius	1828
vanadium	Sefström	1831
cerium	Mosander	1839
lanthanum	Mosander	1839
didymium	Mosander	1841
erbium	Mosander	1843
terbium	Mosander	1843
scandium	Nilson	1879
holmium	Cleve	1879
thulium	Cleve	1879

the earth. A detailed knowledge of minerals was promoted by the new political atmosphere of utilitarianism. There was much more to be done than traditional iron-making or copper-smelting, as a whole range of metals were obtained from native minerals, not least from the great mines at Dannemora. Precious metals used for coinage had to be prepared and then assayed, and a number of men got valuable chemical experience at the Mint in Stockholm. In the late seventeenth century the Board of Mines established a *laboratorium chemicum*, though its most useful work was done a few decades later.[7] Such experience was not unique to Sweden but the ethos and some of the techniques were rarely found elsewhere. It was utilitarian, practical and empirical and did not encourage the spirit of theoretical enquiry until the time of Berzelius.[8]

One example of this work was the isolation of what was thought to be a new earth (since shown to be a mixture of several rare earths). Named yttria it came from a mineral brought from Ytterby near Stockholm. It was isolated by J. Gadolin (1760–1852). He became professor at his native Åbo, in Finland, the then home of the University of Finland, founded in 1640 and relocated in 1828 in

[7] S. Lindqvist, *Technology on Trial: The Introduction of Steam Power into Sweden, 1715–1736* (Uppsala: Almqvist & Wiksell, 1984) pp. 97–9.

[8] A. Lundgren, The new chemistry in Sweden: the debate that wasn't, *Osiris* **4** (1988) 146–68.

Helsinki. He is one of several Finnish men of science who did memorable work in chemistry while Finland was still part of Sweden.

Much of the chemical work was quantitative

Quality control on either minerals or precious metals must involve quantitative analysis.[9] In the eighteenth century the 'quantifying spirit' was more conspicuous than before, showing itself in such diverse fields as statistics, engineering, cosmology and the chemistry of Lavoisier.[10] The use of the balance was developed with more rigour, and in Sweden several important developments were initiated. Bergman, for example, stressed the need for pure reagents and for ensuring that products estimated (by weighing, for instance) were well-defined compounds.

Much experimental work centred on the use of the blowpipe

This technique dates back at least to the seventeenth century. It was described in detail by Johann Kunckel (seventeenth century), was used by Swab in 1738 and by Rinman in 1746 for the examination of antimony and of the ores of tin, respectively. Cronstedt showed how minerals could be fused in its flame if mixed with substances like soda or borax, and this greatly extended its usefulness. So did the technique of placing the substances to be examined on a charcoal block, introduced by Engeström. Still further developments were introduced by Gahn. In 1779 Bergman published a comprehensive account of the technique of blowpipe analysis. It was later – if rather anachronistically – called 'the chemist's stethoscope.' It was largely through this equipment that many of the new elements were discovered in Sweden in the eighteenth and early nineteenth centuries.

Thus chemistry in Sweden depended greatly on the rich mineral resources of the country. In the late eighteenth century it was also favoured by a combination of other circumstances, social, political and even religious. Uppsala, in particular, became a great international centre attracting scholars from all over the continent. Yet in the early nineteenth century the physical sciences in Sweden went into a steep decline. There were many reasons. One was the isolation of Sweden during the Napoleonic wars (the so-called 'Continental system'); it was especially hard on the chemists who (for example) found the utmost difficulty in ascertaining details of the revolutionary ideas of Dalton's atomic theory. By the same token

[9] A. Lundgren, The changing role of numbers in 18th-century chemistry, in T. Frängsmyr, J. L. Heilbron and R. E. Rider, eds., *The Quantifying Spirit in the 18th Century* (Berkeley: University of California Press, 1990) pp. 245–66.

[10] J. L. Heilbron, Introductory essay, in Frängsmyr *et al.*, *op cit* (9), pp. 1–23.

English chemists such as Davy were hindered in their efforts to obtain scientific literature originating in Sweden. A further consideration was the strong influence of the Romantic movement, as represented by the German *Naturphilosophie*. Only botany was to benefit from such ideas, and the tradition of Linnaeus continued to thrive. Chemistry, on the other hand, did not react to Romanticism as Davy had done, and Berzelius thundered his denunciations against its imprecise and woolly approach to nature. Perhaps the final luminary in Swedish chemical science before the Treaty of Vienna (1815) was A. G. Ekeberg, mentioned above, who taught chemistry and wrote poetry at Uppsala. He worked on mineral analyses, discovering tantalum in the process. And in 1800 he examined a dissertation on mineral water by a hopeful young medical student at Uppsala; the candidate's name was J. J. Berzelius and he became one of the truly great figures of chemical history. He was professor of medicine (and then chemistry) at Stockholm from 1807 to 1832, and from 1818 secretary of the Royal Swedish Academy of Sciences.

The chemistry of Berzelius

It is impossible in a few words to do justice to a man who, it has been said, bestrode like a Colossus the whole of European chemistry for the first half of the nineteenth century.[11] He has recently attracted attention as an exemplar of Enlightenment science.[12] Other studies have focused on his extensive correspondence, especially with Trolle-Wachtmeister (1782–1871),[13] and even on his travels in France.[14] We can briefly summarise his main contributions.

Classification and nomenclature

Following in the steps of Ekeberg and others, Berzelius published in 1811 'An essay on chemical nomenclature' in which he proposed an abandonment of trivial names (like 'blue vitriol') and a replacement by systematic names, in Latin, with a comprehensive classification of all materials.' Three years later he made this the basis also of 'A scientific system of mineralogy.' Berzelius the systematiser[15]

[11] J. E. Jorpes, *Jac. Berzelius, his life and work* (trans. B. Steele) (Stockholm: Almqvist & Wiksell, 1966).
[12] E. M. Melhado and T. Frängsmyr, eds., *Enlightenment Science in the Romantic Era: The Chemistry of Berzelius and its Cultural Setting* (Cambridge: Cambridge University Press, 1992).
[13] J. Trofast, *Excellensen och Berzelius* [His Excellency and Berzelius] (Stockholm: Atlantis, 1988).
[14] C. G. Bernhard, *Through France with Berzelius: Live Scholars and Dead Volcanoes* (Oxford: Pergamon Press, 1989).
[15] E. M. Melhado, *Jacob Berzelius: the Emergence of his Chemical System* (Madison: University of Wisconsin Press, 1981).

was to dominate European chemistry for the next two decades in a manner irresistibly reminiscent of Linnaeus.

Discovery of elements

Berzelius published a book on blowpipe analysis in 1820.[16] With the aid of this improved technique several new elements were discovered and much other progress was made in understanding the chemical nature of minerals, including some by Berzelius himself. He discovered selenium in the dust from a lead chamber plant (1818), zirconium by potassium reduction of potassium zirconium fluoride (1825) and thorium from the Norwegian mineral thorite (1828).

Atomic theory

Berzelius had espoused a version of Dalton's atomic theory by 1812 (despite the great difficulties in communication outside Sweden).[17] He differed from Dalton in his supposition that all atoms were spherical and of the same size, in his introduction of 'modern' symbolism for atoms, in his rules of chemical combination, and in his 'theory of volumes' (by which atomic weights were calculated from gaseous volumes before and after a reaction).[18] Having developed a new method of organic analysis by 1814 he wondered if atomic theory applied to organic compounds. Within a year or two he was convinced that it did. However, by then atomism was becoming inextricably linked in his mind to an electrochemical view of matter.

Electrochemical theory

According to this theory chemical combination was essentially electrical in nature, with elements being either positive (hydrogen and metals) or negative (the rest).[19] His great synthesis of electrochemistry and atomism came in 1819 as his *Essai sur la Théorie des Proportions Chimiques, et sur l'influence chimique de l'Électricité*.[20] It led to a new understanding of acids, bases and salts

[16] J. J. Berzelius, *Om blåsrörets användande i Kemien och Mineralogien* [*On the Use of the Blowpipe in Chemistry and Mineralogy*] (Stockholm: 1820).

[17] A. Lundgren, *Berzelius och den kemiska atomteorin* [*Berzelius and the Chemical Atomic theory*] (Uppsala: Uppsala University, 1979).

[18] C. A. Russell, Berzelius and the atomic theory, in D. S. L. Cardwell, ed., *John Dalton and the Progress of Science* (Manchester: Manchester University Press, 1968), pp. 259–73.

[19] C. A. Russell, The electrochemical theory of Berzelius – I: Origins of the theory; II: An electrochemical view of matter, *Ann. Sci* 19 (1963) 117–26; 127–45.

[20] C. A. Russell, Introduction, commentary and notes to Berzelius' *Essai sur la Théorie des Proportions Chimiques, et sur l'influence chimique de l'Électricité* (1819) (New York: Johnson Reprint Corporation, 1972).

and enabled explanations to be given of electrolysis and many other phenomena. For he postulated an electrochemical series, where more positive elements replaced those lower down the scale and in this way was able to account for displacement phenomena, relative ease of combustion and much else. He did not, however, recognise the possibility of combination between atoms of identical polarity, so would not allow Cl–Cl or H–H or any diatomic molecule of an element. Also, his atomic weights were sometimes double their recognised value today. But the biggest threat to his theory came from organic chemistry.

Theory of organic compounds

From Berzelius' electrochemical dualism came a radical view of chemistry generally. In organic compounds a negative atom would be linked to a positive radical (such as alkyl). This proved to be a fertile suggestion and led to the discovery of cacodyl and the isolation of what were thought to be alkyl radicals (actually their dimers). It also, led, ultimately, to a theory of valency.

Where electrochemical dualism led Berzelius into deep trouble was in the matter of chlorination of aliphatic acids on the α-carbon atom. The production of, for example, chloracetic acid led apparently to replacement of a hydrogen atom (positive) by a negative atom, chlorine. No amount of intellectual gymnastics could eliminate the problem and it was on this rock that the whole electrochemical scheme was in great danger of foundering.[21] The notion that hydrogen (or chlorine) could have negative *and* positive characteristics was foreign to the whole of Berzelius' thinking. He died before the question could be resolved, though his legacy lived on in Kolbe, Frankland and their intellectual successors.

The historical question has been well debated as to whether organic chemistry and inorganic chemistry were really two separate sciences, and if not whether it was better to derive laws of organic chemistry from those of inorganic chemistry, or vice versa. It may seem a fairly sterile argument, but central to it was Berzelius' insistence that organic chemistry is best comprehended by *analogy with inorganic*. Only after the failure of dualism as applied to carbon compounds did the French chemists invert the analogy and derive inorganic laws from organic!

Finally, it is to Berzelius that we are indebted for the discovery of pyruvic acid, the term 'protein' and the concepts of catalysis and isomerism.

For 25 years his ideas dominated European chemistry. They were disseminated by innumerable papers in the journals and by his books, notably the *Jahresberichte* and his *Lärbok i kemien*, the latter being available in six European

[21] J. H. Brooke, Chlorine substitution and the future of organic chemistry, *Stud. Hist. Phil. Sci.*, **4** (1973) 47–94.

languages. The process was further facilitated by research students or collaborators from overseas. They included C. G. Gmelin, E. Mitscherlich, Heinrich Rose, Gustav Rose, Gustav Magnus, C. F. Sobrero and Friedrich Wöhler. This devoted band, together with more distant admirers (like Frankland and Kolbe), ensured that, despite all their reverses, his theories have become a permanent part of chemical theory.

Swedish chemistry in the Berzelius tradition

It is a curious fact that chemistry in Sweden never again attained the height to which Berzelius had brought it. In part this must be due to the concerted opposition to his theory by chemists from other countries, most notably France. In so far as the idol could be seen to have fallen his inspirational legacy is likely to have been severely diminished. Yet the general decline of Swedish chemistry after Berzelius must also be seen as part of a wider decline of science generally within the country. There were no extensive coal deposits, and thus no large coal-tar industry, to promote the growth of synthetic organic chemistry as they did elsewhere in Europe. The early attempts at popular science education did not fulfil the hopes of the 1830s, partly because of editorial reluctance to change the format of the periodical *Läsning för Folket*, the journal of the new Swedish Society for the Diffusion of Useful Knowledge.[22] An important means of chemical education was thus lost. However, it is important not to belittle the achievements that were made and to recognise that, by any standards, Berzelius must have been a very difficult act to follow.

His research students naturally included a number from Sweden itself, several of whom did useful work within Berzelius' lifetime and beyond. They included J. A. Arfwedson (1792–1841) and N. G. Sefström (1787–1845), whose discoveries in Berzelius' laboratory respectively of lithium (1818) and vanadium (1830) have already been mentioned. Berzelius' assistant, and successor in 1832, was Carl Gustaf Mosander (1797–1858). He became custodian of the Academy's mineral collection in 1828 and there, in his own laboratory, continued the long tradition of mineral analyses. In 1842 he announced his discovery of two additional elements in 'cerium,' which he named lanthanum [= hidden] and didymium [= twin]. The next year his studies of the rare mineral discovered by Gadolin, yttria, disclosed two further elements, named by him as erbium and terbium. Subsequent work demonstrating the composite nature of his 'didymium'

[22] P. Sörbom, *Läsning för Folket: Studier tidig Svensk folkbildningshistoria* [*Reading for the People: Studies in Early Popular Education in Sweden*] (Stockholm: P. A. Norstedt, 1972).

and 'erbium' does not diminish the value of Mosander's laborious and careful investigations.

Another of Berzelius' pupils, who became his devoted disciple, was L. F. Svanberg (1805–78), a former military engineer. He worked with Struve on molybdenum. In 1853 he succeeded to Berzelius' chair at Uppsala and thereafter abandoned experimental work, concentrating his energies on raising the standards and status of chemistry. His was the impetus that led to the creation of a new Institute of Chemistry at Uppsala in the 1860s. He was the brother of A. F. Svanberg (1806–1857), professor of physics at Uppsala who studied polarisation and other aspects of electric cells. Both were sons of Jön Svanberg, astronomer to the Academy, and a predecessor to Berzelius as its secretary. Two of L. F. Svanberg's pupils went on to discover further elements. L. F. Nilson (1840–99), professor of analytical chemistry at Uppsala, showed that the ytterbium described by Marignac in fact contained another element which he named scandium (1879). Later, often in collaboration with O. Pettersson, he studied compounds of thorium, indium and platinum, and also determined the atomic weight of beryllium. Scandium was quickly recognised by P. T. Cleve as the eka-boron predicted by Mendeléeff.

Cleve himself (1840–1905) taught chemistry at the Stockholm Technical Institute, and eventually succeeded Svanberg at Uppsala. He worked with Wurtz in Paris on platinum ammines, studying their isomerism, and on his return to Uppsala he reverted to his country's propensity for studying the rare earth elements and, in 1879, discovered two new elements in Mosander's 'erbium,' which he named holmium and thulium. By now a new tool for chemical investigation had appeared in the form of spectroscopic analysis. Cleve pursued the spectroscopic study of the rare earth elements, following the work begun at Uppsala by the professor of physics, A. J. Ångström (1814–74), and continued by his successor Robert Thalén. By this means N. A. Langlet, working in Cleve's laboratory, discovered helium independently of Lockyer and Frankland. Eventually Cleve succeeded to Svanberg's chair in Uppsala.

In 1884 Cleve was examining a PhD dissertation at Uppsala by a young man named Svante Arrhenius (1859–1927). Disenchanted by the poor teaching of chemistry at Uppsala, Arrhenius had turned to physics and chose to work under E. Edlund (1819–88) at Stockholm. His thesis concerned electrolytic conduction, introducing the notion of 'activity' which amounts to the modern 'degree of ionisation.' However, his examiners were not impressed and he only just managed to pass. In the next five years he toured Europe, meeting and working with many of the leaders of the new physical chemistry. His associations with Kohlrausch, van't Hoff and Ostwald were particularly fruitful and in due course he was able to enunciate much more clearly his doctrine of electrolytic dissociation. In 1896 he measured infra-red radiation at night, concluding that for each doubling of carbon dioxide concentration in the atmosphere there would be a mean global rise in temperature of 5 °C. He was thus the first to recognise the enhanced greenhouse effect of global warming. Never as much appreciated in his own country as abroad, he managed to obtain, with some difficulty, a teaching post at the Stockholm Technical High School. Recognition came slowly, but in 1903 he was given the Nobel Prize in chemistry. He later became a successful popular writer on science, especially with his *Kemien och det moderna livet*.[23]

An important nineteenth century figure who helped to promote Berzelius' ideas on organic and inorganic chemistry was C. W. Blomstrand (1826–97).[24] After receiving a Berzelius scholarship at the Academy of Sciences at Stockholm, he became lecturer and eventually professor at the northern university of Lund. Although he worked in such diverse fields as metal ammines and mineralogy, Blomstrand was best known for his support of Berzelian dualism and its application to the new doctrines of valency. He opposed the advocates of unvarying valency (such as Kekulé) and proposed his own 'chain theory' of combining atoms, as in $Cl_2Pt(Cl\text{-}Cl\text{-}K)_2$, though he used a non-linear structure for benzene diazonium chloride that is close to that adopted today. His *Der Chemie der Jetztzeit*[25] of 1869 was a not unsuccessful attempt to prolong the influence of Berzelius into the later nineteenth century.

Towards the end of the nineteenth century Swedish chemistry was still preoccupied with the perennial themes of analysis and practical utility. The immense economic importance of the mining and metallurgical industries helped to account for this chemical conservatism. Considerable practical skills were fostered, but there was some reluctance to explore too deeply some of the new areas of theoretical chemistry.

One of the most important figures in Swedish science was, of course, Alfred Nobel (1833–96). He had little formal education, but his revolutionary studies with nitro-cellulose and nitro-glycerine transformed the European explosives industry.[26] However, most of his important chemical work was done overseas, in Britain (particularly at Ardeer in Scotland) or at his international headquarters in Germany or (later) France. His great legacy to Sweden (where he was once

[23] S. Arrhenius, *Kemien och det moderna livet* [*Chemistry in Modern Life*] (Stockholm: H. Geber, 1919).

[24] G. B. Kauffman, Christian Wilhelm Blomstrand (1826–1897): Swedish chemist and mineralogist, *Ann. Sci.* **32** (1975) 13–37.

[25] C. W. Blomstrand, *Der Chemie der Jetztzeit vom Standpunkte der electrochemischen Auffassung aus Berzelius Lehre entwickelt*, (Heidelberg: C. Winter, 1869).

[26] On Nobel see, *e.g.*, N. Halasz, *Nobel, a Biography* (London: Robert Hale, 1960); R. Trotter, ed., *The History of Nobel's Explosives Company Limited and Nobel Industries Limited, 1876–1926*, (London: ICI, 1938); and J. E. Dolan and S. S. Langer, eds., *Explosives in the Service of man: the Nobel Heritage* (Cambridge: Royal Society of Chemistry, 1997).

bankrupt) was his endowment of research institutes, while the world will always remember him for his posthumous provision of the Nobel Prizes.

Apart from studies of mineral products the chief preoccupation was agricultural chemistry.[27] The career of Nilson is instructive. By exchanging an analytical professorship at Uppsala for the chemistry chair at the Royal Academy of Agriculture in Stockholm in 1883, he reverted to one of his earliest interests: agricultural chemistry. Here, amongst much else, he studied the composition of soils and fertilisers and greatly improved the agricultural productivity of his native Gotland.

Perhaps the most significant chemical work performed in connection with plant products has been in the area of terpene chemistry. Here the outstanding figure has been A. O. Aschan (1860–1939). As professor at the University of Helsinki he properly belongs to Finland rather than Sweden, but the old ties remained and much of the Swedish tradition continued, not least in chemical work on tree products.[28] Aschan was notable not only for his researches on camphor and turpentine but also for his studies of the stereochemistry of alicyclic rings. Like many Scandinavian chemists he had studied in Germany, under Wislicenus and Baeyer.

The twentieth century

During the twentieth century, there has been little institutional support for Swedish chemistry. The scientific endeavours of Swedes have been in other directions, largely determined by the special features of their geographical situation. Institutes have been established for astrophysics and for geophysics; there is a general research station at Abisko in N. Lapland and a marine biological station at Kristineberg; other institutions exist for mathematics and for ecology and taxonomy. There has been no equivalent to the Chemical Society in Britain or the Royal Institute of Chemistry, chemical affairs still being covered by the Royal Swedish Academy of Sciences. The nearest to a chemical research association appears to have been the Swedish Institute for Metals Research, founded at Stockholm in 1919. This state of affairs of course reflects the Swedish industrial bias, with wood and mineral products to the fore. The situation remained substantially the same until after World War II. Since the 1970s, however, the mining industry has declined and the fastest growing sector of manufacturing industry is in fact chemicals and pharmaceuticals (9 per cent of total exports in 1992). They include

[27] A. Lundgren, Den Svenska kemin runt sekelskiftet [Swedish chemistry at the turn of the century], *Kemia-Kemi* **18** (1991) 1019–23.

[28] T. Enkvist, *The History of Chemistry in Finland 1828–1918* (Helsinki: Societas Scientiarum Fennica, 1972).

a range of new drugs, including the local anaesthetic xylocaine, discovered in Sweden in 1948.[29]

Nevertheless, Swedish chemistry has been enriched by progress in the two 'hybrid' subjects, physical chemistry and biochemistry, though professorships in these areas took many years to be established. Not until 1905 was Arrhenius appointed director of the Nobel Institute for Physical Chemistry (not a university). A Nobel Institute for Theoretical Physics has since been founded but not (despite much argument) one for chemistry. Biochemists had to wait until the 1930s for their first professorial chairs.

Yet despite the formidable difficulties there has been a revival of Swedish chemistry in our own century. It is well illustrated by the work of two other winners of the Nobel Prize in chemistry.

Theodor Svedberg (1884–1971) studied chemistry at Uppsala and, in 1908, submitted a doctoral thesis more readily received than that of his countryman Arrhenius. He showed how to prepare colloidal solutions of metals by using high frequency electric currents to effect dispersion. In these circumstances the moving particles were visible through an ultra-microscope. From this he was led to observe the familiar phenomenon of Brownian motion which he interpreted as a result of bombardment by moving molecules. His observations were in accord with the predictions of Einstein on the basis of kinetic theory. Even the arch-sceptic Ostwald was persuaded. Svedberg's book *The existence of molecules* (1912) brought these researches to the attention of a wider public.[30] In the same year he accepted the chair of physical chemistry at Uppsala and began to study methods of separating out colloidal particles by gravitational force. His ultra-centrifuge (1922 on) subjected them to an acceleration over a million times that of gravity. It enabled him to ascertain particle size of large molecules such as proteins.[31] By then his laboratory was a world-leader in such research; not since the days of Linnaeus had so many travelled from abroad to study at Uppsala. In 1926 he was awarded the Nobel Prize. His later work was in atomic research at the Gustaf Werner Institute of Nuclear Chemistry.

Svedberg's Swedish students included Arne Tiselius (1902–1971), whose doctoral thesis related to the moving boundary method for studying the electrophoresis of proteins. It was a natural development of his director's work. Having returned to Uppsala after a year at Princeton, he extended the study of electrophoresis to achieve separation of proteins and developed the technique of adsorp-

[29] Svenska Institute Fact Sheets, March and November 1994.

[30] A. Lundgren, The ideological use of instrumentation: The Svedberg, atoms and the structure of matter, in S. Lindqvist, ed., *Center to Periphery: Historical Aspects of 20th century Swedish Physics* (Canton MA: Science History Publications, USA, 1993), pp. 327–46.

[31] B. Elzen, The failure of a successful artefact: the Svedberg ultracentrifuge, in Lindqvist *op. cit.* (30), pp. 347–77.

tion analysis (a form of chromatography). He was appointed to a new chair in biochemistry at Uppsala, paid for by private donation. His electrophoretic and adsorption methods were worked out with extreme care and proved invaluable in the study of proteins and other large organic molecules, by indicating whether or not a sample is homogeneous and identifying specific molecular species.[32] He was awarded the Nobel Prize in 1948. A later extension of the adsorption method to the rare earth elements by Tiselius' British student L. M. Synge brought together two of the most distinctive characteristics of Swedish chemistry dating from at least the eighteenth century: a profound interest in minerals and a commitment to quantitative measurement. By applying physical methods to chemical problems Swedish scientists embraced the new physical chemistry at a time when many of their European colleagues were much less enthusiastic.

Berzelius would have been proud of them.

[32] L. E. Kay, The intellectual politics of laboratory technology: the protein network and the Tiselius apparatus, in Lindqvist, *op. cit.* (30) pp. 398–423.

X

The archives of Sir Edward Frankland: Resources, Problems and Methods

COLIN A. RUSSELL and SHIRLEY P. RUSSELL

The periodical *History of Science* opened auspiciously in 1962 with an article by L. Pearce Williams on 'The physical sciences in the first half of the nineteenth century: problems and sources'.[1] He criticized the nearly exclusive reliance on printed sources then quite common in studies of Victorian science,[2] concluded that much remained to be discovered and closed his paper with these words:

> What papers exist in private hands can only be guessed. I know of a trunk in an attic containing unpublished letters from Darwin, Huxley, Kolbe, Pasteur and a host of others. They are, unfortunately, not available to the scholar and there must be hundreds, if not thousands, of such boxes scattered through America and Europe. As they are discovered, catalogued and made available to scholars, the shape of nineteenth-century science will gradually lose its blurred outlines and the origins of modern science will become clear.[3]

The subject of this paper bears an uncanny resemblance to the contents of the trunk described by Williams, for three of the authors mentioned are indeed represented in great abundance, and the fourth (Pasteur) has one letter to his credit. Since that time further discoveries have revealed material even richer than anyone had envisaged. In terms of sheer scale the present archive is probably one of the largest relating to any scientist to have remained totally hidden until only a few years ago and now to have been systematically explored. However, the lessons that have been (sometimes painfully) learned during its discovery and exploration are applicable to many other manuscript collections in private hands that still await the attention of historians of science.

The purpose of this paper is to give a brief account of the diverse origins of this material and to indicate some of the ensuing problems and their solutions. But first the originator of the archive needs a brief introduction.

1 L. Pearce Williams, 'The physical sciences in the first half of the nineteenth century: problems and sources', *History of Science*, (1962), **1**, pp. 1–15.

2 Ibid., p. 8. One of us (C.A.R.) was not unfairly singled out for special opprobrium; his reply was subsequently published as C. A. Russell, 'The electrochemical theory of Sir Humphry Davy-III: the evidence of the Royal Institution manuscripts', *Annals of Science*, (1963), **19**, pp. 255–271.

3 L.P. Williams (op. cit.), p. 13.

I SIR EDWARD FRANKLAND (1825–1899)

Frankland[4] was born at Catterall near Garstang, Lancashire, and became a pharmacist's apprentice in Lancaster.[5] His early scientific abilities were recognized by some medical friends in the town, and at the end of his apprenticeship it was through their influence that he gained entry to the chemical community in London. Having worked with Lyon Playfair, a rising figure in metropolitan science, he proceeded to take a Ph.D. in Marburg under R. W. Bunsen. Shortly after his arrival back in England he was appointed to the first chair in chemistry at Owens College, Manchester (1851), returning in due course to London (St Bartholomew's Hospital in 1857 and the Royal Institution six years later). In 1865 he replaced A. W. Hofmann as Director of the Royal College of Chemistry and spent the rest of his professional life in that institution or its successors at South Kensington.

Frankland's scientific work placed him in the front rank of British chemists of the last century.[6] He was founder of the now immensely important science of organometallic chemistry.[7] From his research into such novelties as dialkylzinc and tetra-alkyltin compounds he came, still in his early twenties, to a recognition of what we today would call valency (valence); the very terminology of a chemical 'bond' is his, as was much of its subsequent study. In chemical education[8] he was a leading pioneer, advocating radically new methods of laboratory teaching, introducing the modern system of structural (graphic) formulae, and issuing widely used textbooks for the rising class of artisan students.[9] He did work of fundamental importance in physical chemistry (combustion and flame), organic chemistry (acetoacetic ester syntheses) and biochemistry (metabolism and respiration). Yet by his contemporaries he was known above everything else for his development of elegant methods of water analysis, thus enabling him to monitor the provision of (relatively) pure domestic water supplies to the rising population of urban Britain. For this he received a knighthood in 1897.

To the social history of British science Frankland is of extreme importance. The longest serving member of the X-Club,[10] he was engaged in dialogue with the leading

4 Numerous obituary notices appeared, particularly by H. McLeod (*Journal of the Chemical Society*, (1905), **87**, pp. 574–590) and J. Wislicenus (*Berichte*, (1900), **33**, pp. 3847–3874). Also available is the second edition of his autobiographical *Sketches from the Life of Sir Edward Frankland*, edited and completed by his daughters M.N. W[est] and S.J. C[olenso], Spottiswoode & Co., London, 1902. A complete list of obituaries, together with an account of the problems of constructing a Frankland biography, may be found in note 5 (below). The fullest recent survey of Frankland's whole life and work is by W.H. Brock, *DSB*.

5 An account of the first twenty or so formative years of Frankland's life is given in C.A. Russell, *Lancastrian Chemist: the Early Years of Sir Edward Frankland*, Open University Press, Milton Keynes, 1986. This is the first volume of a two-volume biography currently in preparation.

6 His most important scientific papers are collected in his *Experimental Researches in Pure, Applied and Physical Chemistry*, van Voorst, London, 1877.

7 C.A. Russell, 'Edward Frankland, founder of organometallic chemistry', *Chemistry in Britain*, (1982), **18**, pp. 737–738.

8 See his *How to teach Chemistry*, London, 1875.

9 These include *Lecture Notes for Chemical Students*, London, 1866 and later editions, and (with F.R. Japp) *Inorganic Chemistry*, London, 1884.

10 See for example R. Macleod, 'The X-Club, a social network of science in late Victorian England', *Notes and Records*, (1970), **24**, pp. 305–322; J.V. Jensen, 'The X-Club: fraternity of Victorian scientists', *British Journal for the History of Science*, (1970), **5**, pp. 63–72; Ruth Barton, '"An influential set of chaps": The X-Club and Royal Society Politics 1864–65', *British Journal for the History of Science*, 1990, **23**, 53–81.

statesmen of Victorian science, particularly in Britain. He became President of the Chemical Society and Foreign Secretary of the Royal Society. More significantly, perhaps, he was founding President of the world's first professional association of scientists, the Institute of Chemistry, in 1877.[11]

In view of his stature in British chemistry it is surprising that no extended biography has ever appeared (other than his own fragmentary *Sketches*). Two reasons have been * suggested.[12] One is that Frankland was illegitimate and under severe constraints not to identify his father (a distinguished lawyer). Hence enquiries on any aspects of his past were brusquely rebuffed during his lifetime and care was taken to keep all papers from prying eyes. The other reason is that, until now, manuscript evidence has simply not been available. That considerable amounts of material were rumoured to exsist merely deterred would-be biographers still further. Only when these could be examined *and made available to public inspection* could any responsible attempt be made to produce a much-needed full biography.[13] This gave added urgency to the task. Moreover, as will be shown, the material goes far beyond the needs of a biography and should serve a wide range of scholarly interests.

II LOCATING THE ARCHIVES

The present archive consists chiefly of four deposits of Frankland's papers, all in private hands. There are also about sixteen much smaller collections in universities, Record Offices etc., mainly in the U.K. Altogether the papers open to public access amount to less than six per cent of the total currently located. By his first marriage Frankland had four surviving children. After his death, and considerable legal wrangling about his papers, the collection was divided among them (the two daughters by his second marriage apparently receiving none). The recipients were as follows:

Margaret Nannie Frankland, 1853–1941

Frankland's eldest child married a Harley Street physician, Samuel West, in 1881 and, with her sister Sophie, edited their father's *Sketches*. Her portion of the family papers passed eventually to her granddaughter who is now the owner of a considerable volume of Frankland correspondence (mainly of a family nature), photographs, autograph books, portraits etc. This is the **MNW-1 Archive**.

From 1864 Margaret Frankland assiduously kept a private diary, usually producing one bound volume a year. Although not strictly part of the Edward Frankland papers these diaries are an invaluable source of information about her father and many of his scientific friends. They are now lodged with her grandson's widow, and, together with one or two letters, constitute the **MNW-2 Archive**.

11 C.A. Russell, N.G. Coley and G.K. Roberts, *Chemists by Profession*, Open University Press, Milton Keynes, 1977.

12 C.A. Russell, op. cit. (5).

13 These strictures would hardly apply to a first volume of such a biography, dealing with the early years (C. A. Russell, op. cit. (5)), as only external archival sources would be relevant; there is almost nothing extant of Frankland's own correspondence until he reached his twenties.

* **Additional Note [2009].** My own biography of Frankland appeared in 1996. Please see note 1 in the Preface.

Frederick William Frankland, 1854–1916

Edward Frankland's elder son, though destined for a scientific career, was compelled by ill-health to forsake chemistry for mathematics.[14] He emigrated to New Zealand, becoming in due course Government Actuary. He took with him a number of his father's letters. These, with his own papers and nine diaries and journals (1865–1874), are now in the Manuscripts Department of the Alexander Turnbull Library in Wellington.[15] We have not yet seen these but in due course hope to incorporate information about them in the final Archive Index. They will be referred to as the **FWF Archive**.

Sophie Jeannette Frankland, 1855–1936

Frankland's second daughter married in 1880 Frank E. Colenso, son of John Colenso, the controversial Bishop of Natal. Her share of the Frankland archives passed to her daughter Eothen, and thence to the latter's niece. After *her* death the papers were in imminent danger of destruction, but about half of them were rescued by friends who realized their importance. This **SJC Archive** includes correspondence in German between Sophie's mother and her family following her marriage to Frankland, and also letters to Sophie during the terminal illness of her mother at Davos and following her own marriage.[16]

Percy Faraday Frankland (1858–1946)

The only one of Frankland's children to achieve chemical distinction, Percy became first Professor of Chemistry at Birmingham.[17] In 1882 he married the bacteriologist Grace Coleridge Toynbee (1858–1946), their only child being the novelist and landowner Edward Percy Frankland (1884–1958). To Percy came the vast majority of his father's private papers, and these are now owned by his elder grandson. Although stored for many years in several locations in northern Britain they are now securely housed in a bank vault. Designated the **PFF Archive,** this collection comprises nearly eighty-five per cent of all known Frankland manuscripts in private ownership and an even higher proportion of those of direct scientific interest.

Most of this information was unavailable at the beginning of the search. The only fact certainly known was the identity of Percy Frankland's grandson. Approaches to him led to a series of conversations in which it emerged that deposits of papers might also be held in Margaret's branch of the family. Information about the manuscripts in New Zealand emerged after informal conversations in England with Sir Owen Woodhouse and Dr

14 See for example *Insurance Record,* 4 August, 1916; and G.H. Scholefield (ed.), *Dictionary of New Zealand Biography,* Wellington, 1940.

15 Information from Dr Garry J. Tee, Department of Mathematics and Statistics, University of Auckland, and from The Rt Hon. Sir Owen Woodhouse, Law Commission, New Zealand.

16 Some of this material has been used in a paper by F.P. Campbell, 'The meteorological researches of Sir Edward Frankland, KCB, FRS', *Weather,* (1987), **42,** pp. 383–387.

17 See W.E. Garner, *Obit. Not. Fellows Roy. Soc.,* (1947), **5,** pp. 697–715; G.L. Geison, *DSB;* C.A. Russell, *Chemistry in Britain,* (1977), **13,** pp. 425–427.

Garry Tee. The SJC Archive came to our attention after an approach by the owner on a quite different matter.

III CONTENTS AND CONDITION

It is possible here to give only a brief and necessarily selective impression of the wealth of material that has come to light. For Frankland's reluctance to throw anything away posterity has much to be grateful.

First, and most obvious, are many hundreds of *letters*, most of them to Frankland himself but many from him to members of his family. Amongst the in-letters from well-known figures in Victorian science are:

18 from the Duke of Argyll,
23 from Brodie,
11 from Bunsen,
14 from Crookes,
29 from Darwin,
14 from Duppa,
15 from Hofmann,
19 from Hooker,
48 from Huxley,
16 from Bence Jones,
67 from Kolbe,
10 from Liebig,
16 from Lister,
14 from Lubbock,
44 from Nasmyth,
30 from Playfair,
20 from Spencer,
24 from Stokes,
39 from Tyndall.

There are over ten letters each from lesser-known persons (such as H. Grubb, H. Couper Key and George With) concerning Frankland's amateur passion for astronomy. Nearly all his leading scientific contemporaries are represented by some letters as (for example) nine each from Abel, Faraday and Odling. There are over 360 family epistles. Also, several of Frankland's letters *to* other people outside the family have ended up in the collection, as two to his old friend J. C. Nicholson.

Quite apart from letters that Frankland received many of his own *copy-letters* have fortunately survived. These incude a complete copy-letter book known as the Imperial Copying Press which used an oiled sheet to produce copies of a singularly vile legibility. Its 130 pages record letters from the Owens College period between 1853 and 1856. Throughout his later life Frankland made and kept copies of important out-letters, on family crises, his frequent involvement in litigation, contested matters of scientific priority, referees' reports on the work of others, and so on. He soon abandoned the dubious advantages of the Imperial Copying Press and wrote most of these copies by hand, very occasionally employed an amenuensis and eventually succumbed to the charms of the Victorian typewriter (though long after Stokes had become a scientific pioneer in this method of communication).

It would be pleasant to report the discovery of long-lost *diaries*. There is a typescript copy of one kept by Frankland for the years 1848–49[18] which records his months at Queenwood College and his adventures in Paris at the height of the 1848 Revolution. However, that appears to be the only diary of Frankland himself to have survived and it is probable that he did not keep one thereafter. His two sons kept diaries. One by Fred from 1865 to 1874 exists in New Zealand.[19] A diary of Percy from 1868 to 1878 survives and (as has been mentioned) his sister Margaret kept a diary continuously from 1864 until the 1930s. These private documents provide rare insights into the interior worlds of Frankland's children; at moments of deep emotion (like the death of their mother Sophie, Frankland's first wife, and his subsequent courtship and marriage) Margaret relapses into German, the native language of her mother. At happier times they record family holidays, concerts, music lessons, school, the social life of mid-Victorian London, a ceaseless stream of visitors and their (not always flattering) impressions of their father's scientific friends. Alone amongst the children of that first marriage Sophie is not represented by any surviving journals.

The social history of Victorian science is of course illustrated by these diaries and by much of the correspondence. It is also served by *minute books* (as of the famous X-Club) and a whole *file of documents* on the early years of the Royal College of Chemistry. Even such *trivia* as formal invitations to dinner and plans of table seatings at the Athenaeum or elsewhere have much to say about Frankland's perceptions of his scientific colleagues. The archive is a rich source of information about institutions like the Royal School of Mines, the Royal Institution, the Institute of Chemistry and of course the Royal and Chemical Societies.

One of the most important sets of documents are the *lecture notes* of Frankland's own teaching. Courses for which complete or extensive notes survive include the following, for some of which the dates are as yet conjectural:

History of Chemistry, Lancaster, 1846
Agricultural Chemistry, Kings Somborne, 1847
Inorganic Chemistry, Putney, 1850
Technical Chemistry, Manchester, 1853–4
Organic Chemistry, probably Manchester, *c.* 1856
Inorganic Chemistry, London, probably 1860s
Metalloids, London, probably 1860s
Organic Chemistry, London, probably 1860s

When all these are properly dated, and other fragments included, it will be possible to analyse in detail not only the progress of Frankland's chemical understanding but also

18 The diary was copied in typescript at the instigation of John Tyndall's widow, Louisa, in November 1902; other copies of this document exist elsewhere e.g. at the Royal Institution.

19 Information from Dr Garry Tee.

of his pedagogic ideas. This will be enriched further by comparisons with his printed textbooks[20] and notes of his Manchester lectures taken by Frank Spence.[21] In addition there are notes of lecture courses given by others, including Playfair's lectures on chemistry in Putney (1846) and Gerling's lectures on physics in Marburg (1848).

It must be a matter of disappointment that there does not appear to be a complete set of Frankland's *experimental notebooks*. There are plenty of jottings on scraps of paper or at the back of other notebooks, and a few slightly more extensive accounts of laboratory work. One wonders where the bulk of Frankland's scientific observations were located and whether any records of them still remain to be discovered. The great exception is the immense amount of material on water analysis, and specially his water analysis notebooks with hundreds of pages awaiting inspection.

Despite the varied conditions under which they have been stored nearly all these documents are in surprisingly good condition, with very little fading or foxing. Exceptions include some of Kolbe's correspondence, usually written on extremely thin paper and apparently with poor quality ink. The Manchester copy-letter book is in a bad state, though this is more likely to be an inherent problem of the system than a function of storage conditions. Sometimes a large number of related documents were stuffed into one envelope (apparently by Frankland himself). Under those conditions there is a tendency for the envelope to tear and the contents to be more creased and torn than usual. It would be greatly to the advantage of posterity if letters etc., could be properly stored but this is not a likely prospect at present. The more the documents are examined the more this problem will be exacerbated and the more rapidly they will deteriorate, and for this reason alone urgent steps had to be taken to deal with the twin problems of recording and access.

N.B. Section IV of this article, entitled 'Exploitation and Access' (pp. 181–5) has been omitted for the purposes of this volume.

ACKNOWLEDGMENT

We are glad to thank The Open University for the award of an Honorary Research Fellowship to one of us (S.P.R.) and its Research Committee for generously providing both microfilm and computing equipment. We are grateful to our colleagues in Research Group and Department for encouragement and advice, and especially to Dr Peter Morris (Royal Society/British Academy Research Fellow) for help with the computing programme. The world of scholarship has no inalienable right to any material held by private individuals and we are deeply grateful to all the owners for their constant patience and kindness and for permission to film and study the valuable documents in their possession.

20 See note 9.
21 In the archives of Laporte Industries Ltd, Widnes. See P.J. Morris and C.A. Russell, *Archives of the British Chemical Industry, 1750–1914,* BSHS, Stanford in the Vale, 1988, p. 178.

XI

Chemical Techniques in a Preelectronic Age: The Remarkable Apparatus of Edward Frankland

One of the masters of chemical technique in the nineteenth century was Edward Frankland (1825–1899).[1] The illegitimate son of a noted lawyer, he was raised by his mother in Garstang and then in Lancaster, England. After attending a number of schools that mostly failed to provide any kind of scientific education, he was apprenticed to a pharmacist in Lancaster as a substitute for the medical education that his mother could not afford. Here the foundations were laid for neatness and scrupulous accuracy in dispensing, and the continuous use of his manual skills brought them to a high level of proficiency. Meanwhile, he learned the rudiments of scientific theory from popular lectures, at the Mechanics Institute and in a laboratory equipped for young would-be students by a philanthropical doctor in the town.

Through the good offices of this doctor and the local member of Parliament, Frankland then moved to London to work in the chemistry laboratory of Lyon Playfair. Here he began his lifelong friendship with another assistant, Hermann Kolbe, and worked (still with Playfair) at the Putney College of Civil Engineering. After a preliminary visit to Bunsen's laboratory at Marburg during the summer of 1848, he returned to England in some haste, having received a call to teach at Queenwood College in Hampshire, a strange "progressive" institution that left him astonishingly free to develop practical teaching of science in almost any way he liked. Within a year, he was back at Marburg, this time staying for the year needed to start and complete a doctoral degree. This was granted in 1849, largely on the basis of his discoveries of a new kind of compound that he called "organometallic" and of a sub-class known as *dialkylzincs*.

Currently, this is one of the most important branches of chemistry, founded by Frankland and since employed in a multitude of organic syntheses. The great reactivity of these compounds opens up endless possibilities for alkylation and other reactions, whereas the study of their structure is important for both organic and inorganic chemists. In

Publisher's Note: This article was originally published as Chapter 12 in *Instruments and Experimentation in the History of Chemistry*, eds F.L. Holmes and T.H. Levere (Cambridge, MA, 2000). The original numbering of the figures has been retained.

fact, Frankland's organozinc compounds were too reactive, for which reason the full exploitation of organometallic compounds had to await the discovery of their more amenable analogs, the Grignard reagents, in 1900.

After a brief spell back at Putney College, he obtained the first chair in chemistry at the new Owens College in Manchester. He was there from 1851 to 1857. He spent the rest of his career in London, serving at a variety of institutions (sometimes several at once), including St Bartholomew's Hospital, the Addiscombe Military Academy, the Royal Institution, and the Royal College of Chemistry (later Imperial College).

In this chapter, several of Frankland's technical innovations are described. Notably, they sprang from a wide range of chemical interests: civil and military; synthesis and analysis; academically obscure and prominently visible; above all, pure and applied. This breadth of interest was rarely encountered elsewhere at that time and may go some way to explaining why he was so successful. A final section attempts to draw some further general conclusions both as to the reasons for his success and as to the origins and reception of his techniques. An interesting enquiry would be how far they apply to scientific procedures in general. Meanwhile, the following are some of the techniques by which he transformed the face of chemical laboratory practice in the years before black boxes, automation, and other by-products of the electronic age.

Use of the Reflux Condenser

In 1846, Frankland and Kolbe wanted to test the theory that alkyl cyanides were nitriles of carboxylic acids. This was part of a wider hunt for "radicals" that they (and relatively few others) believed to be constituent parts of organic molecules. What was required was a demonstration of the presence of "alkyl" radicals in a range of compounds, including the famous aliphatic acids. Were their nitriles in fact just alkyl cyanides? If so, the acids too must contain alkyl. To establish this hypothesis, they subjected ethyl cyanide (bt 97°C) to boiling, concentrated aqueous potash, a reaction that we now know requires many hours to complete. At first, they seem to have used a distillation apparatus, repeatedly returning the distillate to the retort until no odor of the cyanide was perceptible.[2] After evaporation, followed by acidification of the solid residue, the contents of the flask were redistilled. The

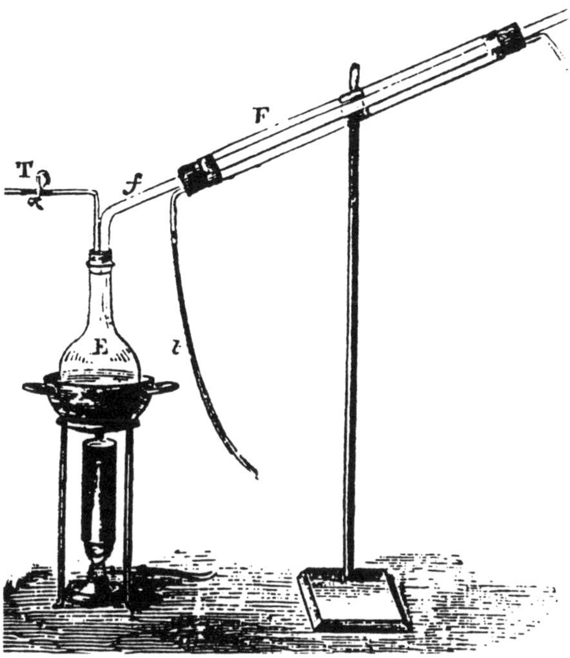

Figure 12.1
An early illustration of a Liebig condenser in the semireflux position, enabling volatile reagents to be recondensed and returned to the flask. [From E. Frankland, *How to Teach Chemistry: Hints to Science Teachers and Students*, ed. G. Chaloner (London: J. & A. Churchill, 1875).]

distillate showed all the characteristics of aqueous propionic acid, thus confirming their theory as to the nature of nitriles.[3] However, this method was not applicable to the more volatile methyl cyanide (bt 82°C), and here Frankland employed a "simple contrivance" consisting of a Liebig condenser whose lower end was bent downward to an obtuse angle, thus enabling the condensed nitrile to return repeatedly to the flask (figure 12.1).[4] This "condensed refrigerator" he claimed to be the first use of a condenser in the reflux position.[5] He had in fact been anticipated by Mohr 10 years previously,[6] but was apparently unaware of that. One wonders why such an extremely simple device had not been in common use for a long time. It probably reflects as much the primitive state of organic manipulation in the early nineteenth century as Frankland's skill at laboratory techniques.

Employment of a reflux condenser is now so commonplace that finding an area of organic chemistry to which it is never applicable is difficult. Subjecting volatile liquids to this technique permits heating them without intermission for an indefinite period of time, the low-boiling components being continuously returned to the flask for further reaction.

It has been employed for simple reactions, such as ester formation and hydrolysis; acylation of amines; aromatic nitrations; preparation of acetoacetic ester; and use in syntheses, the Wittig reaction, the Friedel Crafts reaction, and a host of others. By this means, Grignard was able consciously to follow Frankland's example and discover the alkylmagnesium halides named after him, by heating magnesium, alkyl halide, and ether under reflux conditions. In conjunction with the later Soxhlet extractor, the technique opened up a powerful but simple method of using volatile solvents to extract organic materials from natural products as leaves, bark, and the like.

Heating Under Pressure

At an early stage of his chemical career, Frankland was infected with a passion to isolate the radical "ethyl." It was all very well to know that it might be present in acids, nitriles, and so on. The next step must surely be actually to isolate it, to collect it in some way, and to analyze it eudiometrically (by explosion with excess oxygen). Such thinking was inescapably in the spirit of Berzelius, whose radical theory of organic chemistry was still surviving despite several devastating blows from its opponents. In Kolbe he found a kindred soul, and the latter was in fact able to produce what appeared to be a hydrocarbon "radical" from his electrolysis of fatty acid salts.

While still at Queenwood College, in 1847/48, Frankland made the momentous decision to try to release "ethyl" from ethyl iodide by treating it with the very electropositive metal potassium. A strong evolution of gas ensued when the temperature approached the boiling point of ethyl iodide. It seemed hopeful, but the liquid naturally tended to boil away, so he tried again, this time in a closed vessel to prevent ebullition. The vessel was a glass tube sealed at each end. However, the gas produced was not "ethyl" but a strange mixture that seemed to contain hydrogen and "methyl" (perhaps better called *ethane*). So, he determined to replace the vigorous reagent potassium with the less electropositive element zinc, thus hoping to avoid any decomposition of his expected ethyl.

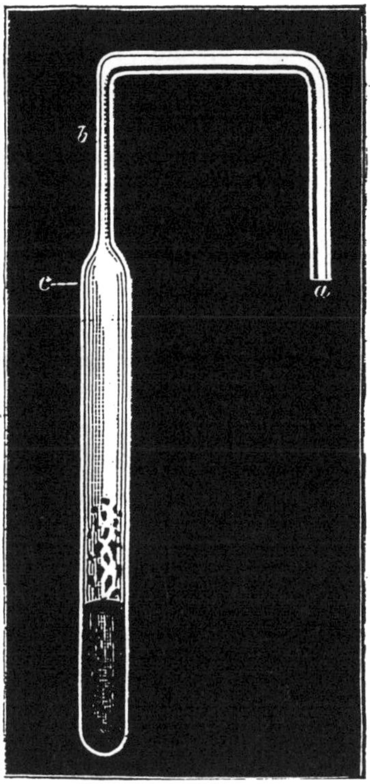

Figure 12.2
Reaction vessel for allowing zinc and alkyl iodides to react. After introduction of the reagents, the tube was heated locally at *c*, and the upper part drawn out into a narrow tube (*b*), which then was sealed at *a*. [From E. Frankland, *Experimental Researches in Pure, Applied and Physical Chemistry* (London: van Voorst, 1877), 69.]

This time he heated the reagents together from the beginning in a sealed tube and made the wholly unexpected discovery of the remarkable zinc alkyls. To prepare these, he took a tube, some 30 mm long and made from hard Bohemian glass. Metal was introduced, and one end was sealed. It was then alternately warmed and cooled, with the open end dipping into ethyl iodide, thus introducing the liquid into the tube. The latter was first evacuated of air by a pump, and the open end then was sealed in a flame (figure 12.2). The sealed tube was then placed in an oil bath and heated for several hours, almost the first time such a technique had been used. Because his only eudiometer at Queenwood had exploded, the tube had to be left unopened and was in fact transported to Marburg. Later, in the following year, he opened the tube under water with spectacular and unexpected results: "A greenish blue flame, several feet long, shot out of the tube."[7]

Several years later, in Manchester, he determined to exploit the reactivity of the zinc alkyls to the full. However, severe limits restricted the amounts that could be made by heating sealed tubes in an oil bath, with the constant danger of explosion. At approximately this time, he developed a lifelong friendship with engineer James Nasmyth. Out of their discussions emerged designs for digesters capable of conducting chemical reactions under considerable pressure. Using iron digesters supplied by Nasmyth, he found that the sealed tubes of reagents were kept from bursting by the external pressure of steam in the vessel.[8] This latest Mancunian technology of high-pressure vessels led to the production of diethylzinc on a large scale. Figure 12.3 illustrates one such apparatus made of copper.

This was another example of Frankland's technical versatility. With the product, he examined many new reactions of diethylzinc (e.g., that with nitric oxide in the apparatus in figure 12.4, where a piston within a condensing syringe [D] forces the gas into the digester). This semiautomatic apparatus was necessary, as the reaction could take several weeks for completion.

Such digesters proved to be of great value to other chemists, including Hofmann[9] and Williamson. Apologizing for a delayed delivery date to the latter, Nasmyth expressed the hope to Frankland that he would "keep a keen look out to commercialise ... and turn them to profitable results" for then, "with such powerful tools as money can furnish," he would be able "to set out as pioneer of that glorious band who are soon in the van of discovery."[10] He was unlikely to have turned away from such a prospect.

The use of sealed tubes is extremely simple, needing only a hard-glass tube and a steam bath or (for higher temperatures) an oil bath. On the other hand, it does require a certain confidence and skill in glass blowing to ensure that the seal is able to withstand the great internal pressures generated.

The procedure enjoyed a certain vogue in the late nineteenth and early twentieth centuries, although the inherent dangers of heating under great pressures deterred many and encouraged a search for alternative techniques. It has the great advantage of enabling liquids to be heated considerably above what would be their boiling temperatures at normal pressures, with no interference from atmospheric contamination or need to employ a reflux condenser. The technique

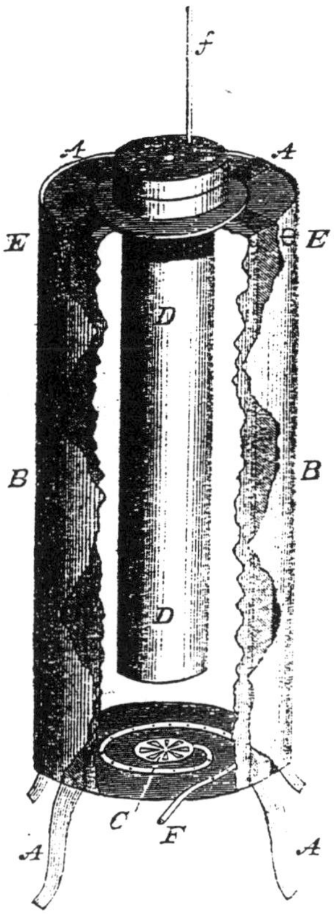

Figure 12.3
Reactions under pressure. A sealed copper digester is placed in the copper oil-bath (*DD*), heated by a gas burner (*f*), and enclosed in a sheet-iron cylinder (*BB*). [From E. Frankland, *Experimental Researches in Pure, Applied and Physical Chemistry* (London: van Voorst, 1877), 194.]

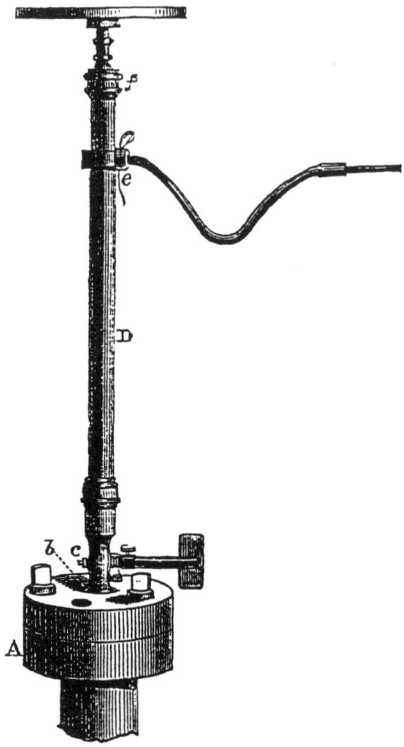

Figure 12.4
Apparatus for introducing a gas (such as nitric oxide) into a copper digester containing dialkylzinc. *A* is the head of the copper digester, *D* is a syringe with a tightly fitting piston, and *c* is a stopcock. [From E. Frankland, *Experimental Researches in Pure, Applied and Physical Chemistry* (London: van Voorst, 1877), 213.]

was formerly employed in the Carius method for estimation of halogen. It has been used synthetically for various condensation reactions (as that between aldehydes and active methyl compounds as 2- and 4-methyl quinolines).

Gas Collection

As part of his quest for "radicals," Frankland wanted to prepare "ethyl" from ethyl cyanide and potassium. Assuming it was gaseous, he devised the apparatus in figure 12.5 to collect the gas in bulk and also in separate sample tubes.

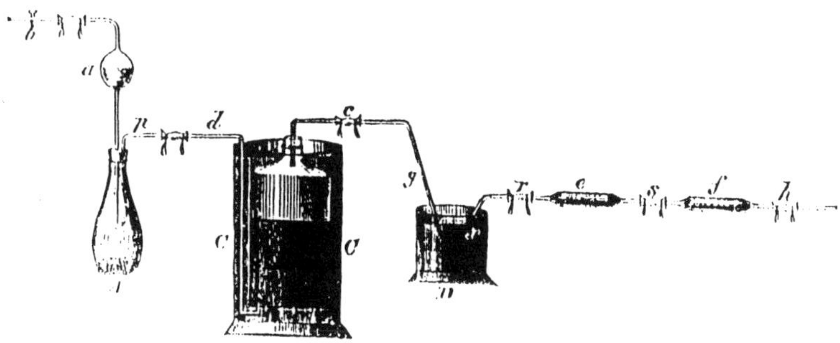

Figure 12.5
Production of "ethyl" (butane) by the action of potassium on ethyl cyanide. See text for details. [From E. Frankland, *Experimental Researches in Pure, Applied and Physical Chemistry* (London: van Voorst, 1877), 50.]

Potassium was in flask *A*, ethyl cyanide in bulb *a*, and mercury in *D*, and the whole apparatus beyond *B* was filled with water. The nitrile was introduced by opening tap *b* and, after displacement of the air, the hydrocarbon gases were collected in *B*. Sample tubes *e* and *f* were filled by opening *h*, thus allowing gas from *B* to displace water in the tubes, which then were isolated, removed, and stored.

The technique was chiefly noteworthy for the added precautions to expel air before collection, to remove the nitrile (by standing the gas in the water in *B* [see figure 12.5] for some hours) and (later) to dry the gas samples and to equilibrate with atmospheric temperature before weighing. The preparation of large volumes of gases is no longer a major feature of organic chemical practice, chiefly because the hydrocarbon products are far better obtained from petroleum and partly because they are no longer identified with "radicals." Further, it cannot be claimed that the specifics of this apparatus were of importance to posterity, but their employment emphasized in general terms several paramount needs in organic chemical practice, especially those of product purification and accurate measurement of volumes and weights.

Distillation in Inert Atmospheres

Distillation in inert atmospheres, commonplace today, makes one of its earliest appearances with Frankland's apparatus for distillation of the highly reactive dialkylzincs under hydrogen (figure 12.6). *D* is a

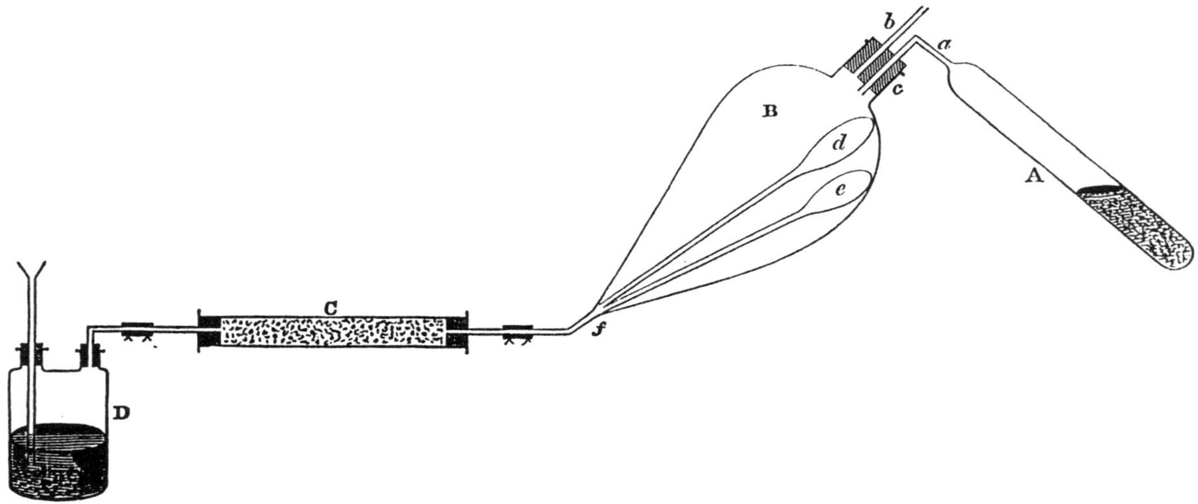

Figure 12.6
Distillation of the highly reactive dialkylzincs under hydrogen. See text for details. [From E. Frankland, *Experimental Researches in Pure, Applied and Physical Chemistry* (London: van Voorst, 1877), 174.]

hydrogen generator, *C* is a calcium chloride tube, *A* is the container for dimethylzinc, and *B* is a receiver containing several bulbs that can be filled with the condensed liquid. After flushing the apparatus with hydrogen, the tubes at *b* and *f* were sealed, and *A* was warmed while *B* was cooled. The entire operation must have been a most hazardous procedure, though the danger was perhaps less than allowing the organometallic compound to have access to either air or water. Probably in organometallic chemistry this procedure is still most employed, though the inflammable hydrogen is now more commonly replaced by pure dry nitrogen or (occasionally) argon.

WATER ANALYSIS

Frankland's development of new methods of analyzing water supplies brought him greatest fame in his lifetime and also performed his most direct service to the community. The problem was complex as, in the 1860s, the microbiological causes of waterborne diseases were ill understood. The approach of Frankland was to measure nitrogen concentration as an indication of previous sewage contamination and also to determine the ratio of organic and inorganic nitrogen in the water.

The old methods all left much to be desired, and his collaborator, H. E. Armstrong, observed of Frankland:[11]

> Thorough in everything he did, he soon became dissatisfied with the methods, especially the determination of organic impurities indicative of sewage contamination. He decided to revise them all. At the close of the summer session of 1866, he did me the great honour to propose that I should carry out the work for him. The method of combustion analysis *in vacuo* we devised was made known, the following year, in a lecture by Frankland at the Royal Institution; the work generally was described at the Chemical Society in February 1868.

The biggest problem, apparently intractable, was that "No process has yet been devised by which the amount of organic matter in water can be even approximately estimated."[12] After nearly 2 years' work, they found an answer.

The new method was a development of Liebig's classic technique for estimating carbon and nitrogen in organic compounds. The essential feature of this variation (figure 12.7) was oxidation of the evapo-

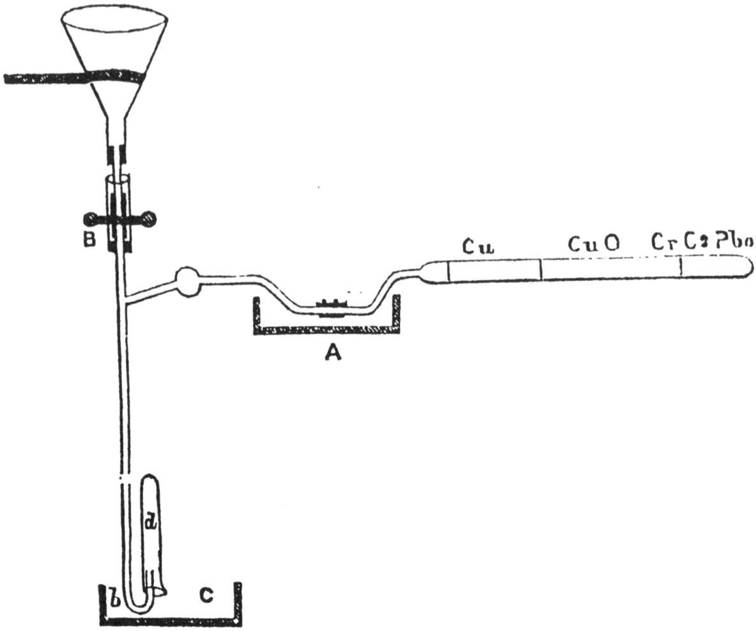

Figure 12.7
Estimation of carbon and nitrogen in water residues, with oxidation by lead chromate and measurement of the volumes of the resultant gases, CO_2 and N_2. [From E. Frankland, *Experimental Researches in Pure, Applied and Physical Chemistry* (London: van Voorst, 1877), 574.]

rated residue with lead chromate in an evacuated tube, the resultant gases (CO_2 and N_2) being estimated volumetrically. A refined form of apparatus for the latter procedure was on general sale (figure 12.8). To avoid interference from carbonates, the preliminary evaporation was first conducted with "sulphurous acid," thereby expelling CO_2, and also liberating nitrogen from nitrates and nitrites (a "remarkable reaction" that "could scarcely have been predicted"). The evacuation was effected, before and after combustion, by the mercury fall-pump described by Sprengel in 1865.[13]

While this work was in progress, an alternative attack on the problem had been developed by Frankland's former student, J. A. Wanklyn (now professor of chemistry at the London Institution) and two of his collaborators.[14] They treated the water first with alkali to expel ammonia from ammonium salts and urea and second with alka-

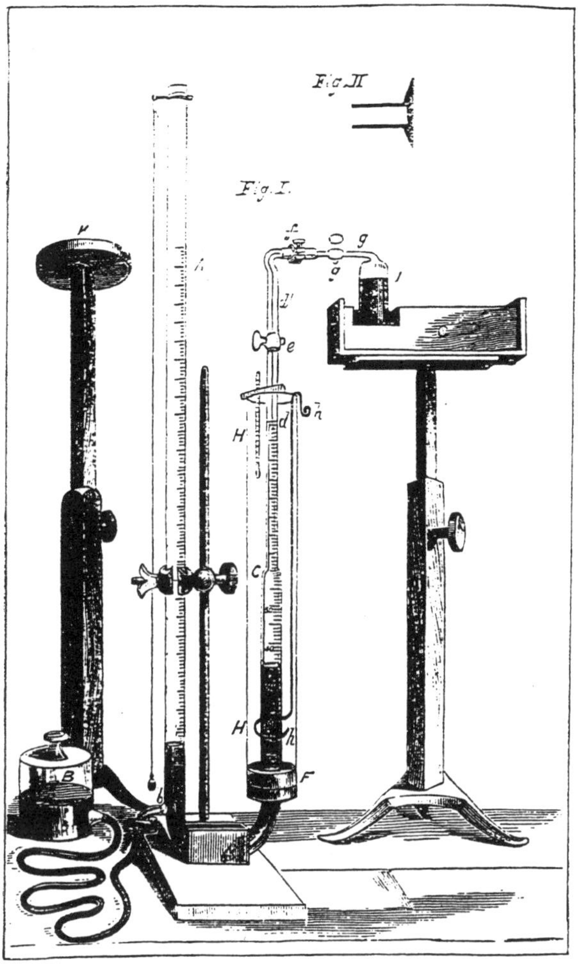

Figure 12.8
Modified apparatus for "analysis of gases incident to water analysis." [From E. Frankland, *Experimental Researches in Pure, Applied and Physical Chemistry* (London: van Voorst, 1877), 594.]

line potassium permanganate, claiming the organic material to be quantitatively oxidized to ammonia, which may then be estimated colorimetrically using Nessler's solution. This method was clearly simpler than that of Frankland and Armstrong but was less accurate. They pointed out that in Wanklyn's technique, even for albumen, oxidation is never complete,[15] and Wanklyn later admitted that only two-thirds of the organic material is so oxidized.[16]

In retrospect, clearly the Frankland technique had several advantages. First, it was extremely accurate. To read Frankland's papers on the subject is to be transported into a world of painstaking attention to detail, of careful standardization of reagents and calibration of apparatus, of blank control experiments, of rigorous procedures for quantitative transfer, of flawless vacuum techniques, even of meticulous instructions for "reading" Nessler tubes. It is a world largely forgotten today, except perhaps by those with long memories stretching back into a preelectronic era. Second, the method was capable of amendment and adjustment as experience demanded. Those (including Frankland's critics) who saw this as a sign of inadequate technique completely missed the point that *all* scientific procedures need constant updating, though often only in minor matters of detail. It was, as Odling said, "a new and very refined process of water analysis."[17]

However, it also had a number of disadvantages. First and foremost among these was the sheer difficulty of the techniques involved. For a Frankland or an Armstrong, this was not a problem, nor was it likely to be so for the numerous assistants whom they had trained. However, for lesser mortals, it presented serious difficulty, though this can be exaggerated. American chemist J. W. Mallet, writing many years later and after some simplifying improvements, described it as "a process of great delicacy, and quite satisfactory in its details with proper precaution and in trained hands."[18]

Two further and related problems were the time taken for a complete analysis (up to 2 days) and the financial cost (in terms of time, apparatus, and chemicals). Finally came the critical question as to what exactly was being measured. One analyst put his finger on the problem thus:

> Most chemists are, we believe, convinced that, assuming the organic matter to be once inside the combustion tube free from an admixture of nitrates, its carbon and nitrogen can be estimated with an extraordinary degree of accuracy, by means of Frankland and Armstrong's

> process. The real questions, however, are, firstly: does the process enable us to get the organic matter dissolved in a litre of water into our combustion tube undiminished in quantity, and freed from the large excess of nitrates, with which it is often associated, and, secondly, can we make accurate allowance for any ammonia which may be present in the water?[19]

By comparison with the Frankland-Armstrong process, the alternative introduced by Wanklyn and his colleagues was both faster (taking perhaps 3 hours) and cheaper. On the other hand, it suffered from the defect that albumen is not quantitatively decomposed into ammonia under the test conditions and moreover, that urea, supposedly eliminated as ammonia by preliminary treatment with sodium carbonate, does not give 100% conversion. For these reasons, Wanklyn introduced his correction factor of two-thirds. This stratagem, together with his failure to conduct his own tests of the rival process, led to much acrimony in the pages of *Chemical News*.[20] The results obtained by this method were almost always lower than those obtained by Frankland and raised the critical question as to whether albuminoid substances were in fact the right subjects for analysis.

By 1872, some claimed that most leading analysts had rejected Frankland's process in favor of Wanklyn's, their number including Thomas Way, Angus Smith, W. A. Miller, A. Voelcker, and Henry Letheby. Crookes, the editor of *Chemical News*, remarked, "[w]e scarcely know a single chemist of reputation who approves of Dr. Frankland's water analysis."[21] Four years later, Frankland himself conceded the popularity of the ammonia process.[22] Nevertheless, one should not exaggerate the monopolistic position claimed for the latter.

Thus, the Nottingham Public Analyst, E. B. Truman, assured Frankland in 1874 that he was using his method for "hundreds of analyses."[23] J. W. Thomas of Cardiff wrote that, though he had temporarily gone over to Wanklyn's method to keep fees down, he hoped to return to the Frankland method as soon as possible,[24] and Odling told Frankland "you may always refer to me as a chemist habitually using your process of water analysis," even though they might occasionally disagree over the conclusions.[25] An undated note in Frankland's handwriting discloses that his process was in use by Bischof, Campbell Brown, H. Brown, Donkin, Hill, Moulting, Odling, O'Sullivan, Russell, Tate, Thomas, Truman, and Williamson.[26] A copy-letter to Sir Hugh Owen, also undated but probably dating in the early 1890s, names Hill, Mills, Odling, and Tidy as users of his own process that, however, is "tedious, laborious and costly." However, the ammonia process, though "utterly untrustworthy," is used by "the profession at large."[27]

In the end, Frankland's water analysis techniques were replaced by those of microbiology. Their influence on subsequent chemistry was trivial. Yet they had an importance for Victorian society that is difficult to exaggerate. Despite the protestations of the chemists employed by the water companies, it seems very likely that urbanized Britain might have suffered something approaching a pandemic through waterborne diseases. Only Frankland's method was capable of identifying water that had been in contact with previous sewage contamination, and this was the critical question.

True, his conclusions were sometimes over-severe (as when nitrate concentrations were abnormally high through seepage of innocuous fertilizers), but it was preferable to err on the side of caution. Water condemned by Frankland was at best a risky commodity and at worst a lethal cocktail of virulent microorganisms. Understandably, his water analysis caused Frankland to became best known by the public at large.

The development of these techniques affected the chemical community in a variety of ways. First and most obviously, they divided it. Chemist was ranged against chemist as testimonies were given, often in a court of law, for and against an individual water company. Of course, the reason was that no real agreement existed as to the true etiology of the outbreaks of typhoid, typhus, cholera, and the like. Second, they were intimately connected with the rise of the chemical profession in Britain. As chemists performed chemical analyses of waters that had once been the domain of doctors, engineers, or even the Astronomer Royal, so their professional awareness slowly increased. It is no coincidence that the first president of the new incarnation of this awareness, the Institute of Chemistry, should be none other than Edward Frankland himself.

Combustion Apparatus

By the early 1860s, a renewed interest was aroused in combustion generally. Once it had been at the very heart of Lavoisier's revolution, but its prominence had receded with the advent of electrochemistry and the growth of organic chemistry. Now the demands of applied chemistry were bringing it back into fashion. After a successful ascent of Mont Blanc in 1859, Frankland conducted a series of experiments on combustion of candles at low and high altitudes. Possibly through his connection with a military college at Addiscombe, he turned his attention to the variation of burning times for shell fuses at different altitudes (and therefore pressures). It had been shown that, unlike Frankland's candles, time fuses took much longer to burn at high altitudes than at low levels.[28]

To examine this phenomenon, he constructed an apparatus that embodied many familiar features; a large iron cylinder with an air pump at one end and at the other a 6-foot-long gas pipe connected to a mercury pressure gauge and capable of receiving the fuse. An iron ball was suspended by a thread at the far end of the fuse, so when combustion eventually reached that end, the thread would burn, and the ball would drop onto an iron plate below. Seemingly this was a rare case in Frankland's work where the critical signal was acoustic. Again, it was an astonishingly simple solution to a complex problem. He concluded that each diminution of 1 inch of barometrical pressure causes a retardation of 1 second in a 6-inch or 30-second fuse. The relevance of this work to gunnery soon became evident to the British army, especially as they operated on mountainous terrain. Yet his discovery of a relation between pressure and combustion rate was to foreshadow the programs of combustion research conducted by the oil companies before and after World War II. Ironically true is that crucial to the complex mechanisms were the formation and reaction of exactly those transient entities sought so diligently but never obtained by Frankland: the alkyl radicals.

CONCLUSION

Even a cursory examination of Frankland's new techniques reveals several obvious reasons for their success.

Availability of Materials

When Frankland returned to England in the summer of 1848, he brought with him, for the benefit of the students at Queenwood, a very large quantity of apparatus and chemicals made in Germany. What could he have imported?

First was Bohemian glass, a commodity prized for its use in chemical manipulations. It is low in lead and low in soda, having a higher melting temperature than the usual alternative, soda glass. To manufacture it into apparatus, a blowpipe was of course required. From this material he constructed his sealed tubes, for example. Another commodity was *caoutchouc* (rubber) that formed the essential joining between adjacent glass vessels. It was made by fastening a sheet around a glass tube and keeping it in boiling water for some time.[29]

For consumable raw materials, Frankland usually made his own chemicals, as ethyl iodide, acetoacetic ester, and the like. He used his considerable network of contacts to acquire all kinds of raw materials, as in the case of 6-inch fuses provided by his friend, F. A. Abel, at Woolwich Arsenal.

Technology Transfer

Frankland was a master at transferring a technique from one field to another. He seems to have learned these techniques from people rather than from books.

His use of the water-cooled condenser was probably learned from his brief sojourn at Liebig's laboratory at Giessen in 1848. To turn its end through an angle of perhaps 120 degrees was a simple adaptation of apparatus from one role (distillation) to another (refluxing). To turn Nasmyth's massive digesters to his own purposes of heating under pressure was not merely the effective invention of the pressure cooker but a successful example of scale-up technology. Learning from Kolbe, and still more from Bunsen, of the latter's techniques of gas analysis, enabled Frankland to develop his own technique of eudiometry, and he was making such equipment by the early 1850s. In 1853, he and W. J. Ward devised a compact apparatus incorporating several features of an apparatus of Regnault (such as water cooling of the eudiometer tube).

However, in the apparatus for water analysis, Frankland most obviously brought together a number of hitherto unrelated techniques, especially two. First, the estimation of carbon and nitrogen was basically an adaptation of the classic technique of Liebig as applied to organic compounds in relatively large quantities. Second, on the scale of water analysis developed by Frankland, the quantities were far smaller, and the apparatus had to be evacuated of air before use. His genius lay in a development recorded by Armstrong: "I recollect, when he suggested that we should make the attempt, that he referred me to the account given by Graham of the use he had made of the Sprengel mercury pump in studying the diffusion of oxygen through india-rubber."

Experimental Skill

One cannot read about Frankland without a growing awareness of his astonishing personal skills at the bench. Armstrong did not doubt their origin: "His real education was probably obtained during his apprenticeship in a local druggist's shop. The work he did there taught him to use his hands." Indeed, the tidiness and self-discipline in a pharmacist's profession gave invaluable lessons in the general manipulation of chemicals, crucially important for anyone working in synthetic organic chemistry.

Writing of his work with Frankland on water analysis, Armstrong observed,

> The experience was invaluable, as I was thrown much on my own resources, though sufficiently aided at critical times ... The task was not entirely simple. A few days later, seeing that I was in difficulties at the blowpipe table, he came to my aid and made the first Sprengel fall-tube for me. We were the first after Graham to use the pump. If the traditions of the early Frankland school had been kept alive and the few students with hands [*sic*] had been regularly tutored in the methods of gas analysis, the vacuum combustion process would now be a preferred process, I believe.[30]

Frankland himself asserted that he acquired considerable skills in glass blowing at Marburg, so much so that he promised to return to Giessen to use his glass-blowing skills for the benefit of that famous laboratory.

Favorable Social Environment

A survey of the conditions in which Frankland worked leaves no room for doubt as to the favorable environment in which he did his best work. In his early years, as at Queenwood, the conditions were not naturally propitious, though it was one of the few schools in the country to take science at all seriously, and Frankland was given a largely free hand as to how he worked. Also, through his drive and ambition, he made the best of what he had. The apparatus he invented was of the simplest kind, and he had ample space in which to use it. In the German universities of Marburg and Giessen he was, of course, in contact with a tradition far more favorable to scientific enquiry than in Britain. In Manchester, on the other hand, conditions were much less satisfactory (heavy teaching load, poor salary, etc.), and research took second place to more remunerative consultancy work.

He came into his own in London, with the unparalleled traditions and good facilities of the Royal Institution, and there he did his best work in organic synthesis. When later he took up his post at the Royal College of Chemistry and shortly afterward became heavily involved in water analysis for the government, he was in a position to make his conditions favorable. Thus, he was rarely in the teaching laboratory; he relieved himself of as many routine chores as possible and eventually set up his own laboratory. Success bred success; by the end if his life, he was in receipt of a very large annual income and well able to spend money on expensive equipment. As was supposed to be the case of Humphry Davy, the very elitism of his social position led to an elitism of his science. Consequently, one could say that his techniques were not applicable to ordinary people. One of them put it thus:

> The Frankland process is quite within the reach of the manipulative skill of any fairly-trained chemist, but it requires practice, and probably pretty constant practice. It cannot be taken up off-hand, and even tolerable results obtained at once. From the hands of a person without proper laboratory training, its results are utterly valueless. It is hence better adapted to regular use in the examination of many samples of water in a large public laboratory than to an occasional use by a private individual.[31]

In other words, it was not even financial but professional elitism that led to a rejection of his methods. For the president of the world's first institute for professional scientists, that is irony indeed.

Two final considerations may place Frankland's technical advances in the widest context. The first concerns the surprising fact that, today, most of this is completely unknown to modern chemists and, in fact, to not a few historians. The apparent eclipse of such a reputation has much to do with Frankland's life style, preference, and extreme reluctance to promote himself in a public way. These have been discussed elsewhere.[32]

Additionally, Frankland never had the good fortune to have a technique, a reaction, or a piece of apparatus named after him. Though generations of chemists have heard of Kipp, Mohr, Victor Meyer, Liebig, Hofmann, and other inventors of laboratory ware, Frankland has not been so commemorated. One reason is that some of his techniques hardly outlived their inventor, but another is more curious. His effective introduction of the reflux condenser and sealed glass tube was so simple, so obvious in retrospect, and so widely applicable that most chemists used them almost as second nature, scarcely bothering to wonder where they came from. They could not even be ordered from the laboratory stores by name, for each was a totally ordinary piece of equipment used slightly differently or adapted by simple glass blowing on the bench. If one *ordered* a "sealed tube," it would be of no use!

Much the same fate befell what was once known widely as "Frankland's notation," though it was again so obvious, so simple, and so universal that it became simply "graphic formulae." We frequently use such devices to this day. Thus, in one sense, Frankland was a victim both of his own success and of the elegant simplicity of his procedures. However, that should not distract historians from a proper evaluation of work that transformed the face of chemistry, particularly the realm of organic syntheses. Despite his past obscurity, it is entirely justifiable to hail Frankland as one of the great founders of synthetic organic chemistry.[33]

The second point is more subtle. The very fact of Frankland's extraordinary experimental success, coupled with the very limited nature of his own chemical training, seems to have stirred him to the invention of one further technique, though this was not chemical but pedagogic: the revolutionary idea that students should learn their chemistry not just from books but from lecture-demonstrations and personal laboratory experience themselves. Of course, he was not the only one to have such ideas, but he was one of the few to translate them into effective practice. His opportunity came when, in 1865, he succeeded Hofmann not only to the chair at the Royal College of Chemistry but to the post of examiner to the Department of Science and Art. Here he had a unique opportunity to fashion the curricula, examination papers, and chemical education of thousands of young people who studied in the evenings for their own improvement.[34]

To bring his dream to reality, he instituted in 1869 a scheme of teacher training in which members of the profession were welcomed for one week to the laboratories of the Royal College of Chemistry. He also wrote a book, *How to Teach Chemistry*, which featured 109 experiments that should be demonstrated to the students.[35] Some of these involved the techniques he had developed: reactions in inert atmospheres, refluxing, gas collection, and so on. Moreover, so popular was his scheme that a number of chemical suppliers competed with each other to provide the apparatus "recommended by Dr. Frankland."

His is a remarkable story. Hugely important advances were made in teacher training and in laboratory instruction. Frankland's name was everywhere in the popular literature of chemistry. His distinctive techniques were no longer the prerogative of an elite of rich chemists but the experience of the common man. In both this democratization of chemical practice and in the introduction of important new procedures, chemistry in Britain gained a very great deal.

Notes

1. Much of this material will be found in my book, *Edward Frankland: Chemistry, Controversy and Conspiracy in Victorian England* (Cambridge, UK: Cambridge University Press, 1996).

2. E. Frankland, *Autobiographical Sketches from the Life of Sir Edward Frankland*, 2nd ed., ed. MNW and SJC (London: privately published, 1902), 169.

3. E. Frankland, "On the Chemical Composition of Metacetonic Acid, and Some Other Bodies Related to It." *Mem. Chem. Soc.* 3(1847): 386–391; *Experimental Researches in Pure, Applied and Physical Chemistry* (London: van Voorst, 1877), 29.

4. E. Frankland, *Experimental Researches*, 41.

5. Ibid., 29.

6. C. F. Mohr, *Annalen* 18(1836): 232.

7. E. Frankland, *Sketches*, 186.

8. E. Frankland, "Researches on Organometallic Bodies—Zincethyl," *Philosophical Transactions of the Royal Society of London* 145(1855): 259–275.

9. Hofmann to Frankland, 28 April 1855 (Raven Frankland Archive, Open University microfilm 01.03.0274).

10. Nasmyth to Frankland, 22 January 1856 (Raven Frankland Archive, Open University microfilm 01.02.0279); regarding Nasmyth (1808–90) see S. Smiles, ed., *J. Nasmyth, An Autobiography* (London: 1885), and *DNB*.

11. *Chemistry & Industry*, 53(1934): 462; the paper was read on 16 January and further discussion ensued on 6 February 1868.

12. E. Frankland and H. E. Armstrong, "On the Analysis of Potable Waters," *Journal of the Chemical Society*, 21(1868): 87.

13. H. Sprengel, *Journal of the Chemical Society* 18(1865): 9–21.

14. E. T. Chapman, M. H. Smith, and J. A. Wanklyn, *Journal of the Chemistry Society* 20(1867): 445–454.

15. E. Frankland and H. E. Armstrong, "On the Analysis of Potable Waters," *Journal of the Chemistry Society* 21(1868): 83.

16. J. A. Wanklyn, *Journal of the Chemical Society* 20(1867): 591–595.

17. W. Odling, testimonial for Armstrong, 1 December 1870 (Imperial College Archives, Armstrong papers, first series, 335).

18. J. W. Mallet, Supplement no. 19, *National Board of Health Bulletin* (Washington, D.C., 27 May 1882), 4.

19. Anonymous, *Analyst* (note 39).

20. For example, editorial comments in September and October 1868, *Chemical News* 18(1868): 151, 153, 165.

21. Editorial, "Water Analysis," *Chemical News* 25(1872): 157.

22. E. Frankland, *Journal of the Chemical Society* 29(1876): 825–851, on 847.

23. E. B. Truman to Frankland, 25 June 1874 (Raven Frankland Archive, Open University microfilm 01.04.0409).

24. J. W. Thomas to Frankland, 22 December 1876 (Raven Frankland Archive, Open University microfilm 01.04.0415).

25. W. Odling to Frankland, 27 April (Raven Frankland Archive, Open University microfilm 01.04.0418).

26. Frankland, undated memorandum (Raven Frankland Archive, Open University microfilm 01.04.0421).

27. Frankland to H. Owen, n.d. (Raven Frankland Archive, Open University microfilm 01.07.0848). As far away as New Zealand and as late as 1896, Dunedin Public Analyst A. G. Kidston-Hunter was arguing the diagnostic importance of albuminoid nitrogen and ammonia in water, while earlier estimations of oxidizable organic matter probably employed permanganate oxidation. If so, Frankland's techniques do not appear to have made much progress in that part of the Empire [R. J. Wilcock, "Water Chemistry," in *Chemistry in a Young Country*, ed. P. P. Williams (Christchurch: New Zealand Institute of Chemistry, 1981), 195–205].

28. J. Mitchell, *Proceedings of the Royal Society* 7(1855): 316–318.

29. E. A. Parnell, *Applied Chemistry* vol. 2 (London: Taylor & Warren, 1844), 249n.

30. H. E. Armstrong, *Pre-Kensington History of the Royal College of Science and the University Problem* (London: Royal College of Science, 1920), 7.

31. J. W. Mallet, *National Board of Health Bulletin*, suppl. 19, 27 May 1882, 4.

32. C. A. Russell, *Edward Frankland*, chapter 17 and elsewhere.

33. G. Porter, "The Chemical Bond Since Frankland," *Proceedings of the Royal Institution* 40(1965): 384–396, on 387.

334

34. C. A. Russell, *Edward Franklin*, chapter 10.

35. E. Frankland, *How to Teach Chemistry: Hints to Science Teachers and Students*, ed. G. Chaloner (London: J. & A. Churchill, 1875).

BIBLIOGRAPHY

Brock, W. H. "Edward Frankland," in *Dictionary of Scientific Biography* (New York: Scribner's, 1970–80).

Frankland, Edward. *Autobiographical Sketches from the Life of Sir Edward Frankland*, 2nd ed., MNW and SJC, (London: privately published, 1902).

Frankland, Edward. *Experimental Researches in Pure, Applied and Physical Chemistry* (London: van Voorst, 1877).

Russell, Colin A. *Edward Frankland: Chemistry, Controversy and Conspiracy in Victorian England* (Cambridge, UK: Cambridge University Press, 1996).

Russell, Colin A. "Edward Frankland," in *New Dictionary of National Biography* (in preparation). Several of the letters and other manuscript documents cited in this chapter are in an archive privately owned by Mrs. Raven Frankland and are at present accessible only in microfilm form at the Open University. Other deposits of Frankland papers, though immensely important, are not directly relevant to the present subject.

XII

KEKULÉ AND FRANKLAND: A PSYCHOLOGICAL PUZZLE?

8.1 A Tale of Two Chemists

Historians of science have habitually, and rightly, concentrated much attention on communication between scientists, whether through correspondence, by published papers in established journals, or by personal contact in the context of scientific of other institutions. Comparatively little attention has hitherto been paid, at least in a systematic way, to the inverse problem, that of scientific *non-communication.* Only where scientific progress has been obviously impeded by isolationism has much interest been shown in the phenomenon. A classic case is that of Berzelius, unable to make contact with English chemists during the Napoleonic Wars for political as well as linguistic reasons, and at a later stage isolated from his contemporaries by virtue of his rigid and unpopular doctrines (1). In the case of Kekulé there was one contemporary whose apparent failure to make or maintain any kind of contact is a major historical mystery. This was the English chemist [Sir] Edward Frankland [1825-1899](2) whose interests so overlapped with those of Kekulé that their failure to communicate raises several important questions not only about the individuals concerned [or rather *not* concerned] but also about the intellectual and social structures of the 19th century chemistry. One observer who knew both men well, H. E. Armstrong, went so far to comment:

> Kekulé was in London in 1854 and consorted with Williamson, who like himself was under the enthralling influence of Gerhardt. Had he consorted with Frankland, a man infinitely in advance of himself and most other chemists as a worker, his attitude would not have been so independent. I have not been able to discover that Kekulé made the least attempt to exchange views with Frankland, having joined another camp. We are in face of a psychological puzzle (3).

Although Armstrong's remarks about a 'psychological puzzle' provide a convenient title for this chapter they must not be allowed to set the agenda. It by no means follows that:

a) A failure to meet or consult in 1854 was a deliberate policy of either man;
b) An initiative failure then or later was the responsibility of Kekulé only;
c) 'Having joined another camp' was a significant cause;
d) The best categories of explanation were [or are] psychological.

We may begin with an enquiry into the extent of the estrangement between Kekulé and Frankland. At first sight it seems to have been almost complete. In view of his acquaintance with both chemists Armstrong's testimony is not likely to have been seriously inaccurate. Frankland's own autobiographical reminiscences (2) confirm that impression, Kekulé hardly crosses their pages. But the most compelling evidence comes from the large collection of Frankland correspondence recently discovered (4). Out of several thousand items there are no copy-letters at all to Kekulé, and only two from Kekulé to Frankland; these both date from 1863 and relate to affairs of the Chemical Society, of which Frankland was then Foreign Secretary (5). Through the rest of the correspondence there are only about 20 incidental references to Kekulé, mostly unflattering particularly in the letters to and from Kolbe .

Yet caution is necessary. The absence of surviving correspondence does not prove that such did not once exist. Moreover, the evidence from Armstrong is late, 35 years after Frankland's death, and could not have been easily contradicted. In any case he was known as a passionate supporter of Frankland, even calling his own son Edward Frankland Armstrong , and wished to present his former mentor in the best possible light. So his evidence must be judiciously evaluated. It is possible to overstate their insulation from each other. In Frankland's own correspondence there is one reference (6) to a meeting between himself and Kekulé, at an evening at Wurtz's house in Paris in 1867. And Anschütz (7) claims that in 1854 Kekulé 'became acquainted with the discoverer of the first organometallic compounds, Edward Frankland' who, he says, 'sometimes came to London.' Yet for all his biased and rumbustious defense of Frankland, Armstrong had correctly perceived something strange and important about Kekulé's inertia from that time, an attitude that cried out for rational explanation.

Not only did they rarely [if ever] meet or correspond, they showed to each other's work what can only be described as a studied indifference. For example the sole reference to Kekulé in Frankland's collected scientific papers is a rather smug note that Kekulé had accepted Frankland's view of the reaction between triethoxyboron and zinc diethyl (8); he could have said so much more about their common interest in valency. In the same way Kekulé studiously ignored Frankland's role in elucidating that theory, claiming priority for himself and never mentioning the English chemist in that connection. As early as 1864 he claimed:

> Unless I am mistaken, it was I who introduced into chemistry the idea of atomicity of the elements (9).

He had already acknowledged debts to other workers [Williamson and Odling, (10) and Wurtz (11)] but never to Frankland. The same three chemists were honorably mentioned in the historical section of Kekulé's *Lehrbuch der organischen Chemie,* (12) but again the name of Frankland is conspicuously absent.

As the years passed their failure to relate becomes even more obvious. But it was a failure that lacked a certain symmetry. On Kekulé's part there was little doubt, as his Memorialist concedes, that he 'paid little attention to Frankland's theoretical views' (13). This is all the more remarkable in the light of Anschütz's conclusion that in the later 1850s Frankland's works were well known in Kekulé's laboratory at Heidelberg (14).

Frankland, on the other hand, did admit in 1877 a debt to Kekulé for the 'laws of atomicity' (15). And in his *Lecture Notes* Frankland gave early prominence to that other great achievement of Kekulé, the cyclic structure of the benzene molecule (16). But the overall impression remains of a pair of chemical giants, so antipathetic to one another that for most practical purposes the other did not exist. Later, as we shall see, their hostility boiled over into bitter recrimination, publicly or in private. Only towards the very end of their lives was anything like a *rapprochement* attained. It is a sorry tale, to be sure, but one that conceals several issues of great importance to any understanding of the course of 19th century chemistry. In the rest of this chapter some attempt will be made to identify the underlying causes. The most blatantly obvious was the question of priority in the discovery and early application of the principle of valency (17).

8.2 Disputes Over Priority

By the late 1870's, a few chemical authors were turning their attention to the historical development of their subject, with special reference to the related themes of atomism, valency and structure theory.

8.21 Debate with Wurtz

An early essayist in this field was Adolphe Wurtz. He had already made several references to Kekulé's contributions (18), stressing also the work of Williamson and Odling. As in his later *History of chemical theory* of 1869 (19) he overlooked Frankland's name. Then in 1878 he brought out a new book and sent a copy to the English chemist: *La théorie atomique* (20). Frankland's reply has survived as a copy-letter (21) and offers an illuminating commentary on his attitude to the whole Kekulé business. It is worth extensive citation. He begins:

> Pray accept my best thanks for a copy of your excellent book 'Théorie Atomique', several chapters of which I have already read with much interest. It will doubtless prove a most useful help to students from the classes with which even the most abstruse chemical laws are expounded. [In] your chapter on the definition & historical development of the doctrine of atomicity I think you have doubtless quite unintentionally done me some injustice. Whilst you described the part which Williamson, Odling, Kekulé and others have taken in this all-important doctrine, no mention is made of what I have always thought was the first definite announcement of the law in a paper communicated by me to the Royal Society in May 1852 and printed in their Phil. Trans.

He went on to explain how he had come to propose a saturation capacity for metals on the basis of the behavior of some of the new organometallic compounds which he had prepared. 'Stannous ethide', for instance, will take up 'an additional atom of oxygen', but no more. As he said, such saturation capacity, or 'limitation of the atom-fixing power of elements appears to me to be the essence of the law of atomicity.' After referring to a paper of his to the British Association (22) in 1855 he continues:

> But it was not until the following year that the law of atomicity was applied by Kolbe and myself to carbon compounds, in a paper which, by mistake, appears in Kolbe's name only ("Experimental Researches" p. 148). Throughout the paper, however, the views advocated are spoken of as ours jointly *"wir sind der Ansicht"* etc. This paper is dated Dec. 1856, about 6 months before Kekulé wrote his first word about atomicity (See Ann. Pharm. 1857, 104, 132) and nearly 2 years before the appearance of his celebrated paper on carbon compounds.

He refers Wurtz, first to the critical notes in his own *Experimental Researches of* 1877 where he explicitly claims 'the discovery of the law of atomicity' (15), and then to Naumann's *Handbuch der allgemeinen und physikalischen Chemie* of the same year (23). This latter work cited with approval an address by Erlenmeyer 'On the task of chemical education' (24), which included the following remarks:

> Following the assumption by Williamson of monobasic and dibasic compound radicals in his 1851 paper on the constitution of salts, the regularity in the combining-ratios of different elements was first recognized by Frankland in the year 1853* and explained in terms of their saturation capacity. [Frankland spoke of triatomic and pentatomic groups]. In 1854 both Odling and Kekulé formulated such saturation laws for compounds and additionally for elementary radicals. In the year 1857* Kolbe had adopted Frankland's way of thinking, and further extended it to a number of carbon compounds (Ann. Pharm. 101, 257), in that he derived the latter from anhydrous carbonic acid (C_2O_4). In the same year Kekulé (Ann. Pharm. 1857, 104, 132) carried the idea of Types, merely anticipated by Gerhardt, with fuller clarity into a way of representing the chemical combination of compounds, because he attributed Gerhardt's Types to mono-, di- and triatomic elementary atoms, and in a certain sense Kolbe's carbonic acid Type to a tetratomic carbon atom. In the above case, and still more in a paper appearing in 1858 (Ann. Pharm. 106, 129), Kekulé gives us to understand that he

regarded it as necessary for representing the constitution of chemical compounds to bear in mind the constitution of radicals and to extend this even to the elements. Nevertheless he still held to fixed Types.

Frankland took exception to the attribution of the 1856/7 (25) paper to Kolbe alone [a natural mistake since the title had accidentally omitted his own name (26)]; and he objected to the two dates indicated above by asterisks (*), which should be respectively 1852 [though Erlenmeyer was actually citing the slightly later German edition and gave the reference (27) which Frankland omitted to quote] and 1856 [which was when Frankland said they *wrote* the paper (25)]. But he added that 'otherwise Erlenmeyer's statement is a very fair history of the development of the doctrine of atomicity.' He concluded his letter:

> In thus bringing under your notice the discovery of the essential principle of the law of atom-fixing power, *Sättigungscapacität*, or atomicity of elements, and, in conjunction with Kolbe, the first application of this law to compounds of carbon, I do not forget how much the present developments of this law owe to other chemists, especially to Kekulé and Cannizzaro. Indeed until the latter had placed the atomic weights of the metallic elements on their present consistent basis the satisfactory development of the doctrine was impossible. Whilst therefore I should not be surprised to hear different estimates put upon the value of the respective contributions of different chemists to the present condition of the law of atomicity, I do not think that my papers of 1852 and 1856 can be fairly altogether ignored in a history of this great chemical law.

8.22 Intervention by Kolbe

There the situation might have remained, as just a private correspondence between colleagues, but for two events which brought affairs into the open. The first of these was an intervention by one of chemistry's most famous [or notorious] polemicists, Hermann Kolbe. Ever a critic of what he saw as attempts to import mysticism into chemistry, Kolbe had long had Kekulé in his sights and had been engaged in verbal trench warfare with him for many years. Acrimonious attacks on chemical opponents had enlivened the pages of Kolbe's *Journal für praktische Chemie* for over a decade, and recently had concentrated on the formulae and structures suggested by Kekulé himself (28). Having failed to convince many of his contemporaries of the justice of his claims, Kolbe decided in 1881 to set the record straight by separately publishing his own historical account: *Meine Betheiligung an der Entwickelung der theoretische Chemie* (29). He wrote to his old colleague, Frankland, telling him of his intention [and suggesting he might like to provide an English translation](30). Frankland's reply (31) conveyed more than polite interest in Kolbe's venture, promising 'to take steps to have the book translated into English' when all was complete. He remarked:

> In regard to atomicity, I have always thought that Kekulé has behaved very ungenerously towards both of us in assuming the authorship of atomicity without a word of acknowledgement in respect of what we had previously published. I fully appreciate the services which Kekulé rendered to chemistry by his subsequent developments of the doctrine, but he has no claim whatever to the authorship of the theory or its application to carbon compounds and I am very glad to hear that you are about to put the matter in its real historical light.

Kolbe quickly responded with the first two sections of his *Betheiligung* and another letter (32), full of stinging accusations against Kekulé who had been guilty of 'such insolence' against them. Despite some notable experimental work 'Kekulé is an uneducated man; he cannot think or write logically.' The Foreword to his essay did not mince words either, speaking of 'all the crooked opinions and misrepresentations with which Kekulé's *Lehrbuch* is replete,' and alleging that Kekulé had partly concealed the work of Frankland and Kolbe, 'partly distorted it and partly appropriated it for himself.' Yet even Kolbe had second thoughts on one phrase. He had intended to announce that 'averse in principle to priority disputes I have hitherto kept silent on each distortion of the truth', but in the corrected page-proofs he replaced the inflammatory last phrase *Verunstaltungen der Wahreit* by the more innocuous 'commotion' [*Treiben*] (33).

In his letter Kolbe went on to address his friend as 'the principal founder of exact chemistry through your discovery of the saturation capacity of the elements' (32). He followed this with another epistle (34) containing the flattering suggestion that the *Betheiligung* should appear under their joint names. Not unnaturally Frankland read all this with 'much interest', agreeing with what Kolbe said about Kekulé and Baeyer [though not about Dumas, Wurtz, and Gerhardt]. His name could appear in the pamphlet's title 'provided that you state clearly in the preface that I have nothing to do with the authorship and that you alone are responsible for the contents.' He added this revealing comment:

> I never knew the former [Kekulé] had actually claimed the authorship of atomicity. That was *Frechheit & Unverschämmtheit* [audacity and effrontery](35).

Nor, indeed, was he to know that Kekulé had reiterated and amplified his own claims in a 'History of valency theory', submitted to *Annalen* as part of a paper on *Carboxytatronsäure* (36). He was dissuaded from publication by the editor, Volhard, and the essay remained unpublished until 1929 (37).

The last surviving letter from this exchange in 1881 is Kolbe's reply conceding that his name alone should appear on the title page, but wondering what Frankland was so worried about (38). Further abuse of the French chemists followed, particularly Dumas ['more shameless and impudent than Kekulé']. Their correspondence was nearly at an end (39) but Kolbe had succeeded in fanning the flames of hostility between Frankland and Kekulé, and confirming in his English colleague a sense of genuine grievance. Kekulé had not merely ignored him; he had stolen his discovery.

Yet still Frankland was not to be goaded into public claims, even though Kolbe had complained of his silence and urged that he 'enter the battle' (40). That silence was immensely significant, as we shall see. Another decade was to pass before one final stimulus initiated another brief exchange of letters and very nearly [but not quite] triggered a public response.

8.23 Recognition in the 1890's

Early in 1892 the Leipzig organic chemist Ernst von Meyer wrote to Frankland about the English translation (4l) of his *Geschichte der Chemie* (42). Though well received in England the book had brought a 'very angry' response from Kekulé over his part in the discovery of valency, and Meyer asked for Frankland's opinion (43). At last justice had been done and the contributions of all had been fairly evaluated. This at least was the opinion that Frankland expressed in a long and cordial reply (44); von Meyer had got it right:

> There is no longer any doubt that Kekulé was anticipated by 5 years in respect of Valency and by 6 months in respect of the application of the theory of carbon compounds.

Regarding Kekulé's counter-allegations [in 'very questionable taste'] Frankland remarked that if he still 'asserts in print his claims to the authorship of atomicity and its application to carbon, I shall certainly no longer remain silent but will defend Kolbe's and my position.' This however did not happen; as von Meyer noted some months later 'Kekulé has so far said nothing in public; from now on he is certainly silent' (45). And now, of course, the need for recriminations had diminished. Kolbe had been dead some years; Kekulé had made conciliatory noises in his famous Festschrift speech (46) though still maintaining his own priority; his life was nearing its end and, as Kolbe had long ago remarked (34), he was old and reluctant to fight. So yet again Frankland kept his own counsel. When, in 1897, he received an open letter from Anschütz (47), appealing for support to erect a monument to Kekulé, Frankland promised a subscription but wished to disassociate himself from the public appeal because of 'one or two statements ... to which I could not conscientiously subscribe'(48). These doubtless included the ascription to Kekulé of 'the laws of valency and atom-linking, and the views we hold today on the structure of carbon compounds.'

However vindication was to come in the form of Japp's Memorial Lecture for Kekulé (13). In reply to an appreciative letter from Frankland, Japp observed that 'of late years, German chemists have shown themselves more just than formerly, in discussing your claims' (49). He noted how Anschütz, in a new edition of Richter's *Organische Chemie,* said that Frankland enunciated the theory of valency 'for the elements of the nitrogen group' (50), and that Königs [another pupil of Kekulé] said much the same in an obituary notice for his teacher. Though unintentional, this was still an 'unjust limitation,' of Frankland's achievement, probably derived from Ladenburg's *Entwicklungsgeschichte der Chemie* (51), but 'an improvement on their former treatment.' There was still something to fight for, but only in his autobiographical *Sketches* did Frankland attempt a further printed vindication of his claim (52). In a book intended for the amusement of his family, this manifesto consisted mainly of extensive reproduction of passages from his *Experimental Researches* of 1878. In fact there was little new to say.

8.3 Questions of Chemical Ideology

So far it may seem that the estrangement between Kekulé and Frankland was little more than a squalid dispute about priorities. If so it is a common enough occurrence in human affairs [including science] and calls for no special comment. However its maintenance over several decades appears to betoken something more than mere pique, or even deeply hurt pride. This impression is confirmed by the protagonists rhetoric, suggesting a high wall of misunderstanding between them, and the adoption of mutually incompatible viewpoints. Only in these terms does their long hostility even begin to be explicable. In fact they found themselves on opposite sides of three major conflicts of 19th century chemistry. The first was about atomic weights.

8.31 Atomic weights

When Frankland and Kolbe, separately (53) or together (54) presented the early papers that seem, at least in retrospect, to have laid the foundations of valency theory they did so in terms of the old atomic weight system (C = 6; O = 8). Although a common practice at the time, when atoms and equivalents were often undifferentiated, it greatly offended Kekulé, the convener of the Karlsruhe Conference. He wrote:

> Frankland ... spoke now of atoms, now of equivalents, whereas valency theory assumed a sharp difference between the two ideas. Frankland argued with a false atomic weight for oxygen ... The law refers not to atoms but equivalents. It is right for equivalents but false for atoms, and therefore false in the form arrived at by Frankland (55).

Kekulé's contention was therefore not that Frankland was using the wrong atomic weights but that *he was not using any at all.* It is idle to argue from hindsight whether or not Kekulé was right to dismiss Frankland out of court so brusquely. The fact is that he did it.

These words of Kekulé were never read by Frankland, [though others such as Baeyer made essentially the same point (56)] but he was in little doubt of at least this reason for his rejection. As he wrote to Kolbe in 1881:

> In our application of the theory to carbon compounds, it was no doubt unfortunate for us that we were still hampered with the old atomic weights of C & O. It is owing to this circumstance alone, in my opinion, that our ideas were never properly recognized & appreciated; but this unfortunate circumstance obviously gives no right to others to appropriate our ideas, applying them to revised formulae and then re-issue them as their own new & original inventions (57).

Writing ten years later in similar vein to von Meyer, he adds the revealing admission, not only that others were alienated, but that *they themselves were inhibited* by their own misjudgment:

> One of the chief causes of the neglect with which the views of Kolbe and myself were treated was the blunder we made in adhering too long to the small atomic weights of C and O. This is the only reason why we did not develop the valency of C and *die Verkettung der Atome* years before they occurred to Kekulé (58).

Yet we may well ask if this is a sufficient explanation. Perhaps the fairest contemporary assessment offers a clue to an additional element in the ideological division. F. R. Japp, like Armstrong, had worked closely with both men, and he suggested the following:

> A discovery, made by an adherent of the radical theory and correctly formulated by him in terms of the old equivalents, does not

> become the property of the first adherent of the type theory who happens to translate it into the new molecular weights (59).

Japp's reference to 'the type theory' was not merely incidental. It may offer an additional insight into the Frankland/Kekulé feud: was this estrangement also a legacy of the war between radicals and types that had so divided the previous generation of organic chemists? It seems probable that it was.

8.32 *The Theory of Types*

As is well known, valency emerged as a joint offspring of the radical and type traditions of organic chemistry, the one stretching back to Berzelius and ultimately Lavoisier, the other owing its lineage to Gerhardt, Laurent, and Dumas. Kekulé's great achievement was commonly held to be the establishment of the marsh gas type (CH_4), analogous to ammonia, water and hydrogen chloride:

$$\mathrm{H\ Cl} \qquad \left.\begin{array}{l}\mathrm{H}\\ \mathrm{H}\end{array}\right\}\mathrm{O} \qquad \left.\begin{array}{l}\mathrm{H}\\ \mathrm{H}\\ \mathrm{H}\end{array}\right\}\mathrm{N} \qquad \left.\begin{array}{l}\mathrm{H}\\ \mathrm{H}\\ \mathrm{H}\\ \mathrm{H}\end{array}\right\}\mathrm{C}$$

thus effectively recognizing a tetravalent carbon atom. Whether Kekulé actually *intended* to place the marsh-gas type alongside the other three has been questioned (60), and Frankland considered (61) that he and Kolbe had arrived at essentially the same conclusion [via radical theory] six months before Kekulé (62). Yet the fact remains that in Kekulé's eyes the theory of valency was indissolubly linked to the theory of types, so there was no room for pretenders who founded their ideas on such heretical concepts as outdated radicals. Even though Frankland had used the symbolism of types in his classic hope that type and radical theories might be reconciled, Kekulé remained adamantly opposed to his English rival. But there was more to the story than simply a partisan refusal to recognize merit in another viewpoint.

The theory of types showed remarkable resilience and survived, at least in vestigial forms, long after valency had rendered it redundant as an explanatory tool. Yet it must have seemed that Frankland was intent on undermining the very foundations of type theory and thus of Kekulé's chemical philosophy. Such a program would be very unpopular with all who owed allegiance to Kekulé. Yet it is doubtful if Frankland intended a sustained campaign; his private papers and publications suggest a gradual disenchantment.

A letter of 1860 to Lyon Playfair (64) explained his ideas of saturation and unsaturation in terms of *biatomic* [divalent] and *quadratomic* [tetravalent] *double atoms* of carbon, as in C_2Cl_4 and C_2O_4; 'all organic compounds of the C_nH_{n+l} and C_nH_{n-1} families are constructed upon these types', as for instance *trichlorhydrin* [1, 2, 3-trichloropropane] and *chloride of allyl* [3-chloro-1-propene]:

$$C_2\left\{\begin{array}{l} C_4H_5 \\ Cl \\ Cl \\ Cl \end{array}\right. \qquad \text{and} \qquad C_2\left\{\begin{array}{l} C_4H_5 \\ \\ Cl \end{array}\right.$$

In other words, he attributed the unsaturation of alkenes and kindred compounds to change in valency of the carbon atoms, but still used the language of types to express it. Next year he wrote to Cahours expressing glowing appreciation of Dumas' types (65). He continued to use type formulae until 1866 (66), but shortly afterwards abandoned them for the now familiar system of linear *bonds* [the very term being Frankland's gift to chemistry (67)]. Taking up an idea of the young Scottish chemist A. Crum Brown, he wrote to him applauding his new symbolism and declaring 'the water-type, after doing good service, is quite worn out' (68). A letter by return of post from the delighted Scotsman gave details of an Edinburgh firm of printers who could make castings for printing the new formulae, in 'two sizes of circles' to contain the atomic symbols (69). So, when Frankland's textbook *Lecture Notes for Chemical Students* appeared shortly afterwards, it was adorned with the new formulae, and introduced with the words:

> I have entirely adopted the graphic notation of Crum Brown which appears to me to possess several important advantages over that first proposed by Kekulé (70).

By a truly remarkable misfortune even the *glyptic formulae,* based on ball-and-stick models, were wrongly attributed to Kekulé in the second edition of Schorlemmer's *Rise and Development of Organic Chemistry* (71) As Frankland asked the editor/translator 'are we not indebted to Crum Brown rather than to Kekulé for graphic and glyptic formulae?' (72)

If type notation was out, its theoretical underpinning was in considerable danger. Frankland did not bother to attack type theory further in print; he simply threw it away as the husks of an outworn dogma, no longer of service to chemistry. His private papers reveal his attitude clearly enough. Kolbe suggested why Wurtz, Williamson and Odling may have overlooked their work on atomicity: 'it was the death blow to the types upon which so much of their reputation had been built.' Their continued belief in types 'betrayed a complete ignorance of the first principles of chemical combination' (73) A later [undated] scrap of paper in Frankland's writing contains the sentence: 'No one having even a conception of valency could have imagined such things as double or treble water types or mixed types' (74).

Whether or not Kekulé knew of such sentiments it is not hard to understand how his adherence to the theory of types so greatly diminished him in the eyes of Frankland.

8.33 Structure Theory

The so-called 'structure theory' of organic chemistry did not emerge overnight, nor can any one man be said to have been its founder, including Kekulé (75). In its developing phase there was plenty of room for ambiguities, to which some further misunderstandings between Frankland and Kekulé can be plausibly attributed. Two things are clear: that Kekulé was commonly regarded as the champion of structure theory (76), and that Frankland did not respond to the clarion call to declare himself a devoted supporter. Frankland's own views were, to say the least, at times inconsistent. On the very basic question 'who founded structural chemistry?' he gave conflicting answers. To von Meyer he announced in 1892 that 'I have always considered Kekulé to be the founder of *Strukturchemie'* (77), but to Ost he confided in 1884 that Kolbe was, 'if the title can be assigned to any one man, the father of *Strukturchemie'* (78). Such inconsistencies may be partly explained on the assumption that Frankland wrote what he thought his correspondents wanted to read, or less cynically that his own views changed over the eight intervening years. However, he could be using the same terms with different shades of meaning, and this seems to have been the case.

For Frankland 'structure theory' seems primarily to have meant the abundant use of graphic, or constitutional, formulae as in his *Lecture Notes.* Receipt of this book triggered a hostile response by Kolbe, who felt the graphic formulae were expendable, doubtful and positively misleading to students who might infer too much from them (79). Yet in Frankland's view they enable 'constitution to be expressed in the simplest and clearest manner' and are 'merely symbolic expressions of atomicity' (80). When, years later, Kolbe sought to enlist his support in a crusade against *Strukturtheorie* Frankland was horrified, and emphatically stated his view that

> Chemistry owes its progress from empiricism to exact science entirely [so far as theoretical conceptions are concerned] to Strukturchemie. Without Strukturchemie there is no science of chemistry (81).

Nevertheless he had long regarded the ideas on chemical constitution associated with Gerhardt, Kekulé, Erlenmeyer and others as 'absurd', as worthy of Kolbe's 'severest criticism', and greatly inferior to his own system using 'the very neutral word *bond*' (82).

For Frankland, as for Kolbe, a difference from the *Strukturchemiker* of the Kekulé school was that 'they deny the actual existence in compounds of compound radicals.' For them methyl, ethyl etc. were only ideal entities of great convenience (83). This insight informed their early work, was totally absent in Kekulé, and continued to insulate both of them from the Kekulé school.

Finally, as Frankland pointed out, structural formulae do need to be correct. Being a *Strukturchemiker* does not guarantee infallibility. Thus, he agreed with Kolbe's criticisms of structures suggested by Baeyer who 'either cannot write clearly or does not think it worth his while to write clearly' (84). Similarly, although in his *Lecture Notes* Frankland used Kekulé's hexagon formula for benzene, he admitted privately that he preferred Ladenburg's prism formula, but had been over-persuaded by his assistant Japp, 'a fanatical disciple of Kekulé' (85).

In all these ways it is clear that Frankland's view of Kekulé's chemistry was highly ambivalent, with suspicion mingled with a somewhat qualified admiration. But perhaps the most persuasive explanation for their alienation lies in the personal circumstances of each man.

8.4 Chemists in Context

Nothing is easier than to construct elaborate explanatory structures in history when the simplest and most obvious explanations lie to hand, unnoticed and overlooked by virtue of their very simplicity. While matters of chemical theory and chemical priority were unquestionably important elements in the Frankland/Kekulé relationship, a glance at their personal circumstances will immediately suggest others, perhaps even more important.

If their paths seldom crossed in the 1850's that in itself does not to bear a necessarily sinister interpretation. Thus consider the remarks by Armstrong about their failure to communicate in 1854 (3). Armstrong's complaint of a failure by Kekulé to contact Frankland in 1854 overlooks the latter's almost continuous absence from London. He was then struggling hard to fulfill his responsibilities as the first Professor of Chemistry at Owens College, Manchester, and visits to the capital were a rare luxury. In the unexpurgated first edition of his *Sketches* [London, 1901] Frankland remarks that 'after April, 1854 there were two babies to attend to, and we were, therefore, less mobile than hitherto'; the only visit to London recorded was for a lecture which he gave on 'Artificial light' at the Royal Institution, his spare time being devoted to showing his German sister-in-law the new Crystal Palace (86). Moreover it is not at all clear why in the early years Frankland *should* have sought out a relatively unknown research student, four years his junior. However, quite the same cannot be said of Kekulé, for Frankland's paper of 1852 did raise fundamental questions about chemical structure [though not by that name] which were later to concern Kekulé so deeply and which might even in 1854 have seemed relevant. Frankland's 1852 paper does not seem to have aroused much comment anywhere for some considerable time [except for the striking discovery of organometallics], so Kekulé may have been quite typical in failing to notice its general importance.

After Kekulé's return to Germany he gradually built up an international reputation as an organic chemist, particularly as a theoretician. He did little work outside organic chemistry. His influence was consolidated and extended by his *Lehrbuch der organischen Chemie* which began to appear in 1859. Above all his insights into the structure of benzene [from 1865] made an indelible mark on the whole science. Frankland, by contrast, spent his Manchester years [1851-1857] on a different intellectual plane. He wrote a few papers on a subject that was not obviously connected with the classical organic chemistry then taking shape in Germany: organometallic chemistry (87). Judging by copy-letter books that have survived from that period(88) most of his scientific activities were connected with technical consultancies of a bewildering variety, doubtless to supplement his stipend as a professor. Not until he was well established in London, in the 1860's, did Frankland return to serious research on organic chemistry, when he worked on aliphatic esters, acids and ketones, often in collaboration with B. F. Duppa (89). From then onwards he became preoccupied with the scientific establishment (90), the Rivers Commission and a whirl of social engagements befitting an upwardly mobile, ambitious, metropolitan chemist. By 1870 he had virtually finished original investigations in organic chemistry, and by 1877 the publication of his *Experimental Researches* was a public signal that academic research was largely over. In von Meyer's opinion Frankland's 'retirement' from organic chemistry by 1870 was a major explanation for the neglect of his work in the priority disputes about valency (91). Not since his Marburg days had Frankland belonged to a large and active school of organic research. Kekulé, on the other hand, was now at the head of a growing school of organic research, though his own active research days were numbered owing to ill-health and premature aging. If the paths of the two men rarely crossed in a literal sense, this is more profoundly true in the realms of ideas and cultural values.

It is tempting to perceive nationalistic overtones in all this. In a sense that is quite justified. Kekulé did personify the archetypal German professor: secure in a university supported by the state and respected by the local community; the father-figure to generations of loyal research assistants; dedicated to research that was both deep and narrow, of which organic chemistry was the perfect paradigm; and patriotically convinced of the superiority of German science. Similarly Frankland may have seemed [especially to Kekulé] an embodiment of the very opposite characteristics: an academic whose teaching often left little time for research, and whose research interests would cover a wide range of subjects. Unlike his German counterpart an English scientist had to devote much of his energies to persuading society of the value of his work, cajoling government into providing more money and to supplementing his low academic income from consultancies which would often dominate his working days. All this was certainly true of Frankland and may help to explain something of the cultural antipathy between the two men. It would almost have been possible for Kekulé to regard Frankland as something of a renegade from the noble science he once espoused.

Yet this is all too simple. Frankland was far from hostile to German values. His first wife was German, his own Ph.D. was from Marburg, and he sent his son Percy to study at Würzburg under Wislicenus. His letters from the vitriolic Kolbe, though admittedly idiosyncratic, for ever eliminate the myth of a monolithic German chemical culture. Kolbe derides as 'chemical nonsense' addresses given by Baeyer and his master Kekulé (92), and chides Frankland for sending Percy to study under Wislicenus ['no longer an exact scientist but a *Naturphilosophe,* a metaphysician'](93). Kolbe's dislike of Kekulé, which may well have infected Frankland, spilled over into his even greater dislike of all things French after the Franco-Prussian war of 1870. This was specially true of the type theories of Dumas. Gerhardt, Laurent and Wurtz, and what now alarmed him was 'the spread of this dogmatism to Germany, in the person of Kekulé and his structuralist school' (94).

So perhaps Armstrong was right after all. If one German was guilty of selling out to the French he was hardly likely to be *persona non grata* to Frankland merely because of his nationality. Perhaps the best understanding will be reached when the psychological factors have been properly assessed. Psychohistoriography is a notorious minefield for the unwary but, in conclusion, a few fairly obvious remarks may be made. On Kekulé's side there is little doubt that he revelled in polemics. As Farrar has observed he was not a particularly good lecturer or bench chemist (95). This fact, combined with a remarkable critical acumen, would naturally lead to both a defensive attitude towards what he had accomplished, and an aggressive stance in public about his priority claims. Kekulé's training as an architect probably endowed him with a better than average grasp of spatial relationships and thus a greater commitment to structure theory. Given the misunderstandings over what constituted the true foundation of valency theory, the failure of Frankland to produce anything comparable to the benzene theory, his early withdrawal from academic research, his known friendship with the arch-enemy Kolbe, and the simple fact that they hardly ever met, it is not surprising that Kekulé saw no need to make advances of any kind in Frankland's direction.

From Frankland's perspective matters stood very differently. Throughout his life Frankland possessed a superb experimental technique, probably a legacy of his pharmaceutical apprenticeship. Armstrong said he never knew anyone with greater experimental ability (96). In this respect he differed from Kekulé, just as he did in a lack of visionary tendencies. It stood him in good stead in the multitude of practical tasks to which he was drawn in later life by a combination of financial ambition, government pressure and social conscience. Deeply sensitive about his own position in society he was far less secure than Kekulé, and indeed he sprang from much humbler origins, having also to guard the secret of illegitimacy (97). His natural shyness deterred him from public argument, as he often admitted in private: 'I have always been very reluctant to enter upon controversies and have no doubt suffered in consequence' (98), though he persuaded himself that 'by *Privatebriefe* I have obtained what would probably not have been got by any amount of publication' (99), but he was not readily deflected from a social ambition, reflecting his humble origins and evident on a hundred pages of his letters and autobiographical notes. If Kekulé wished to dominate the European chemical scene Frankland needed to be appreciated by the social and political leaders of his own country, not least the rising hegemony of scientific naturalists. And so he strove to be the typical late Victorian English gentleman, with a country seat, well travelled, given to hospitality in the Athenaeum. If Kekulé was to be ennobled by a grateful Emperor who had once been his pupil, to Frankland came the ultimate accolade of a K.C.B in the Queen's Jubilee Honours List.

Was it then Frankland's Englishness that served to distance him from Kekulé? That is a dangerous assumption for it hypostases a quality that is in essence indefinable. Moreover, there is always the hazard of generalizing on the basis of the individual, and in that individual and his context lie the best explanations of his behavior. One final illustration must suffice.

After the demise of a distinguished scientist it fell to the President of the Royal Society to make appropriate remarks in his next Annual Address. Following Kekulé's death the President, Lord Lister, sought relevant facts about him. He wished to know whether Kekulé was of comparable stature to Fizeau who had also died, and 'whether his work was of such world-wide importance as to make it desirable for me to refer to him specially in my address' (100). Remarkably, the informant to whom he wrote was Frankland. He requested him to 'ask some one to furnish me with some details concerning him [Kekulé]', and this was duly done. Lister's address then included the following remarks:

> The death of August Kekulé will be felt as a severe loss to chemical science all over the world. Not only did his great activity

> in original research enrich organic chemistry with many new and interesting compounds, but his announcement of the tetradic valency of carbon, and, especially, his theoretical conception of the benzene ring, gave an impulse to the study of structural chemistry which has introduced order into the vast array of organic compounds, both of the alcoholic and aromatic types and has not yet fully expended itself (101).

If any statement may be said to represent a typically English view of one man's work, that must surely be the one. But its most interesting feature is not obvious from the text. Frankland's response to Lister was to write a note himself; perhaps he saw this as a last opportunity to set the record straight? Even more remarkable is the fact that, with the addition of a short sentence about Kekulé's Copley Medal, *Lister's text is identical with the note furnished by Frankland* (102). Through no less a mouthpiece than the President of the Royal Society Frankland was at last able to deliver a public evaluation of his great colleague and rival, August Kekulé.

References and Notes

Nearly all the Frankland papers are on microfilm at the Open University, U.K. They are referred to (a) by the name of the collection, and (b) by an unambiguous 8-digit number designating their location on an Open University microfilm. The first part of this number denotes their provenance, the second the number of the film, and the third the first frame on which the document appears. For further details see note 4 (below).

1. Lundgren A, *Berzelius och den kemiska atomteorin,* Uppsala Academy of Sciences, Uppsala, 1979

2. *Sketches from the Life of Sir Edward Frankland,* second edition, edited and completed by his daughters M. N. W[est] and S. J. C[olenso], Spottiswoode & Co., London, 1902; Brock WH, 'Edward Frankland ', *Dictionary of Scientific Biography;* Russell CA, *Lancastrian Chemist: the Early Years of Sir Edward Frankland,* Open University Press, Milton Keynes, 1986 (the first volume of a two-volume biography currently in preparation)

3. Armstrong HE, Oration at the Lancastrian Frankland Society, Lancaster Guardian, 26 January, 1934

4. Russell Colin A and Russell Shirley P (1990) The Archives of Sir Edward Frankland: Resources, Problems and Methods, Brit J Hist Sci, 23: 175

5. Letters from Kekulé to Frankland, 18 January, 1863 and 23 December, 1883,

Raven Frankland Collection [O.U. microfilm nos. 01.02.0081 and 0080]

6. Letter from Frankland to his wife, 1 May, 1867, reproduced in Frankland (note 2), p. 400

7. Anschütz R (1929) *August Kekulé,* Berlin, Verlag Chemie, vol. i, pp. 39: 41

8. Frankland E (1877) *Experimental Researches in Pure Applied and Physical Chemistry,* London, van Voorst: p. 283

9. Kekulé FA (1864) Compt rend, 58: 510-514

10. Kekulé FA (1857), Ann Chem u Pharm, 104: 129-150

11. Kekulé FA (1859), letter to Wurtz, 15 Feb., 1859, reproduced in Anschütz (note 7), vol. i: p., 147

12. Kekulé FA (1881), *Lehrbuch der organischen Chemie,* Erlangen, Enke

13. Japp FR (1897) 'Kekulé Memorial Lecture' *Chemical Society Memorial Lectures* No. VII: pp. 118-9

14. Anschütz R (Ref 7), vol. i, p.80. See also Chapter 16 and 17 for other examples of Kekulé's selective memory

15. Frankland E, (Ref 8), p. 154

16. Frankland E (1872), *Lecture Notes for Chemical Students,* 2nd ed., London, Vaan Voorst vol. ii: p. 42.

17. This question has been fully discussed in Russell CA, *The History of Valency,* Leicester University Press, Leicester, 1971, pp. 108-134

18. Wurtz A (1858), Rép. de chimie pure, 1, 24n (an editorial footnote to a paper by Kekulé (pp. 20-24); Ann chim phys, 1859 [3], 55: 400

19. Wurtz A (1869), *Histoire des doctrines chimiques depuis Lavoisier jusqu´á nos jours,* Paris, Hachette English translation by Watts H,(1869) *History of chemical theory from the age of Lavoisier to the present day,* London, Macmillan, 1869

20. Wurtz A (1880), *La théorie atomique,* Paris, Baillière There were several later editions, and an English translation by Cleminshaw E (1880), The *Atomic Theory,* London, Paul

21. Copy-letter (or possibly draft-letter) from Frankland to Wurtz, 25 Dee., 1878,

96

Raven Frankland Collection [O.U. microfilm no. 01.04.0629]. Presumably he felt sufficiently strongly about the matter to spend part of Christmas Day compiling his reply [however that morning the rest of the family went to church and in the afternoon his in-laws were entertained. (Margaret Frankland's Diary, Myra Bucknall Collection); letter-writing may have been a welcome diversion!]

22. Frankland E (1855) Rep Brit Assoc, Trans Sections, 62 [a brief report of his work on diethylzinc]

23. Neumann A (1877) *Handbuch der allgemeinen und physikalischen Chemie,* Heidelberg: Winter pp. 73-4

24. Erlenmeyer E (1871) *Ueber die Aufgabe des chemischen Unterrichts,* Munich: von Bayern p.26

25. Kolbe H (1857) Ann Chem u Pharm, 101: 256

26. Frankland E, (Ref 8): p. 148

27. Frankland E (1853), Ann Chem u Pharm, 85: 329-373 (368); this date, instead of 1852 [note 1887, 53a, below], continues to be perpetuated in German literature, e.g., Neufeldt S, *Chronologie Chemie, VCH,* Weinheim: p. 43

28. Phillips JP (1866), Liebig and Kolbe, critical editors, Chymia, 11: 89

29. Kolbe H (1881) J prakt Chem, 131: 305, 353, 497; 132: 374

30. Letter from Kolbe to Frankland, 11 Feb., 1881, Raven Frankland Collection [O.U. microfilm no. 01.02.1447]

31. Letter from Frankland to Kolbe, 6 March, 1881, Deutsches Museum Archive, Munich, Urkundensammlung Nr. 3570

32. Letter from Kolbe to Frankland, 20 March, 1881, Raven Frankland Collection [O.U. microfilm no. 01.02.1439]

33. Corrected page-proofs marked *Vorwort*, P.1, Raven Frankland Collection [O.U. microfilm no. 01.02.1444]

34. Letter from Kolbe to Frankland 23 March [1881], Raven Frankland Collection [O.U. microfilm no. 01.02.1442]

35. Letter from Frankland to Kolbe, 29 March, 1881, Deutsches Museum Archive, Munich, Urkundensammlung Nr. 3571

36. Kekulé FA (1883), Ann Chem u Pharm, 221: 230

37. Anschütz R (Ref 7), vol. i: p. 554-578. See also Chapter 17 in this book

38. Letter from Kolbe to Frankland, 1 April, 1881, Raven Frankland Collection [O.U. microfilm no. 01.02.1436]

39. The only later letters known to have survived are from 1883: two from Kolbe to Frankland (30 July and 28 October, Raven Frankland Collection (O.U. microfilm nos. 01.02.1533 and 1526]) and one from Frankland to Kolbe (23 September, Deutsches Museum Archive, Munich, Urkundensammlung Nr. 3573)

40. Letter from Kolbe to Frankland, 30 July, 1881, Raven Frankland Collection [O.U. microfilm no. 01.02.1533]

41. von Meyer E (1891) *History of Chemistry from the Earliest Times to the Present Day, being also an Introduction to the Study of the Science,* trans. G. McGowan, London, Macmillan

42. von Meyer E (1889), *Geschichte der Chemie, von der ältesten Zeiten bis zur Gegenwart. Zugleich Einführung in das Studium der Chemie,* Leipzig, Verlag von Veit

43. Letter from von Meyer to Frankland, 18 January, 1892, Raven Frankland Collection [O.U. microfilm no. 01.07.0486]

44. Copy-letter (or draft-letter) from Frankland to von Meyer, 28 January, 1892, Raven Frankland Collection [O.U. microfilm no. 01.01.0852]

45. Letter from von Meyer to Frankland, 10 December, 1892, Raven Frankland Collection [O.U. microfilm no. 01.01.0839]

46. Kekulé FA (1890) Ber Deutsch Chem Gesell, 23: 1304

47. Printed *Aufruf zur Errichtung eines Denkmals für August Kekulé in Bonn,* 1897, signed by Anschütz, Baeyer, Claisen, Glaser, Klinger, Ladenburg, Wichelhaus and Zincke, Raven Frankland Collection [O.U. microfilm no. 01.01.0831]

48. Copy-letter from Frankland to Anschütz, 14 July, 1897, Raven Frankland Collection [O.U. microfilm no. 01.01.0833]

98

49. Letter from Japp to Frankland, 3 March, 1898, Raven Frankland Collection [O.U. microfilm no. 01.02.0675]

50. Anschütz R (ed.)(1899), *Chemie der Kohlenstoffverbindungen* by V. von Richter, e.g. 3rd American edition, *Richter's Organic Chemistry* (trans. Smith EF), London, pp. 36-7; the error was still being repeated in the English edition of 1915 (trans. Spielmann PE), London, Kegan Paul 1915: p. 21

51. Ladenburg A (1888) *Vorträge der Entwicklungsgeschichte der Chemie in den letzen hundert Jahren,* Braunschweig, Vieweg: pp. 245-8, etc

52. Frankland E, (Ref 2): pp. 186-203

53. Frankland E (1852) On a new series of organic bodies containing metals, Phil Trans, [142]: 417; Kolbe H, Ann Chem u Pharm, 1850, 75: 211

54. [Frankland E and] Kolbe H, (1857) On the constitution of fatty and aromatic acids, Ann Chem u Pharm, 101: 256 [Frankland's name being inadvertently omitted]

55. Kekulé FA, (Ref 7)*'History of Valency theory',* in Anschütz: p. 561

56. Baeyer A (1858) Ann Chem u Pharm, 105: 265

57. Letter from Frankland to Kolbe, 6 March, 1881, Deutsches Museum Archive, Munich, Urkundensammlung Nr. 3570

58. Frankland to von Meyer (Ref 44)

59. Japp FR (1898) Kekulé Memorial Lecture, J Chem Soc: 73: 97

60. von Meyer E (1898) *A History of Chemistry* (Ref 41), 2nd ed., p. 309

61. Frankland to von Meyer (Ref 44); c.f. Frankland E and Duppa BF, Synthesis of butyric and caproic ethers from acetic ether, Proc Roy Soc, 14: 191

62. Frankland E and Kolbe H (Ref 54)

63. Frankland E (Ref 53), p. 441-2

64. Copy-letter from Frankland E to Playfair L, 24 June, 1860, Raven Frankland Collection [O.U. microfilm no. 01.04.0596]

65. Copy-letter from Frankland E to Cahours, A.A.T. 1 January, 1861, Raven Frankland Collection [O.U. microfilm no. 01.04.0011]

66. Probably for the last time in his paper on notation, Frankland E (1866) Contributions to the notation of organic and inorganic compounds, J Chem Soc, 19: 372

67. Ref 66: p. 377

68. Letter from Frankland to Crum Brown, 4 June, 1866, reproduced in J Chem Soc, (1923): 123, 3422

69. Letter from Crum Brown to Frankland, 5 June, 1866, Raven Frankland Collection [O.U. microfilm no. 01.04.1266]

70. Frankland E (1866) *Lecture Notes for Chemical Students embracing Mineral and Organic Chemistry,* London,van Voorst: p. v.

71. Schorlemmer C (1894) *Rise and Development of Organic Chemistry,* 2nd ed., ed. and trans. A Smithells, London, Macmillan: p. 117

72. Copy-letter from Frankland to Smithells A, 5 June, 1896, Raven Frankland Collection [O.U. microfilm no. 01.07.05243

73. Letter from Frankland to Kolbe, 19 May, 1881, Deutsches Museum Archive, Munich, Urkundensammlung Nr. 3572

74. Undated note by Frankland, Raven Frankland Collection [O.U. microfilm no. 01.01.0877]

75. Russell (Ref 17): pp. 146-152

76. Thus Kekulé was described in the Appeal for his statue (Ref 47)

77. Copy-letter from Frankland to von Meyer [1892](Ref 44)

78. Frankland E, obituary note for Kolbe accompanying letter to Ost H, 20 December, 1884, Deutsches Museum Archive, Munich, Urkundensammlung Nr. 3576

79. Letter from Kolbe to Frankland, 9 July, 1867, Raven Frankland Collection [O.U. microfilm no. 01.04.1374]

80. Letter from Frankland to Kolbe, 3 December, 1871, Deutsches Museum Archive, Munich, Urkundensammlung Nr. 3566

81. Letter from Frankland to Kolbe, 23 September, 1883, Deutsches Museum Archive, Munich, Urkundensammlung Nr. 3573

82. Letter from Frankland to Kolbe (Ref 80)

83. Letter from Kolbe to Frankland, 21 April, 1871, Raven Frankland Collection [O.U. microfilm no. 01.03.0600]

84. Letter from Frankland to Kolbe (Ref 81)

85. Letter from Frankland to Kolbe (Ref 57).

86. Frankland E, Sketches (note 2), lst ed., 1901:, p. 310

87. Frankland E,, (1855) Researches on organometallic bodies, Phil Trans 145: 259; 1856, 147: 59; Proc Roy Soc, 1857, 8: 502

88. Several copy-letter books from Owens College in Raven Frankland Collection: 1851-2, 1852-3, 1853-6 [O.U. microfilm nos. respectively 01.06.0109-0219, 01.06.0220-0364 and 01.08.0013-0278]

89. A large number of papers from 1863 to 1870, recorded in *Experimental Researches* (Ref 8): p. 299

90. The Chemical Society, The Royal Society, the X-Club, etc.

91. von Meyer E (Ref 60)

92. Letter from Kolbe to Frankland, 27 November, 1878, Raven Frankland Collection [O.U. microfilm o. 01.02.1423]

93. Letter from Kolbe to Frankland, 5 January, 1879, Raven Frankland Collection [O.U. microfilm no. 01.02.1432]

94. Rocke A (1987) Kolbe versus the "Transcendental chemists": the emergence of classical organic chemistry, Ambix, 34: 156

95. Farrar WV (1969) 'Kekulé', *A Biographical Dictionary of Scientists,* ed. T. I. Williams, Black, London

96. Armstrong HE (1933), Proc Inst Chem, 57: 62

97. Russell CA (Ref 2): p. 162

98. Copy-letter from Frankland to von Meyer (Ref 44)

99. Letter from Frankland to Kolbe (Ref 81)

100. Letter from Lister to Frankland, 10 November, 1896, Raven Frankland Collection [O.U. microfilm no. 01.02.0503]

101. Lister J, (1896) Presidential Address to-the Royal Society, 30 November, pp. 4-5, Lister Papers no. 70, Royal College of Physicians, London; see Proc Roy Soc

102. Manuscript note in Frankland's hand, undated, Raven Frankland Collection [O.U. microfilm no. 01.02.0505]

Publisher's Note: This article was originally published as Chapter 8 in *The Kekulé Riddle: A Challenge for Chemists and Psychologists*, ed. J.H. Wotiz (Vienna, IL, 1993). The original numbering of the sections has been retained.

XIII

EARLY CONCEPTS OF AROMATIC SUBSTITUTION

12.1 Introduction

In 1865 the cyclic structure of benzene was established (1). Within fifteen years it had become widely accepted, though pockets of skepticism remained (2). Meanwhile the daunting problems of isomerism in substituted benzenes were gradually solved. At first the now familiar terms *ortho, meta and para* were used with complete arbitrariness. Kekulé, for example, believed that *ortho* compounds were derived from aniline, and *meta* compounds from [m-]dinitrobenzene (3). To establish reliable standards of reference required, in Lachman's words, 'ten years of incessant labour' in what was to be veritably 'a chemical Eldorado' (4). Above all it was the work of Kekulé's former assistant Wilhelm Körner [1839-1926](5) who established the configurations of *ortho, meta and para* derivatives, in essence by examining the conversions [not necessarily in one stage]:

$$C_6H_4X_2 \longrightarrow C_6H_3X_3$$

and separating and counting the numbers of trisubstituted products [2 from *ortho,* 3 from *meta* and 1 from *para*](6).

This classic work, reported to generations of students ever since, raised the further question: *what determines the position of substituents in polysubstituted benzenes?* Logically it is the next problem to solve as again a century of undergraduate instruction has emphasised. However, the historical order of events is not necessarily that which our *post hoc* insights might suggest to be the most logical. Many counter-examples exist to warn us against writing history deductively. Yet in this instance there seems at first sight to be a strong possibility that the historical sequence of events might bear an almost uncanny resemblance to the, logical order, as might have been prescribed for a modern research program. Whether this coincidence is real, and if so some of the implications, will be considered at the end of the present chapter. However, appearances can be deceptive, and an apparent 'linearity' of connected developments raises an intriguing historical question.

It is usually assumed by historians of science that history about successes is one-sided, unreal and open to the ultimate condemnation of being *Whiggish.* What we need, they say, is not a linear account of progress in science but an attempt to explain and understand all that happened in a given period, success, failure, the lot. We must explore the blind alleys as much as the highways of science. Only then shall we really begin to understand the nature of the scientific enterprise. So the question to be addressed may be put in these terms: does the early history of aromatic substitution conform to a *Whig* interpretation of history, or is the linearity of ideas merely an artifact of the historian?

Before an even tentative assessment can be given it is necessary to survey briefly the main developments, to comment in passing on the historical aptness of some modern critiques, and finally to reflect briefly on some implications of the story.

The first stage was the emergence of certain rule-of-thumb generalizations, or empirical rules.

12.2 Early Empirical Rules for Aromatic Substitution

12.21 Empirical Rules Involving Acidity

W. Körner It would have been surprising if Körner (4) had failed to enquire into the cause of the regularities that he was beginning to perceive; preparation of no less than 126 new compounds proved to be a powerful stimulus to enquiry. He wrote:

> When Cl, Br, I or HNO_3 act on Cl-, Br- or I-benzene, on aniline, phenol or toluene, there is always formed as principal product a compound of the *p*-series; at the same time there is formed a compound of the *o*-series in 14-15% of the cases. If the group already present has an acid nature, being CO_2H, NO_2 or SO_3H, the action of the same reagents yields a compound of the 1,3 series as principal product, compounds of the 1,2 series and often also of the 1,4 series also being obtained (7).

The first sentence is pure empiricism. In the second sentence, however, there is a first step beyond the merely empirical in that three groups (carboxyl, nitro, and sulfonic acid) are given the designation 'acid.' While this is obviously and empirically true of the first and third, it is by no means so clearly the case with nitro-groups, which in simple nitro-compounds are chemically neutral by all criteria then used. A former colleague of Körner was to spell this out in a little more detail.

H. Hübner Another pupil of Kekulé, and of Wöhler and Bunsen before him, was Hans Hübner [1837-1884](8). Newly appointed as full professor at Göttingen, he came to similar conclusions to those of Körner, independently and almost at the same time. He wrote:

> In the replacement of hydrogen in the benzene nucleus the incoming negative [acid] substituents at the same time enter the *para* and *ortho* positions with respect to the least negative or acid substituent already present. It follows from this that if an acid [negative] substituent is already present and a second acid substituent

> enters, the latter will avoid the *ortho* and *para* positions as far as possible and enter mainly the *meta* position (9).

Here also the term 'acid' is assigned to some substituents, though with the alternative meaning of 'negative.'

E. Noelting Emilio Noelting [1851-1922](10) had studied at Zurich with Victor Meyer and Kopp and was to make his name as a dyestuffs chemist, first at Lyons and then at Mulhouse. Here he eventually became director of the School of Chemistry, and later was one of the first chemical influences on the young Alfred Werner. In 1876 he expressed a similar generalization to those of Körner and Hübner, though with rather greater precision:

> If a neutral, basic, or weakly acid group, such as CH_3, Cl, Br, I, NH_2, OH, occupies position 1, by the action of Cl, Br, I, HNO_3, and H_2SO_4 the main product will be a *para* compound together with varying but always smaller quantities of *ortho* derivatives. But if the position 1 is occupied by an acid group, such as NO_2, CO_2H, or SO_3H, the action of the above reagents produces mainly a *meta* compound together with small quantities of the *ortho* and *para* series (11).

All these contributions made explicit use of a concept of acidity. In the case of existing substituents CO_2H and SO_3H there is no problem; both carboxylic and sulfonic acids, as their names suggest, display strongly acidic properties in all the conventional ways. But this is not true of the remaining member of the trilogy which, as we have noted, was the most common: nitro. Simple nitro-compounds do not give any of the usual reactions for acids. In what sense, therefore, could all three authors describe the nitro group as 'acid'? The authors offer no explanation themselves so we are left to speculate on possibilities.

(i) Though simple nitro-compounds display no acidic properties, the nitro-group does enhance the 'acidity' of the phenolic function when it is in the *o/p* relation to it. Certainly this was known in 1877 (12). As Kekulé himself observed 'all nitro-derivatives of phenols are decided acids', maintaining this view even in 1882 (13). But the known neutrality of nitrated hydrocarbons (14) renders this hypothesis rather questionable.

(ii) An alternative explanation for the 'acidic' character of the nitro group could lie in the strongly acidic medium from which such compounds were prepared [concentrated nitric and sulfuric acids]. This would be true of the conditions for sulfonation also, though hardly for the introduction of the carboxyl group by the oxidation of the methyl group [though this *could* involve acidic conditions]. But by the same token [basic] amines were prepared by the acidic reduction of nitro-compounds, so this explanation looks implausible.

(iii) Noelting's equating of 'acidic' with 'negative' suggests the most likely provenance of this concept. It comes straight from the pages of Berzelius. For the great Swedish chemist an acid was what we should call an anhydride, usually the oxide of a non-metal, but always something that could combine with a basic oxide to form a salt, and would in aqueous solution give the familiar reactions with indicators etc. In that sense the nitro group may be identified with the molecule NO_2, which in Berzelian terms is truly acidic, being, as we would say, a mixed acid anhydride:

$$Na_2O + 2NO_2 \longrightarrow NaNO_3 + NaNO_2$$
$$H_2O + 2NO_2 \longrightarrow HNO_3 + HNO_2$$

But this cannot apply to COOH or SO_3H *unless they too are considered in pairs*. In those circumstances we have, respectively, oxalic and dithionic ['hyposulfuric'] acids, both of which had long been known (15) and which are clearly defined acids: $(COOH)_2$ and $(SO_3H)_2$. Underlying all these ideas was the clear Berzelian doctrine, derived in its turn from Lavoisier, that 'acidity' is a quality inherently associated with certain elements or groups. Kekulé agreed with him when he observed that 'the entry of an acidic element or radical enhances the acidic nature of a compound' (16).

If Berzelius did provide the basis for this interpretation, several considerations follow. Most obviously, perhaps, it suggests an unexpected longevity for his views, particularly remarkable from a source that might have been expected to have repudiated dualism utterly in the light of unitary and type theories (17). Secondly, it corrects later indictments of Hübner *et al.* in their use of 'the vague terminology of the already discredited electrochemical theories of Berzelius' (18). Criticisms like the following from A. W. Stewart are quite beside the point:

> Obviously if the nitro-group and the bromine atom are to be regarded as belonging to the acidic class, the generalization breaks down at the first test, since a nitro-group directs incoming substituents into the *meta*-position, whereas the bromine atom directs them to the *ortho*- and *para*-positions (19).

In fact Hübner never included bromine as a strongly acidic group (20). Nor in Berzelian terms, could it be possibly so considered. The point was not that the terms were vague but that they were precise. Noelting explicitly includes bromine as a 'neutral, basic, or weakly acid group.' However, like all later commentators, Stewart was perfectly correct to stress the limitations of such generalizing rules, and to indicate exceptional cases, such as the CHO group which even in Berzelian terms is neutral yet it is *meta*-directing. Meanwhile, another approach to the problem was being formulated.

12.22 *Empirical Rules Based Upon Saturation*

H. E. Armstrong In 1887 one of Frankland's former associates, H. E. Armstrong [1848-1937](21) proposed a new rule (22). If a monosubstituted benzene can be represented as C_6H_5-XYn the group will be *o/p* directive provided X is monovalent [in which case n = 0], or that all the Y atoms are monovalent; if any one of them is polyvalent [e.g., if the substituent contains a multiple bond], then *m*-substitution will result. This wholly empirical rule works well for all the substituents so far mentioned [including CHO], but fails with other molecules like azobenzene (C_6H_5-N=N-C_6H_5) and cinnamic acid (C_6H_5CH=CH.COOH), where *o/p* substitution takes place. Moreover, as Armstrong himself was to discover (23), in strongly acid solution amines lose their *o/p* directive character [when the aromatic substrate is of course of the form $ArNH_3^+$].

D. Vörlander Like Armstrong, the Halle chemist Daniel Vörlander [1867-1941](24) declared unsaturation as the condition for *meta*-substitution (25). Given a compound C_6H_5E, the substituent 'element' E will exert different effects according to whether it is saturated or unsaturated:

> The radicals which orient an incoming nitro-group into the *meta*-position are unsaturated at the point adjacent to the benzene ring:
>
> -NO_2, -CN, -CHO, -COOH, -SO_3H;
>
> whilst those groups which favor an *ortho-para* type of substitution have a saturated atom next to the benzene ring:
>
> -Cl, -Br, -OH, -CH_3, -CH_2Cl, -CH_2.COOH.

As in Armstrong's rule, the term 'saturation' required greater precision. Both the amino group and the iodine atom are capable of accepting addenda [in salt formation and oxidation respectively] so the question remains as to the sense in which they could be called 'saturated.' And again there are anomalies, as in the *meta*-substitution in phenylnitromethane. Empiricism was to receive the fullest expression in a third kind of approach.

12.23 The Crum Brown – Gibson Rule

This was an earlier suggestion from the Scottish chemist A. Crum Brown (26), and his assistant J. Gibson (27). Given a compound C_6H_5X, the directive power of X would be determined thus. If the hydride HX were to be incapable of direct oxidation to HOX, then X in an aromatic nucleus would be *o/p* directive; if such oxidation were possible, then X would direct to the *m*-position. The following is a version of an explanatory table provided:

Substituted benzene	*Substituent*	*Hydride of substituent*	*Oxidation product*	*Directive power*
C_6H_5Cl	Cl	HCl	HOCl	*o-p*
C_6H_5Br	Br	HBr	HOBr	*o-p*
$C_6H_5CH_3$	CH_3	HCH_3	$HOCH_3$	*o-p*
$C_6H_5NH_2$	NH_2	HNH_2	$HONH_2$	*o-p*
C_6H_5OH	OH	HOH	H_2O_2	*o-p*
$C_6H_5NO_2$	NO_2	HNO_2	HNO_3	*m*
$C_6H_5CCl_3$	CCl_3	$HCCl_3$	$HOCCl_3$	*o-p*
C_6H_5CHO	CHO	HCHO	HCOOH	*m*
C_6H_5COOH	COOH	HCOOH	$CO(OH)_2$	*m*
$C_6H_5SO_2OH$	SO_2OH	HSO_2OH	$SO_2(OH)_2$	*m*
$C_6H_5COCH_3$	$COCH_3$	$HCOCH_3$	CH_3COOH	*m*
$C_6H_5CH_2COOH$	CH_2COOH	HCH_2COOH	$HOCH_2COOH$	*o-p*

This rule was of rather greater influence than its predecessors. Yet it rested on very insecure foundations. There was an inherent ambiguity in the authors' phrase 'direct oxidation,' though it was usually taken as meaning 'in one experimental operation;' itself a highly ambiguous concept. Moreover, any rule depending on negative evidence stands in peril of contradiction by later discoveries. Today most, if not all, of the 'impossible' oxidations of HX to HOX have been accomplished. At much earlier times several were questioned.

Thus the oxidation

$$CH_4 \longrightarrow CH_3OH$$

was postulated in 1908 by Bone (28) on the basis of a proposal by Armstrong (29). Similarly, Nernst showed the formation of hydrogen peroxide on treatment of water with oxygen (30).

There were also objections based on the occasional predictive failures of the rule. Perhaps the most famous were the *meta*-nitrations of phenylnitromethane (31) and benzonitrile (32), *o/p* substitutions being expected in both cases since neither of the following oxidations was known:

$$CH_3NO_2 \longrightarrow HOCH_2.NO_2$$
$$CH_3CN \longrightarrow HOCH_2.CN.$$

12.3 Mere Empiricism?

Much criticism has been levelled by later chemists at these early generalizations. Adverse comments by Waters (18) and Stewart (19) have already been noted; the former also asserts that all the foregoing rules were 'empirical summaries applicable for the prediction of the structures of the products of substitution reactions', and nothing more; they failed to 'focus attention on the true functional characteristic of a substituent' (33). Yet these *post hoc* criticisms have missed several elements in the early rules that indicate a genuine attempt to transcend 'mere' empiricism and, albeit speculatively, to grasp the theoretical nettle.

Underlying all the admitted empiricism of the various 'rules' was a generalization so quickly accepted, and to us so obvious, that it is easy to overlook it altogether. It was a recognition of the importance of the group(s) already present in an aromatic molecule before a substitution reaction commenced. Orientation depended not so much upon reagent, or solvent or other external conditions, but upon existing substituents. Perhaps the first to voice this with unambiguous clarity was Körner himself, who spoke of the *influenza orientate of* such groups (6). Even though the reason was quite unknown there was some way in which *substituents could guide the course of a subsequent reaction.* That they could exert a *directive influence* was a perception going well beyond an empirical rule-of-thumb.

The same is true of Hübner's account. His reference to 'acid' or 'negative' groups in both molecule and reagent has the clear implication that like groups somehow repel each other in the *o/p* positions, and therefore adopt a *meta*-configuration by default. It was crudely represented by his formula:

This insight was apparently the rationale behind all the generalizations depending on the concept of acidity. It also enshrined a recognition of the importance of a reagent, although at that stage, and for long afterwards, this was seen to be secondary to that of the directive group. As late as 1931 Stewart was able to dismiss this aspect of Hübner's work with the remark 'the nature of the incoming substituent plays a hardly appreciable part in the problem' (34). Today's chemists would place a considerably higher value on Hübner's theoretical insights.

If Körner, Hübner and Noelting have attracted most criticism for being 'mere' empiricists that is probably because their language suggested they were something more [as indeed they were]. Few of their successors attempted to offer any theoretical justification for their own 'rules.' An exception is Vörlander who had previously explored the significance of unsaturation in affecting the reactivity of a hydrogen atom, observing that it is more labile when a double bond occupies a 3:4 position relative to it (35). His own formulae are on the left, some familiar examples on the right [H = hydrogen; E = a non-metallic element]:

1 2 3 4 5
H.E.E:E
	R.C=O
	   |
	  OH
		R.CH=N-OH	R.N=N.OH

H.E:E	CH_2=CH_2		
H.E.E.E:E	CH_3.CR_2.CH=NR		

Here are hints of the importance of unsaturation and its influence on the mobility of hydrogen atoms, as in the benzene ring.

12.4 Towards a Pre-electronic Explanation

If the early writers were in fact tainted by inductivism, an aside by Crum Brown and Gibson offers a suggestive hint of the direction in which some chemists' thoughts were leading them. The authors admitted that their 'rule' is 'not a law because it has no visible relation to any mechanism by which the substitution is carried out' (36). This is noteworthy not only for its honest disclaimer, but also for being one of the first explicit references to the *mechanism* of an organic reaction, even though this was entirely unknown. But even if the word was seldom used there was occasional speculation in what we should call mechanistic terms.

12.41 Some Preliminary Moves

Once again Kekulé set the tone of the debate. He had long held that substitution reactions in general may proceed by addition to form an intermediate (37):

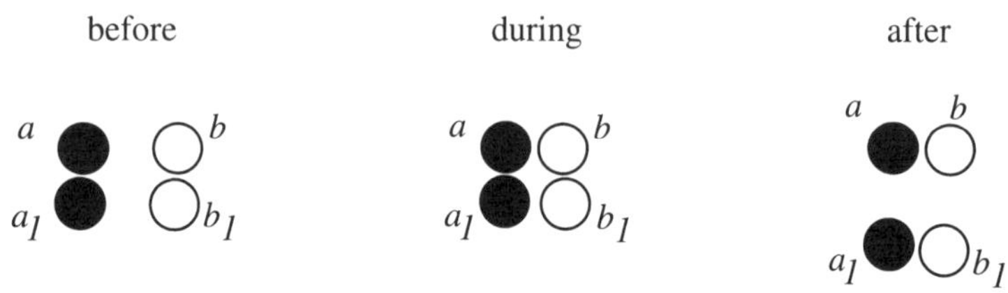

For Partington this implied the modern theory of the intermediate complex (38). Its application to benzene substitution was not fully worked out, but he did use the same underlying idea of weakened affinities in his explanation of *para*-dibromination: atoms near the first bromine atom have their affinity for bromine weakened, so the second bromine atom takes up a position as far as possible from the first (39). A similar general theory of substitution was offered by Armstrong (22), who draws on his own ideas of residual affinity and explicitly cites the intermediate complex theory in Kekulé's *Lehrbuch.* Following the initial formation of an addition compound he believed that para-substitution would result from a subsequent intramolecular change, due to a 'tendency to form symmetrical di-derivatives.'

Yet another move towards mechanism was made by J. J. Blanksma of Amsterdam, who cited Armstrong's theory. He distinguished between direct and indirect substitution, stressing the energy difference between chain- and nuclear-substituted products (40). Nuclear attack gives a mixture of isomers but with *meta* predominating, sidechain attack producing overwhelmingly *o/p* substitution. In the latter case the sidechain (OH, NH_2) could be attacked first, followed by rearrangement to a ring-substituted product (41).

All these ideas were largely incidental to the main thrust of the work described. Several writers, however, made rather more specific attempts to arrive at a theoretical basis for aromatic substitution, most notably Flürscheim and Holleman.

12.42 Flürscheim's Theory

B. J. Flürscheim [1874-1955](42), after studying with Werner at Zurich and Thiele at Heidelberg, settled in England, largely as an independent worker. In 1902 he began to expound what Ingold rightly called 'the first theory [as distinct from a rule] of orientation in aromatic substitutions' (43). Flürscheim drew upon Werner's theory of maximum disposable affinity (44), in which affinity was continuously divisible. Elements [as X, below] with high affinity demand would cause adjacent atoms to exert low affinity on the atoms to which they were attached, and so on. The opposite situation would arise with elements [as Y] whose affinity demands were low. An alternation of weak and strong bonds would appear down a chain of atoms:

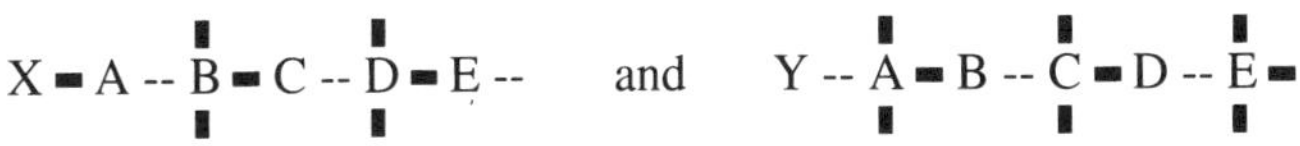

Applying this idea to substituted benzenes he wrote symbols like the following; in the first a chlorine has a stronger affinity than hydrogen for carbon, so there will be an excess of affinity at the *ortho* and *para* positions, and substitution will occur there, while in the second the sulfonic acid group has a weaker affinity so *meta* attack would be expected:

Where an atom attached to the ring was in its highest valency state [as N in NO_2, S in SO_3H, C in COOH] its affinity resources would be overtaxed, its link with the ring would be weak, and *meta*-substitution would be the rule. This offered a rationale for the ideas of Armstrong and, as Ingold recognized, ‘played an important role in the development of the modern theory’ (44). Somewhat similar ideas were published by Obermiller (45) and Tschitschibabin (46). They were severely challenged by Holleman.

12.43 Holleman's Theory

A. F. Holleman [1859-1953](47) had studied at Leiden with Franchimont, then with Bunsen and Baeyer, becoming assistant to van't Hoff at Amsterdam in 1887. Aware of the inconsistencies and flawed predictive power of previous generalizations, he began an extensive programe of research in 1895 (31) separating and identifying isomers from a wide range of substitution reactions. He showed the rarity of pure *o/p* or *m*-substitution, though the former occurs in phenols and the latter with positive charges against the ring [as $C_6H_5NH_3^+$]. He noted the frequent association of oxygen-containing substituents with dominant *meta*-substitution.

Holleman also investigated the sequence of directive powers in substituents. This involved reactions under standard conditions [e.g., nitration with fuming nitric acid at 0°C.]. By using disubstituted derivatives he could compare the effectiveness of two different substituents. Thus:

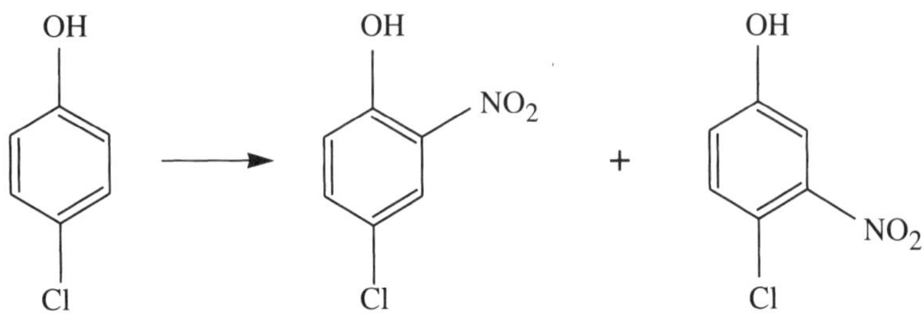

Since the first product exceeded the second in yield it followed that hydroxyl had a more powerful directive influence than chlorine. In this way a whole sequence of directive powers was constructed, and with it came an awareness of the importance of rate of reaction. In comparison with benzene it was generally shown that *o/p* directive groups facilitate substitution while *m*-directive groups retard it. The ring was *activated* in one case, *deactivated* in the other. He came very near to our modern concept of partial rate factors. This and much else was summarized in a book (48) and review article (49). Nor was Holleman alone in such research. Similar studies had been undertaken by F. Kehrmann in 1890 (50), showing that bromination of 4-ethyltoluene predominated at position 2 rather than position 3; he also compared the dinitrations of resorcinol and its monomethyl ether:

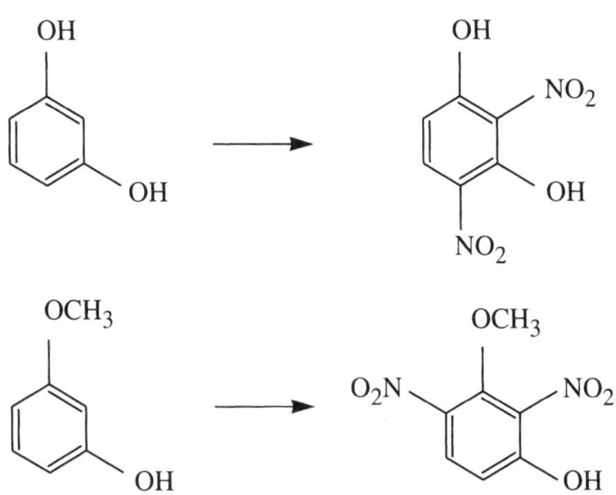

In terms of mechanism Holleman, like Kekulé and Armstrong, proposed an initial addition reaction. This would then be followed by elimination to yield the observed substitution product. His scheme was predicated on two further assumptions. Aware that double bonding affects the reactivity of substituents [as seen in the vastly different mobilities of chlorine in vinyl and allyl chlorides] he argued that the converse should also be true and that substituents should modify the behavior of double bonds, as in the benzene ring. He also assumed the possibility of 1,4 addition, as described by Thiele for conjugated polyenes (51). Thus a reagent HY could add to an aromatic compound C_6H_5X in three ways, the first two being where X increased reactivity of the adjacent double bond, and the third when it decreased such reactivity:

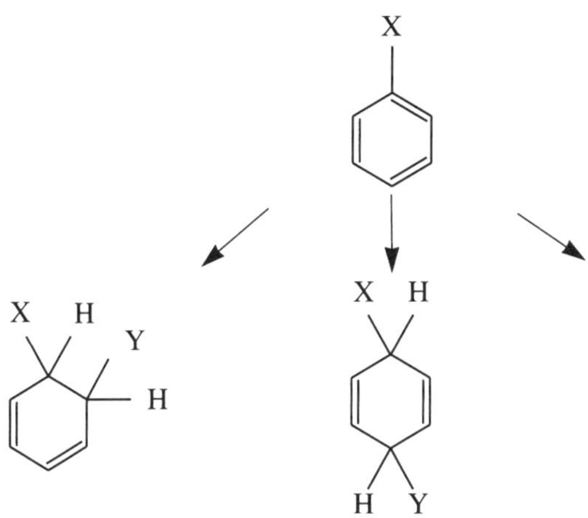

This was the last great scheme before the electronic theory of organic chemistry rendered all its predecessors obsolete. Its weakness lay not in its apparent commitment to a one-stage addition, rarely known now even for alkenes (52), but for a certain circularity of argument:

> The nitro-group is a *meta*-orienting radicle [sic] because it depresses the activity of the adjacent double bond. But how do we know that it does this? Because it orients to the *meta*-position (53).

12.5 An Assessment

With half a century of efforts to bring order out of the chaos of aromatic substitution reactions it is pertinent to attempt a brief estimate of the role of that work in organic chemistry as a whole. The overwhelming impression is that, *at the time,* the concepts of aromatic substitution were of little importance. Only as viewed by posterity can they be seen to have been significant stages in the growth of fundamental knowledge. One or two examples must suffice.

Writing in 1906 the American chemist Ira Remsen posed an 'interesting question':

> Which one of the three possible dibrom-benzenes is formed by direct treatment of benzene with bromine? The answer to the question is equally interesting. The main product of the action is *para*-dibrom-benzene, while there is always formed in much smaller quantity some of the *ortho* product. The reason why these products are formed, and not the *meta* compound, is unknown; nor has any plausible hypothesis been suggested to account for the fact (54).

With such a pessimistic conclusion it is hardly surprising that others gave little attention to the rules of aromatic substitution. Perkin and Kipping in 1911 have a few short paragraphs on the subject, concluding with the Crum Brown-Gibson Rule as an expression of 'general behaviour' (55). The same rule had been the only one cited [in small print] in less than half a page of exposition by Richter on 'regularities in substitution' (56).

There seem to have been several clear reasons for this neglect. In the first place there were so many *anomalies and unexplained contradictions.* At the end of the century Holleman established and codified a number of inconsistencies between the experimental results and the predictions of his predecessors, as well as inconsistencies with each other. He was specially concerned with compounds of the type $C_6H_5CH_2X$ where *o/p* substitution was the norm, but not when X = NO_2. He referred to the earlier predictive rules, some of whose results are summarized in the following table:

Aromatic substrate		$C_6H_5CH_2COOH$	$C_6H_5CH_2NO_2$	C_6H_5CN
Predictions	Körner/ Hübner/ Noelting	meta	ortho/para	ortho/para
	Armstrong	meta	**meta**	**meta**
	Vörlander	**ortho/para**	ortho/para	**meta**
	Crum Brown & Gibson	**ortho/para**	ortho/para	ortho/para
Experimental results		ortho/para	meta	meta

Thus the infrequency of the correct predictions [in bold] was a feature of all six schemes, but especially those based on a concept of 'acidity.' It is hardly surprising that practical men in the laboratory had little time for speculative generalisations and preferred instead to work by rule-of-thumb, thankfully accepting lists of orienting groups but not enquiring too deeply into the reasons behind them.

That, however, is hardly a sufficient explanation. There was also a recognition of *other influences than directing groups alone.* Hübner had acknowledged the importance of the reagent and Holleman was later to stress this more clearly. Another relevant factor was temperature, particularly noticeable in the *o/p* ratio. Kekulé himself had reported that sulfonation of phenol at low temperatures leads chiefly to the *ortho* product, whereas at high temperatures *para*-substitution predominates (57). Nor was this an isolated observation. Goldstein showed that the *o/p* ratio increases with rising temperature in the nitration of phenol (58). and Rosenstiehl had found a similar effect in the nitration of toluene (59). These and other reactions formed part of a discussion on the *o/p* ratio in Erlenmeyer's influential textbook of 1894 (60). That ratio was to be the theme of many enquiries in the new century.

Thirdly, there were many other *distractions* in studies based on benzene and its derivatives. Some of these are apparent in the writings of Armstrong who, though an author of an attempt to codify aromatic substitution, nevertheless seems much more interested in questions as sidechain *versus* nuclear replacements (61). In the same way, perusal of the papers of Baeyer suggests a far greater fascination with the new field of alicyclic chemistry than with aromatic compounds *per se.* It is not hard to multiply examples of this kind. For Kekulé and many of his followers the unresolved question of the fine structure of benzene was as much a deterrent to detailed speculation as it was an embarrassment. How could one theorize about a reaction if the reagent itself was of uncertain nature? Obermiller, for example, preferred the hexagonal formula of Claus (62) to that of Kekulé in his account of aromatic substitution (63).

A fourth consideration was the sheer *complexity* of the very idea of aromatic substitution. The existence of sidechain substitution [especially with halogens] indicated unwelcome complications that might be turned to good effect in a new scheme, but chiefly underlined the overall complexity. There was always the possibility that, for some unknown reason, a rearrangement had taken place. Kekulé, for example, knew that all three sulfonated phenols on alkaline fusion lead to resorcinol (57). As we now know such reactions involve nucleophilic, not electrophilic, replacements, for which all the usual rules for orientation are totally inapplicable, though at the time there was no reason to suspect such complications. Yet nucleophilic substitutions had been known for years, though not by that name (64). Their reversal of even the rules then known was quite inexplicable. Again, some reactions studied in the late 19th or early 20th century are now known to proceed by a homolytic mechanism, so there is small wonder at the confusion caused by a diversity of experimental results, as *ortho-* and *para-* substitutions in nitrobenzene. Thus when Kühling treated nitrobenzene with sodium *p*-nitrobenzenediazotate he obtained mainly 2,4'- and 4,4' dinitrobiphenyl, *i.e., o/p* substitution (65). Similar effects were later found with substituted benzenes in the reaction discovered in 1897 by Bamberger (66):

$$C_6H_5.N(NO).CO.CH_3 + C_6H_6 \longrightarrow C_6H_5.C_6H_5 + N_2 + CH_3COOH.$$

When Bamberger conducted the reaction in the presence of nitrobenzene, for example, he obtained both 2- and 4- methyldiphenyls. Another classic case, mentioned in some perplexity by Holleman (67), was that of the mercuration reactions studied by Dimroth (68)[which can proceed by homolytic *or* electrophilic mechanisms]. Often these involved the decomposition of mercuric acetate initiated by free alkyl radicals from acyl peroxides. Well into the 1930s these phenomena were called 'amphoteric aromatic substitutions' or 'invariable *ortho-para* substitutions', their homolytic character only being recognized in 1934 (69).

Also it has been well observed (70) that historians, though not 19th century chemists, largely confine their attentions to benzene and its derivatives, whereas naphthalene presented a further set of orientation problems, with α- and β- substitution being governed by other factors than merely the nature and location of substituents.

12.6 Conclusion

The account here offered is supremely Whiggish in one sense. It does focus on theories that seem to have borne fruit much later. One can trace a direct genesis of modern electronic theory in nearly all of the stumbling attempts to rationalise phenomena in the later 19th century. In so far as this approach enables us to understand the rise of electronic theories of aromatic substitution it would appear to have some justification. But in another sense recapitulation of the work of Hübner, Crum Brown, Holleman and all the others is very unwhiggish for the simple reason that a 'dead-end' is exactly how their work was viewed by most of their contemporaries. Moreover it can be argued that some of them played no part in a linear chain of influential development from Kekulé and Körner to Robinson and Ingold (71). Perhaps the truth is that Whiggishness is primarily in the eye [or on the pen] of the beholder and very hard to define in precise terms.

A representation of an extreme realist position, making no concessions to Whiggish views of progress, might be like this:

Körner ⟶ }
Hübner ⟶ } —— rules based on acidity ⟶ NIL
Noetling ⟶ }

Armstrong ⟶ }
Vörlander ⟶ } —— rules based on saturation ⟶ NIL

Crum Brown/Gibson ⟶ Rule based on oxidation ⟶ NIL

If the rules were devoid of theoretical assumptions, and their authors innocent of aspirations to mechanistic hypotheses, this stark scenario might well be justified. But, as we have seen, this was not always the case. One concept, in particular, had a continuity of development that was demonstrable in the sense that authors cited their predecessors. This was the notion of residual affinities, with a network of relationships going back to Kekulé. Hatched lines indicate presumptive influence on the grounds of pupil/teacher relationship, full lines explicit citations:

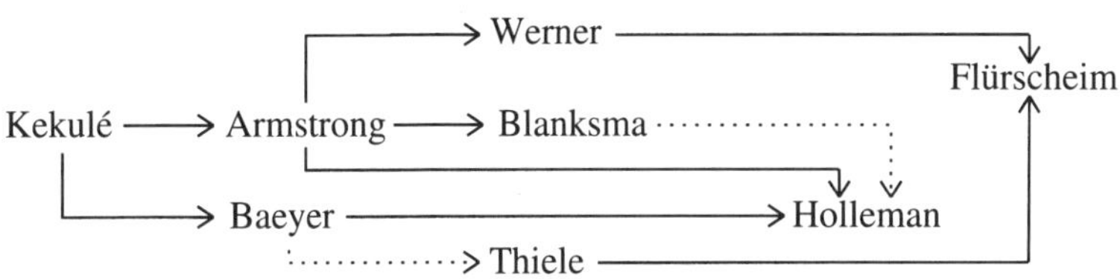

From all this it follows that the views of Flürscheim and Holleman, of great influence later, did connect to previous ideas, if in a rather complicated fashion. Whatever else is not illustrated by these events it is surely clear that chemical theory did progress, but did not do so in a uniformly linear manner. With hindsight we can see that the electronic theory was to give depths of understanding quite unattainable in a pre-electronic age. In arriving at that theory workers like Robinson and Ingold were clearly indebted to Flürscheim and Holleman [though in rather different ways], and thus to *their* predecessors. There was a true progression of thought, even a genuine and sustained attempt to reach out to a theory often just beyond the grasp. In so far as there was some degree of linear development it is pertinent to enquire why.

One obvious cause for the step-wise evolution of a generalized theory lay in one reaction which turned out to be overwhelmingly predominant. Partly because it was easy to conduct, but much more because it opened up an immense field of useful derivatives, especially in the dyestuffs industry, the favored reaction was *nitration.* So true was this that it became [perhaps unfortunately] the paradigm of aromatic substitution, whereas in fact it is neither simple nor typical (72). But it had the great advantage of introducing a substituent that deactivated the nucleus towards further attack of the same kind, thus eliminating in large measure the complications of polysubstitution. It was a powerful tool, even if it came to hand adventitiously.

Another thing that emerges is the largely static and unchanging nature of the general theory of organic chemistry through the later 19th and early 20th centuries. Despite all the setbacks, failures, anomalies, and irregularities chemists continued to synthesize new aromatic compounds by their tens of thousands. Only a paradigm of immense strength and resilience could have allowed that to happen. There can be few other examples in the history of science that so clearly demonstrate the characteristics of normal science.

The slow emergence of coherent schemes capable of accommodating an ever-growing body of empirical facts is an unrivaled model of how science can progress and adapt. For this very reason the story of aromatic substitution can be of rare value to modern chemical students. There is another reason: 'the development of theoretical organic chemistry was intimately concerned with the development of that particular aspect of it concerned with aromatic substitution' (73). Its history shows us all something of the complexities of modern science, of a pathway from the purely empirical to the highly theoretical, of the actual issues involved in this branch of chemistry and of the perennial fallibility of its practitioners.

Publisher's Note: This article was originally published as Chapter 12 in *The Kekulé Riddle: A Challenge for Chemists and Psychologists*, ed. J.H. Wotiz (Vienna, IL, 1993). The original numbering of the sections has been retained.

References and Notes

1. Kekulé FA (1865) Bull Soc Chim Paris, [2], 3: 98; Bull Acad Roy Belg, [2], 19: 551

2. Such was the case of Edward Frankland, in matters of benzene structure a crypto-Ladenburgian (see Section 3.3, and Editor's note: Chapter 9 by Prof. Schiemenz)

3. Kekulé FA (1882) *Lehrbuch der organischen Chemie,* vol. iii, Stuttgart, Enke: pp. 25, 92 [he employed the terms *meta* and *para* in the opposite way to modern usage]

4. Lachman A (1899) *The Spirit of Organic Chemistry,* New York, Macmillan: p. 34

5. Anschütz R (1926) Obituary notice for Körner, Ber Deut Chem Gesell, 59A (1): 75

6. Körner W (1874) Gazz chim Ital, 4: 305

7. *Ibid.*, p. 446

8. Beilstein FK (1884) Obituary notice for Hübner, Ber Deut Chem Gesell, 17 (3): 763

9. Hübner H (1875) Ber Deut Chem Gesell: 8: 873

10. Friedlander P (1922) obituary notice for Noelting, Ber Deut Chem Gesell, 55A **(1):**137

11. Noelting E (1876) Ber Deut Chem Gesell, 9: 1797

12. Mono-, di-, and tri-nitrophenols were described by Körner (1868)[Zeitsch f Chem, 11: 322], though of course picric acid had been known since the 18th-century and was synthesized from phenol by Laurent [Ann Chim, 1841, 3: 195 (221)]. Engelhardt A and Latschinoff P (1870)[Zeitsch f Chem, 13: 225] stressed that the two mononitrophenols that they had studied had the ability to decompose carbonates. They concluded that such compounds are acids but that the influence of one nitro group is not sufficient to make nitrophenols 'completely acidic' [*vollständig sauren*](p. 230)

13. Kekulé (3, p. 41). From the melting temperatures given for the two mononitrophenols [110° and 45°] they must be respectively *ortho* and *para*; the *meta* compound [which does not show enhanced acidity] is not described, though it was prepared by Fittig in 1874 [Ber Deut Chem Gesell, 7: 179]. Kekulé also mentions a dinitrophenol [which must be 2, 4] and picric acid. Thus all reported examples would justify his generalization

14. Engelhardt and Latschinoff (Ref 12: p. 226)

15. Oxalic acid had been known for centuries, its dicarboxylate structure appears first to have been so represented by Frankland E and Duppa BF (1866)[Phil Trans, 156: 309; dithionic acid was discovered in 1819 [Gay-Lussac JL and Welter J J, Ann chim phys, (1819)[2], 10: 312]

16. Kekulé FA (1861) *Lehrbuch der organischen Chemie,* Erlangen, Enke vol i: p. 127

17. However Kekulé did suggest that, in due time, dualism would be fruitful again, 'in a rejuvenated form' [Rectoral Address, 1877, in Anschütz R, *August Kekulé,* Verlag Chemie, Berlin, 1929, vol. ii: p. 913]

18. Waters WA (1950) *Physical aspects of organic chemistry,* 4th edn., Routledge & Kegan Paul, London, p. 463

19. Stewart AW (1931) *Recent advances in organic chemistry,* 6th edn.,Longmans Green London, vol. i: p. 365

20. Even if the 'double group', as in oxalic or dithionic acids, is considered, Br_2 was not notable for its acidic reactions

21. Vargas Eyre J (1958) *Henry Edward Armstrong,* Butterworth, London

22. Armstrong HE (1887) J Chem Soc, 51: 258

23. Armstrong HE and Berry W (1900) Proc Chem Soc, 16: 159

24. Weygand C (1943) obituary notice for Vörlander, Ber Deut Chem Gesell, 76A: 41

25 Vörlander D and Meyer F (1902) Annalen, 320: 122

26. Walker J (1923) Obituary notice for Crum Brown, J Chem Soc, 123: 3422

27. Crum Brown A and Gibson J (1892) J Chem Soc, 61: 367

28. Bone WA and Drugman J (1906) J Chem Soc, 89: 660

29. Armstrong HE (1903) J Chem Soc, 83: 1088; Bone's 'conversion' to Armstrong's idea is recounted in Eyre (21: pp. 201-2)

30. Nernst W (1905) Z Electrochem, 11: 710

31. Holleman AF (1895) Rec trav chim, 14: 121

32. Beilstein FK and Kuhlberg A (1868) Ann Chem u Pharm, 146: 317

33. Waters (Ref 18): p. 464

34. Stewart (Ref 19): p. 365

35. Vörlander D, Ber Deut Chem Gesell (1901) 34: 1632. Meanwhile Thiele had used his famous theory of partial valency to explain both the reactivities of alkenes *and* the relative non-reactivity of arenes, thus providing another important link between the two classes of hydrocarbons [see below, ref. 51].

36. Crum Brown and Gibson (27), p. 369.

37. Kekulé FA (1858) Ann Chem u Pharm, 106: 129-52; *Lehrbuch,* vol- i, (note 16 p. 143)

38. Partington JR (1964) *A history of chemistry,* vol iv, Macmillan, London: p. 538

39. Kekulé FA (1866) *Lehrbuch der organischen Chemie,* Erlangen, Enke vol. ii: p. 553

40. Blanksma JJ (1902) Rec trav chim, 21: 269

41. *Idem, ibid.* (1904) 23: 202

42. Ingold CK Obituary notices for Flürscheim, (a) Nature, (1955) 176: 191; (b) J Chem Soc (1956): 1087

43. Ingold (Ref 42: p. 1088)

44. Russell CA (1971) *The history of valency,* Leicester University Press, Leicester, pp. 201-23

45. Obermiller J (1907) J prakt Chem, 75: 1; (1908) 77: 65

46. Tschitschibabin AE (1912) J prakt Chem, 86: 381

47. For a brief account of Holleman's life see Kubbinga HH (1980), *De benzeentheorie van Cornelis Willem Lely [1882-1936]; hoofdstromen in de Nederlandse organische chemie anno 1922,* Het Waterlant-College, Amsterdam, pp. 46-7; and Wibaut JP, (1954-5) obituary notice for Holleman, *Jaarboek Koninklijke Nederlandse Akademie van Wetenschappen*, pp. 280-287

48. Holleman AF (1910), *Die directe Einführung von Substituenten in der Benzolkern,* Leipzig, Veit

49. Holleman AF (1925) Some factors influencing substitution in the benzene ring, Chem Rev, 1: 187

50. Kehrmann F (1890) Ber Deut Chem Gesell, 24: 130

51. Thiele FKJ (1899) Ann Chem u Pharm, 306: 87 and following papers

52. It was known as long ago as 1936 that nitric acid and sulphuric acids do not add in one stage to alkenes [Michael A and Carlson GH (1935), J Amer Chem Soc, 57: 1268; Michael A and Weiner N, *ibid.*, 1936, 58: 294]

53. Stewart (Ref 19: pp. 382-3)

54. Ira Remsen (1906) *An introduction to the study of carbon compounds, or organic chemistry,* 4th ed., London: pp. 275-6

55. Perkin WH and Kipping FS (1911) *Organic Chemistry,* Chambers, London: pp. 392-3

56. Anschütz R (ed.)(1899) *Chemie der Kohlenstoffverbindungen* by V. von Richter, e.g. 3rd American edition, (1913) *Richter's Organic Chemistry* (trans. Smith EF), Blakiston, Philadelphia: p. 69

57. Kekulé FA (1869) Ber Deut Chem Gesell, 2: 330 (cf. 3)

58. Goldstein M (1878) Ber Deut Chem Gesell, 11: 1943 [brief account]

59. Rosenstiehl A (1872) Ann chim phys, (4), 27: 433

60. Erlenmeyer E (1894) *Lehrbuch der organischen Chemie,* 3rd edn., ed. Meyer R and Goldschmidt H, Leipzig, Winter: p. 219

61. Armstrong HE (1880) *Introduction to the study of organic chemistry,* 2nd edn., London, Longman: pp. 118-20

62. Claus A (1867) *Theoretische Betrachtungen und deren Anwendung zur Systematik der organischen Chemie,* Freiburg, Poppen, p. 207

63. Obermiller J, *Die orientierenden Einfüsse und der Benzolkern,* Leipzig, J A Barth 1909

64. *E.g.,* the replacement of chlorine by OH or NH_2 in picryl chloride (2, 4, 6-trinitrochlorobenzene)[F. Piriani (1854), Compt rend, 39: 852

65. Kühling O (1895) Ber Deut Chem Gesell, 28: 4; (1896) 29: 165

66. Bamberger E (1897) Ber Deut Chem Gesell, 30: 366

67. Holleman (Ref 49): p. 199

68. Dimroth O, Ber Deut Chem Gesell, (1902), 35: 2032, 2853; Ann Chem u Pharm (1925), 446: 148. *C.f.* Klaproth WJ and Westheimer FH, (1950) J Amer Chem Soc, 72: 4461

69. Hey DH and Grieve WSM (1934) J Chem Soc 56: 1797, 1797. *Cf* Hey DH, 'Homolytic aromatic substitution', in Waters WA (ed.) 1959 *Vistas in Free Radical Chemistry,* Pergamon, London, pp. 209-223

70. Ward ER, *Incidents and personalities in the history of aromatic nitration and aromatic nitro-compounds,* privately published, n.d., p. 5.

71. Ingold himself, with uninhibited hindsight, pointed out the kernels of electronic truth embedded in each of the early schemes: the inductive effect in the rules relating to acidity, the mesomeric stabilization of oxidation products behind Crum Brown's rule, and the electromeric effects in unsaturated compounds behind the rules of Armstrong and Vörlander. But none of these features can necessarily imply actual historical connexions [Ingold CK, *Structure and mechanism in organic chemistry,* Bell, London, 2nd edn., (1969), pp. 267-81]

72. However the majority of nitration reactions studied do proceed via the nitronium ion, NO_2^+

73. Hoggett JG, Moodie RB, Penton JR and Schofield K (1971) *Nitration and aromatic reactivity,* Cambridge University Press, p. 3

XIV

August Wilhelm Hofmann – Cosmopolitan Chemist

In the summer of 1862 August Wilhelm Hofmann wrote to his English friend Edward Frankland: "I sincerely hope that you and your family are enjoying the stay in the country. London becomes unbearable. The second great wave of scientific Europeans has just arrived. There is no alternative but flight. I have taken some rooms in the country . . ."[1] At first sight this sounds like a heart-cry from a German emigré weary of his adopted country and longing for home. Closer inspection of their correspondence shows this to have been not the case. Indeed the "great wave of scientific Europeans" was in large measure a consequence of his own endeavours in London, conducted with immense zest and enjoyment. When, several years later, Hofmann was persuaded to forsake England for Germany he told Frankland that "it was only with great pain that I was able to make up my mind to say farewell to the country of my adoption".[2] He was much at home in England, and late in life spoke of England as his "adopted country",[3] claiming to know London better than Berlin even after twenty years here.[4] This remarkable chemist was well described by one of his memorialists as "cosmopolitan in speech, cosmopolitan in sympathies".[5]

Hofmann was first attracted to chemistry by the teaching of Liebig at Giessen (where his father, an architect, had been responsible for enlarging the laboratories in 1839). In 1845, at the age of 27, he was called from Bonn to direct the new Royal College of Chemistry in London, a foundation inspired by the example at Giessen. Instrumental in enticing him to London was the German-born Prince Albert, consort of England's Queen Victoria.

During his 20 years in London Hofmann transformed the face of English chemistry. So quickly did he establish himself that as early as July, 1845, Liebig was doubting whether Hofmann would want to return to Germany;[6] in 1851 Hofmann himself thought he would return one day, though confessed it hard to think of leaving London;[7] and as late as 1863 Liebig wrote to him, "Ich habe es nie verstehen können, daß Sie sich entschlossen haben, England zu verlassen, ein Land, welchem anzugehören ich für ein großes Glück halte, ... um sich in die Misere von deutschen Universitätsverhältnissen zu begeben."[8] There must have been something very special about the Royal College of Chemistry.

Originally a private institution the Royal College of Chemistry became (in 1853) part of the Government's Royal School of Mines.[9] During these years of exile Hofmann raised the College to a "high position among the chemical schools of Europe".[10] His scientific activities have been abundantly chronicled in the Memorial Lectures of the Chemical Society[11] and elsewhere, while these and other aspects of his life may be seen more informally in his copious correspondence, or in diaries like those of his assistant Herbert McLeod.[12] Yet when all has been said the enduring and unique legacy of this "cosmopolitan chemist" lies surely in his international influence, one that certainly transcends England and Germany. How was this realised in practice?

Friendships within the international scientific community

From his earliest days in London Hofmann enjoyed the confidence of many influential people, including the British Royal Family. The Queen and Prince Albert treated with him with great kindness, the latter becoming a personal friend and frequent visitor. This was also true of their eldest daughter the Princess Royal who was to become Empress Frederick of Germany. Hofmann also entered into warm and lasting friendship with many leading chemists. Two examples must suffice.

Hofmann's correspondence with Liebig reveals the close bond of comradeship between pupil and master. During his stay in England Hofmann became a channel of communication between the English and German chemical communities, never more obvious than when acting on behalf of Liebig. From 1848 onwards there were innumerable requests from Liebig for assistance of all kinds. He asked Hofmann to get books, papers, materials, apparatus etc., – even to interview prospective parents-in-law for his daughter![13] Although very much 'his own man' Hofmann was nevertheless the most influential apostle of Liebig's ideas in mid-Victorian Britain.

Much less well-known was his friendship with the Englishman Edward Frankland who became, in Hofmann's own words, his "natural successor".[14] They had known each other since 1849 when Hofmann came to Frankland in Giessen to learn to make sealed tubes,[15] and dined together from time to time over the next 40 years.[16] On behalf of Poggendorff Hofmann wrote for Frankland's biographical details.[17] In informing Frankland of his election to the German Chemical Society Hofmann said he took an "infernal pleasure",[18] though a few days later confessed to "an indescribable pang" over the death of Frankland's wife Sophie.[19] Such deep personal friendship quite naturally sprang from close collaboration in matters of chemistry.

His network of international friendship led Hofmann throughout his life to undertake travel on a prodigious scale – the Victorian precursor of the jet-set.[20] A visit to Paris, meeting Cahours and Thenard, sent him back to London full of new zest for work.[21] W. H. Perkin remarked that when he became President of the Chemical Society "Dr. Hofmann travelled all the way from Berlin in order that he might preside, a practical mark of friendship for his old student, which he would never forget."[22]

Propagation of the Giessen academic tradition in England

Liebig already had a few devoted followers in England but none was more effective in transplanting the whole Giessen model of chemical education than Hofmann himself. Here, at the Royal College of Chemistry, was a new emphasis on practical instruction, on collaborative work in the laboratory and on research in organic chemistry. In the 1830s this branch of chemistry was little studied in the United Kingdom; a would-be researcher had to go to Göttingen or Giessen,[23]

Tab. 1: Hofmann's academic connections: destinations of some of his London students

British		
C. Bloxam	King's College, London	
H. Noad	St. George's Medical School, London	
A. H. Church	Agricultural College, Cirencester	
A. Matthiesen	St. Bartholomew's Hospital, London	
J. S. Muspratt	College of Chemistry, Liverpool	
G. D. Liveing	University of Cambridge	
H. E. Armstrong	Central Technical College, London	
Overseas	from	to
H. How	England	Nova Scotia University
T. Rowney	England	Queen's College, Galway
R. Galloway	England	Royal College of Science, Dublin
J. Volhard	Germany	University of München
C. H. Wichelhaus	Germany	University of Berlin
J. Blyth	Jamaica	Queen's College, Cork
A. A. T. Cahours	France	Ecole Polytechnique, Paris

or, a little later, to Marburg. Since organic chemical work was (and is) crucially dependent on practical instruction there is little wonder that the rise of English organic chemistry can be largely attributed to Hofmann.

Every teacher likes to see his values and methods reproduced in the next generation. In that respect Hofmann was singularly fortunate.

Introduction of English methods to Germany

It may well be wondered if there was any repayment of England's chemical debt to Germany. Apart from his constant communication with former colleagues in Britain Hofmann did import several useful items across the North Sea. At a comparatively trivial level he conveyed a constant stream of British chemicals and equipment to Germany, while he equipped his German laboratories with slate from English quarries.[24]

Much more important was his rôle in founding the Deutsche Chemische Gesellschaft. Berlin was a natural birthplace, for now, as Baeyer observed, the University of Berlin for the first time had a chemical teaching laboratory.[25] Hofmann cited the example of London and sought to establish the new Society on lines of the London Chemical Society. He stressed that what was needed was a *national* society, and that while London could stand for England, Berlin could not stand for Germany. Hence it had to be a German institution, and national and provincial interests were displayed in its full name Deutsche Chemische Gesellschaft zu Berlin.[26]

There was, moreover, another major aspect of Hofmann's international influence that, at least at first, was a transference to Germany of English practice: a deep concern for the application of chemistry.

The widespread application of chemistry

At the Royal College of Chemistry Hofmann did not merely teach or undertake research in pure chemistry. In a long letter of self-defence Frankland claimed "the undertaking of professional work in moderation the general practice in all British chemical schools with which I am acquainted." He went on to draw attention to the many interesting cases in Hofmann's annual reports.[27] That it had long been a particularly English phenomenon is borne out by a remark made by Hofmann in 1851 to the effect that, since he was unable to devote all his time to pure science, a return to Germany did have some attractions.[28]

While in England Hofmann undertook much Government work, including water analysis,[29] advice on lubricating grease for rifles,[30] membership of a small commission (including Faraday) on "Lighting Picture Galleries by gas"[31] etc. In the private sector he undertook beer analysis for Allsopp's Brewery,[32] and tested samples for Darwin[33] (who afterwards lamented his absence in Berlin[34]). He was frequently involved as an expert witness, as in the famous Simpson vs. Holliday case.[35] But above all he played a vital rôle in the infant dyestuffs industry.[36] His philosophy is well-known:

> Though proud of her office as guide of industry, science acknowledges without blushing that there are territories on which she cannot advance without leaning on the strong arm of her powerful companion.[37]

Something of Hofmann's international influence on the chemical industry may be seen from Tab. 2, which applies only to the early years in London.

Tab. 2: Hofmann's industrial connections

Coal tar products	C. B. Mansfield	C_6H_6 and $CH_3{\cdot}C_6H_5$ from Coal-Tar	London
Pharmaceuticals	G. Merck	Papaverine	Darmstadt
Sugar refining	J. A. R. Newlands	Periodic Law	London
Brewing	P. Griess	Diazotisation	Burton-on-Trent
	C. O'Sullivan		Burton-on-Trent
Explosives	F. A. Abel	Nitrocelluloses	Woolwich
Metallurgy	J. S. Abel		Silver Works, Chile
	F. Manning		Hong Kong Mint
	C. Tookey		Hong Kong Mint
	J. Riley		Dowlais Ironworks, Wales
	W. Child		Dowlais Ironworks, Wales
	W. Matthey	Pt Alloys	Platinum manufacturer, London
Calico-printing	B. Cooper		Lancashire
	G. Schlieper		Elberfeld, Germany
Dyestuffs	E. C. Nicholson	H_3PO_4 and Organic Bases	London
	G. Simpson		London
	G. Maule		London
	H. Medlock	Magenta; Arsenate Oxidation of $PhNH_2$	London
	F. Field		London
	J. Spiller		London
	W. Spiller		London
	W. H. Perkin	Mauveine	Greenford
	C. A. Martius	$PhNHCH_3$; Eosin; Naphthalene Dyes	Berlin
	C. H. G. Williams	Pyridines in Coal-Tar	Brentford

In his later years in Germany Hofmann maintained and increased his industrial interest. Thus he became a founding member of the (British) Society of Chemical Industry, even though by then he had been 17 years in Germany. His was the very first name to appear in its illustrious *Journal*.[38] Students from industry flocked to hear him in Berlin. One such was Dan Dawson, a Huddersfield dyestuffs manufacturer, who later made a fortune and built many houses, including one for himself at Milnsbridge, near Huddersfield. His address was Hofmann Street![39]

It is sad to record that after Hofmann went to Germany English research had little relevance to chemical industry, and it is probable that in France Berthelot's views hindered the growth of aromatic chemistry. Thus the leadership in organic chemistry went to Germany.

Effective use of scientific media

For good scientific communication there need to be effective media, and an effective language. Hofmann was concerned with both.

Necessary to all modern science are journals. Although the *Annalen* and the *Annales de chimie* had long dominated the continental scene, Hofmann soon perceived the need for a publication outlet for his English students. He was advised by Sir James Clark against launching a new journal in the crucial early years of the College.[40] Indeed that was the year (1847) in which Hofmann decided to report his work on aniline to the French Academy.[41]

The solution was to be the *Journal of the Chemical Society*, and for years Hofmann was a mainstay of the Publications Committee of that society, guiding and shaping editorial policy until the *Journal* gained its international recognition.[42]

A further necessary medium is textbooks. In 1865 Hofmann published the latest version of his lectures at the Royal College of Chemistry. His *Introduction to Modern Chemistry* is replete with woodcut illustrations and the most up-to-date formulations of gas reactions, using the new atomic weights of Cannizzaro. The book's very thesis is international: Hofmann speaks of new chemical doctrines which have only recently acquired "a logical consistency, and a consequent ascendency throughout Europe".[43] A German edition guaranteed still wider circulation of this important book.

Nor must we forget Hofmann's numerous and eloquent memorial lectures,[44] including the notable Faraday Lecture on Liebig.[45] They are still of value to historians, and it is said that these obituary memoirs were among the "fascinating and exciting reading that stimulated young Willstätter".[46]

Nor was it always a question of printed media. One cannot fail to notice the rise of international exhibitions, an institution of which Hofmann took full advantage, starting with the Great Exhibition of 1851. Thus in 1855 he was called upon to exhibit alkaloids in Paris[47] and he acted as juror at other times. The exhibitions of 1851 and 1862 gave him an opportunity to declaim before a wide public the merits of chemical science.

Development of an international chemical language

Turning to matters of language Hofmann was, as Armstrong remarked, "cosmopolitan in speech".[48] With an enviable fluency in English, and conversational French and Italian, he sought also for the greatest clarity in chemical language.

Advocating the French metric system, Hofmann condemned the varied systems of weights and measures used elsewhere as making "the records of one nation practically a sealed book to the philosophers of another".[49] Yet he also complained of the French reluctance to use the symbol N for nitrogen, seeking the "abandonment of her one discordant symbol".[50] In attempting to internationalize the language of chemistry he was following in the footsteps of Berzelius.[51] In order to rationalise the nomenclature of hydrocarbons he proposed the suffixes -ane, -ene, -ine, -one and -une,[52] the first three of which were accepted by the Geneva Conference of 1892.[53] An even later survival of Hofmann's influence on nomenclature came in 1919 when the American Chemical Society was experiencing a wave of anti-German feeling. It abandoned the term "benzol" (of German origin through Liebig) for "benzene", proposed by Hofmann but apparently acceptable because he was living in England![54]

For a final example of Hofmann's concern for clear scientific communication we go to a popular lecture delivered in 1865 at London's Royal Institution. Here he illustrated a new idea with which chemists were beginning to struggle: the notion that was later to be called (by Frankland) a chemical "bond". So abstract a concept he made real to his audience by a series of models, using balls taken from that most English of games, croquet.[55] He was wrong on some of his structures, had no idea of tetrahedral carbon, and others were to advance the new notation and to make better models to supplement it. But on the central question of what to call the new phenomenon, Hofmann played a vital part. In that same year he called it "quantivalence",[56] a contribution to nomenclature that had international results. The term was employed in England, France and the United States of America for a few years, but in Germany was rapidly shortened to *Valenz*.[57] The almost universal use of valency and cognate terms we owe directly to Hofmann.

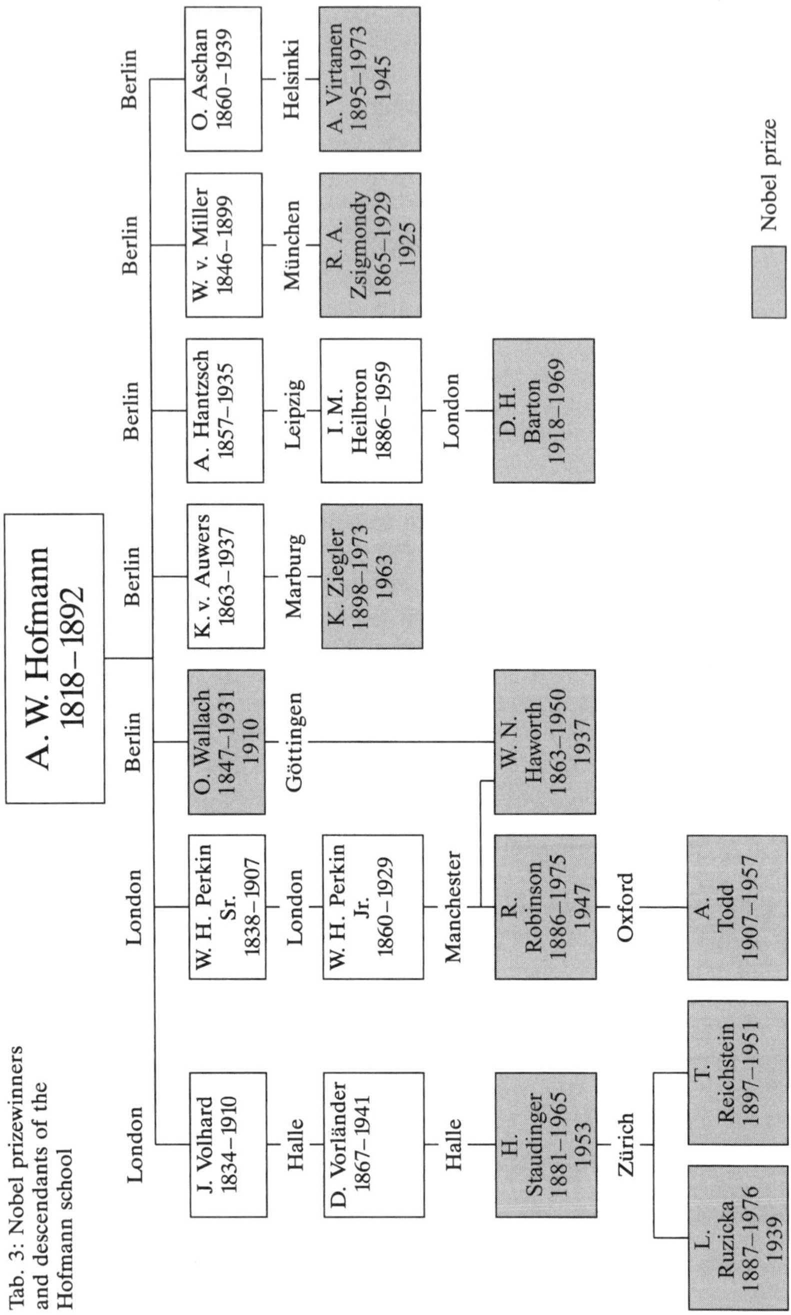

Tab. 3: Nobel prizewinners and descendants of the Hofmann school

Spread of research results

In England Hofmann's academic research in organic chemistry began with his demonstration in 1843 that the same product, aniline, was obtained directly from coal-tar (Runge), the alkaline degradation of indigo (Fritsche) or the reduction of nitrobenzene. From this early work on organic bases he advanced to "diatomic" and "triatomic ammonias", to the phosphorus series of ammonium bases, alkaloids and the whole dazzling range of aniline dyes. By judicious choice of publishing journals he kept an international public aware of his work. Such was its quality that few organic chemists could have been unaware. Similarly his method for vapour densities was soon "in general use".[58]

One of the surest indicators of a chemist's influence is to have a reaction bearing his name. In Hofmann's case there are three such reactions (apparently a record). The Hofmann Exhaustive Methylation procedure was reported[59] in 1851 but not fully appreciated until he published his work on the constitution of piperidine.[60] The Hofmann-Martius Rearrangement (of, for example, methylaniline to 4-toluidine) was discovered in 1871[61] and elucidated shortly thereafter in terms of the hydrochlorides. The Hofmann Reaction (or Degradation, or Rearrangement) involving the hypohalite decomposition of an amide to an amine with one less carbon atom appeared in 1881, when Hofmann showed the intermediate formation of an isocyanate.[62] Each of the last two reflects on a molecular scale one characteristic of its discoverer: a tendency to migrate!

With such an immense reputation it is hardly surprising that, to the end of his days, Hofmann was able to attract chemists of outstanding calibre. One measure of this is the succession of Nobel Prizewinners who trace their descent from perhaps the greatest of them all, August Wilhelm Hofmann.

Thanks are expressed to Mr and Mrs Raven Frankland for permission to cite from documents in their possession (The "Raven Frankland archive" [quoted as RFA]); microfilms of these documents are at the Open University and are quoted here by OU Microfilm numbers.

1 Hofmann to Frankland (1862 Aug 10), RFA; OU no. Z. 01.03.0131.
2 Hofmann to Frankland (1869 [1867?] Apr 22), RFA; OU no. Z. 01.01.0533.
3 F. A. Abel, "The history of the Royal College of Chemistry and reminiscences of Hofmann's professorship," *J. Chem. Soc. Lond.* 69 (1896), 580–596, on p. 595.
4 W. Thorp cited in (anonymous) report of meeting (1892 May 30) "The late Professor A. W. von Hofmann, F.R.S.," *J. Soc. Chem. Ind.* 11 (1892), 485–486, on p. 486.
5 Henry E. Armstrong, "Notes on Hofmann's scientific work," *J. Chem. Soc. Lond.* 69 (1896), 637–732, on p. 729.
6 Liebig to Hofmann (1845 July 26), in: William Hodson Brock (ed.), *Justus von Liebig und August Wilhelm Hofmann in ihren Briefen* (Weinheim 1984), pp. 32–33.
7 Hofmann to Liebig (1851 May 18), ibid., p. 113.
8 Liebig to Hofmann (1863 Nov 14), ibid., p. 197.

74

9 See G.K. Roberts's paper in this volume.
10 Abel (1896), p. 588.
11 *J. Chem. Soc. Lond.* 69 (1896), 575–732; also separately as *Memorial Lectures delivered before the Chemical Society, 1893–1900* (London 1901), pp. 575–732.
12 F. A. J. L. James (ed.), *Chemistry and Theology in mid-Victorian London: The Diary of Herbert McLeod, 1860–1870* (London 1987).
13 Liebig an Hofmann (1848 Oct 26; 1854 Jan 12, etc.), in: Brock (1984), pp. 79–80, 167, etc.
14 Hofmann to Frankland (1868 Feb 19), RFA; OU no. Z. 01.07.0330.
15 E. Frankland, *Sketches from the life of Sir Edward Frankland*, 2nd ed. (London 1902), p. 111.
16 E.g. in Paris (1867 May) at Thenard's and also at Wurtz's, and in Berlin (1890 Sept 7); ibid., pp. 400 and 444.
17 Hofmann to Frankland (1852 Nov 20), RFA; OU no. Z. 01.02.0344.
18 Hofmann to Frankland (1873 Dec 17), RFA; OU no. Z. 01.01.0542. He added that "the election of honorary members is always a source of considerable agitation".
19 Hofmann to Frankland (1874 Jan 11), RFA; OU no. Z. 01.01.0544.
20 Of course he also travelled for other reasons (to exhibitions, for example) or to see his family.
21 Hofmann to Liebig (1850), in: Brock (1984), pp. 92–93.
22 W. H. Perkin cited in "The late Professor" (1892), p. 486.
23 L. Playfair, "Hofmann Memorial Lecture," *J. Chem. Soc. Lond.* 69 (1896), 575–579, on p. 576.
24 Cannizzaro to Frankland (1873 Sept 18), und Duppa to Frankland (1873 Oct 26), RFA; OU no. Z. 01.08.0444 and 01.02.0154.
25 Walter Ruske, *100 Jahre Deutsche Chemische Gesellschaft* (Weinheim 1967), p. 29.
26 Ibid., p. 38. He probably did not know that in that very year England gained its first provincial Chemical Society, that of Newcastle-upon-Tyne.
27 Frankland to T. H. Huxley (1884 June 21), RFA; OU no. Z. 01.03.0477.
28 Hofmann to Liebig (1851 June 11), in: Brock (1984), pp. 115–117. This makes Playfair's assertion that at the Royal College of Chemistry Hofmann "thought of science only, not of practice" border on the incredible; see Playfair (1896), p. 579.
29 W. Farr to Frankland (1865 Oct 5), RFA; OU no. Z. 01.03.0090. Hofmann was reported as saying one should not work for the Government for nothing, so he agreed £100 p. a. for water analysis.
30 Hofmann to Liebig (1853 June 26), in: Brock (1984), pp. 159–161. The grease had to be insect-proof and inodorous when hot.
31 Report of Commission on Lighting Picture Galleries by Gas (1859 July), 106, Session 2.
32 Hofmann to Liebig (1852 Apr 20), in: Brock (1984), pp. 127–129.
33 Darwin to Frankland (1873 July 12), RFA; OU no. Z. 01.08.0486.
34 Darwin to J. S. B. Sanderson (1873 Sept 14), Cambridge University Library, Darwin Collection, ref. 8089 (original at University of British Columbia).
35 Report on court case in *Chemical News* 10 (1864), 48; 11 (1865), 35.
36 Anthony S. Travis, "Science's powerful companion: A. W. Hofmann's investigation of aniline red and its derivatives", *The British Journal for the History of Science* 25 (1992), 27–44.
37 A. W. Hofmann, "Researches on some artificial colouring matters, 1: On the composition of the blue derivatives of the tertiary monamines derived from cinchonine," *Proc. Roy. Soc. Lond.* 12 (1862/63), 410–418, on p. 418.
38 Proceedings of first General Meeting (1881 June 28–29) in *J. Soc. Chem. Ind.* 1 (1881), 1.
39 G. T. Morgan, *Hofmann Memorial Lecture at Imperial College* (London 1936), p. 45–51.

40 Hofmann to Liebig (1847 Feb 4), in: Brock (1984), pp. 66–69.
41 Liebig to Hofmann (1847 Nov 22), in: Brock (1984), pp. 74–75.
42 Robert F. Bud and Gerrylynn K. Roberts, *Science versus Practice: Chemistry in Victorian Britain* (Manchester 1984), p. 66.
43 August Wilhelm Hofmann, *Introduction to Modern Chemistry Experimental and Theoretic* (London 1865), pp. vi-vii.
44 August Wilhelm Hofmann, *Zur Erinnerung an vorangegangene Freunde*, 3 vols (Braunschweig 1888).
45 A. W. Hofmann, "The life-work of Liebig in experimental and philosophic chemistry," *J. Chem. Soc. Lond.* 13 (1875), 1065–1140.
46 Richard Willstätter, *From my Life: Memories of Richard Willstätter*, ed. by A. Stoll, trans. by L. S. Hornig (New York 1965), p. 42.
47 Hofmann to Frankland (1875 Apr 1), RFA; OU no. Z. 01.03.0268.
48 Armstrong (1896), p. 729.
49 Hofmann, *Introduction* (1865), p. 122.
50 Ibid., p. 208.
51 J. J. Berzelius, "Essai sur la nomenclature chimique," *Journal de Physique* 73 (1811), 253–286.
52 A. W. Hofmann in: *Jahresberichte über die Fortschritte der Chemie* 1865, p. 413n.
53 Paul E. Verkade, *A History of the Nomenclature of Organic Chemistry* (Dordrecht 1985), p. 23.
54 Ibid., p. 99.
55 A. W. Hofmann, "On the combining power of atoms [1865]," *Procedings of the Royal Institution* 4 (1866), 401–430, on p. 414.
56 A. W. Hofmann, *Introduction* (1865), p. 168–169.
57 C. A. Russell, *The History of Valency* (Leicester 1971), pp. 85–88.
58 Carl Schorlemmer, *The Rise and Development of Organic Chemistry*, rev. ed. (London 1894), p. 103.
59 A. W. Hofmann, "Researches into the molecular constitution of the organic bases," *Phil. Trans. Roy. Soc. Lond.* 142 (1851), 357–398; also in: *Ann. Chem. Pharm.* 78 (1851), 253–286.
60 A. W. Hofmann, "Ueber die Einwirkung der Wärme auf die Ammoniumbasen," *Ber. dtsch. Chem. Ges.* 14 (1881), 494–496; idem, "Beiträge zur Kenntniss des Piperidins," ibid., pp. 659–672; idem, "Beiträge zur Kenntniss des Coniins," ibd., pp. 705–713; see R. C. Denney, *Named Organic Reactions* (London 1969), p. 194.
61 A. W. Hofmann and C. A. Martius, "Ueber die Methylirung der Phenylgruppe im Anilin," *Ber. dtsch. Chem. Ges.* 4 (1871), 742–748.
62 A. W. Hofmann, "Ueber die Einwirkung des Broms in alkalischer Lösung auf Amide," ibid., 14 (1881), 2725–2736.

XV

THE CHANGING ROLE OF SYNTHESIS IN ORGANIC CHEMISTRY

In 1793 Lavoisier defined chemistry as "the science of analysis".[1] According to a distinguished fellow countryman 67 years later he was wrong. Said Marcellin Berthelot:

> Having completed his analytical work, the chemist proposes to re-compose what he has destroyed; he takes as his point of departure the ultimate degree of analysis, i.e. simple bodies, and he compels them to unite with each other and by their combination to re-form the same natural principles which constitute all material beings. Such is the object of chemical synthesis.[2]

For Berthelot synthesis was more than a programme of research. It was a way of life, pursued with unremitting zeal for over 10 years. However, as we shall see, the very word "synthesis" is loaded with ambiguities and has all sorts of connotations, some of which Berthelot accepted, others he did not. That synthesis *in some sense* was crucial to the development of chemistry in the nineteenth century was, however, beyond doubt. It has been the subject of many extravagant claims and has had spectacular results, some rather more tangible than others. Without it not only organic chemistry but the shape of modern chemical science and indeed European history would not have been the same. Before examining these developments it is necessary to clarify two matters which have much confused the issue in the past.

First, there was the actual meaning of "synthesis". Apart from a general connotation of "building up" it has been used with various shades of meaning[3] as follows:

(1) Formation of a natural organic product from any other simpler natural product.
(2) Formation of a natural organic product from an artificial product.
(3) Formation of a natural organic product from the elements directly.
(4) Formation of an artificial organic product from any other artificial product.
(5) Formation of an artificial organic product from the elements directly.
(6) Formation of an artificial organic product from a natural one of simpler structure.

Also it is well to remember, as Graebe has pointed out, that some very early synthetic work was not regarded at the time as such, but merely as simple preparation.[4] This included Scheele's production of prussic acid (from carbon, carbon dioxide and ammonia, 1783), Wöhler's production of oxalic acid from cyanide (1824), Gmelin's discovery of the peculiar "croconic acid" from potassium and carbon monoxide (1825) and the preparation of alcohol from ethylene by Faraday and Hennell in 1825.

A second confusion often exists between the *intention* and the *actual effect* of synthetic studies. Intentions have ranged from the ruthlessly avaricious to the faintly dotty, from the "let's try it and see" philosophy to a meticulously conceived programme of research. Effects have ranged from precisely nil to something approaching a revolution in chemical theory, from the utterly unexpected to the fulfilment of years of hope and aspiration.

Whether intended or not there seem to be at least five different kinds of outcome attending the rise of synthetic organic chemistry. Chronologically they overlap to some extent but it may be helpful to unweave this tapestry, pull out these five separate strands and look at each one in turn.

1. Synthesis and the Unity of Chemistry

The story of Wöhler's conversion of ammonium cyanate to urea in 1828 is familiar enough. So, unfortunately, is the legion of misrepresentations to which it has been subjected, notably that it bridged the chasm between organic and inorganic chemistry by once and for all laying the ghost of vital force, by synthesizing a genuinely organic product from a non-organic starting material. That this is a misleading construction of events has been made abundantly clear by many scholars since Douglas McKie, who roundly asserted in 1944 that "those who believe that Wöhler drove vitalism out of organic chemistry will believe anything".[5] It is not necessary to rehearse the arguments here, though they have been modified and refined by several historians since.[6] However, one intriguing aspect of the Wöhler myth will emerge at the end of this paper.

It has been cogently argued[6c] that the gap between organic and inorganic chemistry was considerably narrowed in the decades following the urea synthesis, though the latter was hardly responsible. By 1862 another French chemist, Wurtz, proclaimed "there is but one chemistry".[7] If there was indeed substance in Wurtz's proclamation an explanation might be sought in the proliferation of organic synthesis in the previous 30 years. Indeed some protagonists of a new synthetic philosophy went so far as to argue that this was the case (as we shall see). However Dr. John Brooke has persuasively argued that it was not so much synthesis as the establishment and fortifying of a prolonged analogy between organic and inorganic chemistry that really helped to cement the union between them. In fact the time was to come when inorganic chemistry would be indebted to the newer organic branch for analogies and guiding principles. There can be no doubt of the force of the analogical argument. Yet synthesis cannot be neglected altogether. And it now seems probable that it did have a significant effect on the unification of chemistry, though not in the sense once ascribed to Wöhler and later to Berthelot. The explanation may lie in the origins and institutionalisation of synthetic organic chemistry.

In 1825, three years before urea presented itself to the astonished gaze of Wöhler, Justus Liebig became Professor of the University of Giessen. For 28 years he was to supervise a chemical laboratory that was to acquire world reputation for its course of practical instruction and its research output, above all in the field of organic synthesis. With generous funding from the state of Hesse-Darmstadt it became the Mecca for students from all over Europe who wished to learn chemistry and to obtain a recognised qualification. The whole subject was taught, but organic chemistry was pre-eminent. Within five years Liebig had developed his analytical techniques to such a degree that they could be confidently applied to the vast numbers of new compounds he and his students were going on to prepare. His influence was enormous. As time passed his own students went out to different centres of Europe (especially Germany) to continue and extend his method of laboratory training and instruction. Even London was included within his sphere of influence, when Hofmann imported the German methods into the Royal College of Chemistry in 1845. An "apostolic" succession of chemical disciples made sure that continuity was preserved.

That even the most simple organic synthesis demands considerable skill and development of characteristic techniques is obvious from the early literature, inspection of the famous picture of Liebig's laboratory,[8] from a visit to it today, or best of all laboratory experience oneself. The methods derived for preparation, isolation and purification are not acquired in a day and considerable laboratory training is essential. This was available in Giessen and in its later satellites. The whole effect of such chemical education was to produce something of a race apart, chemists whose skills and expertise were really quite different from those who spent their time in mineral analysis and other areas of inorganic chemistry. If this is so it would appear that the development of German synthetic organic chemistry would tend towards differentiation rather than unification of the two branches.

The experience of the young Lancashire chemist, Edward Frankland, is instructive. Going to another German laboratory in the 1840s—to Marburg where Bunsen reigned supreme—Frankland became the very first Englishman to graduate there as a Ph.D. Entering the uncharted waters of what he was later to call organometallic chemistry Frankland became proficient in glass-blowing and became the first chemist to develop the technique of heating materials in a sealed tube. He also pioneered the use of a Liebig condenser in the reflux position and was one of the first to employ ethyl iodide as a laboratory reagent. The products of his research were mostly inflammable gases or liquids and their handling and subsequent analysis required entirely different techniques from those in inorganic chemistry.[9] But it was not only by its demands for laboratory rigour that organic synthesis tended to maintain its practitioners as a race apart. So also was the scale of the problem disclosed afresh by every new synthetic product. Vast numbers of new compounds were piling up on the shelves and in the literature, and problems of isomerism and polymerism were not paralleled in the inorganic branch at that time. In order to encompass these synthetic novelties, it was possible to proceed quite satisfactorily with a minimum of theory. In 1841 Liebig unfavourably compared the French tendency to wonder whether radicals were types with the German dedication to productive experiment in the laboratory.[10]

An interesting confirmation of this effect of synthesis on the unity of chemistry may be seen, ironically enough, in an address by Hofmann to the Royal Institution in 1853 when he was formally arguing that "the separation of chemical science into inorganic and organic is by no means found in nature".[11] Speaking at the very time when he was seeking Government support for the Royal College of Chemistry he, perhaps naturally, had to portray his subject in the best possible light, with all its practitioners united in one cohesive group. Yet in actual fact he of all people was a living demonstration of the separation of organic chemistry. He had produced about 360 research papers of which only a very few were outside that area. Indeed it has been argued that Hofmann, more than anyone else, set the fashion for organic research in England. So it is not surprising that at the end of his lecture, he had to admit: "I am almost afraid, gentlemen, that you will object to me that in denying the distinction of inorganic and organic compounds I lose the very ground upon which I stand".[12] In so far as that was the case it must be attributed to the astonishing success of his research programme which led him, on more than one occasion, to announce to his delighted assistants "Gentlemen, new bodies are *floating* in the air".[13]

Thus synthesis, far from having a unifying effect (as was once alleged for Wöhler's work on urea) actually helped to differentiate the two branches of chemistry and postpone their unification until well after 1860. When that did happen, synthesis was to turn the course of chemistry in another direction altogether. How that was to occur we must now consider.

2. Synthesis and Fundamental Chemical Theory

Reverting again to the Wöhler synthesis of 1828 it may be noted that the modest, genuine, contribution made to chemical theory was a clarification as to what was later to be called "isomerism" and "isomeric change", for ammonium cyanate and urea possessed not merely the same empirical formula but also the same molecular formula. Other examples of isomerism were to be revealed by later syntheses but little real progress could be made in this direction until the whole concept of chemical bonding and chemical structures became clarified. An important step in this direction was associated with another adventure in chemical synthesis. In 1847 Edward Frankland was lured away from his work of mineral analysis at the Geological Survey in London to a three-months' research trip to the University of Marburg, to work under Bunsen. It was here that he acquired a specific objective:

> "the isolation of the alcohol [alkyl] radicals was, at this time, the dream of many chemists, while others doubted or even denied their very existence. I was also smitten with the fever and determined to try my hand at the solution of the problem".[14]

Accordingly, together with his friend Hermann Kolbe he decided to try to extract the ethyl radical from ethyl cyanide by means of potassium metal. The result of the reaction did not give ethyl but rather a complex series of events which included a base-catalysed trimerization of ethyl cyanide to an aminopyrimidine:

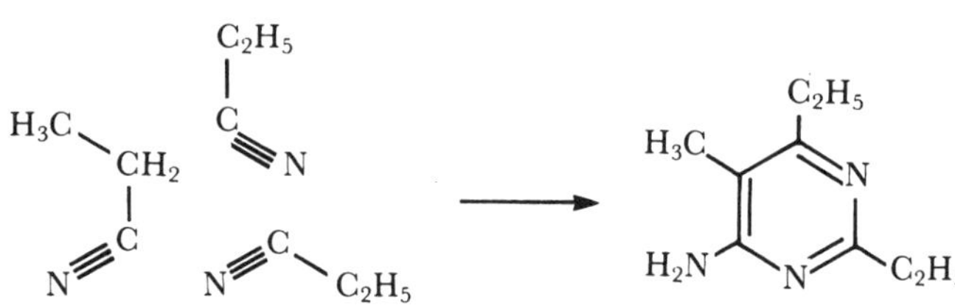

He then tried to make ethyl from the iodide by using potassium. He found (of course) a very vigorous reaction with numerous hydrocarbon products. So he tried the much less reactive metal zinc. The outcome of his bold experiment was an explosion, the evolution of large quantities of gas (in which he thought he had obtained "ethyl"), but (most important of all) a new class of chemical compound. This included ethyl zinc iodide and diethylzinc, the first compounds to which the name *organometallic* was given.

$$
\begin{aligned}
C_2H_5I + Zn &= C_2H_5ZnI \\
2C_2H_5ZnI &= (C_2H_5)_2Zn + ZnI_2 \\
C_2H_5ZnI + C_2H_5I &= C_4H_{10} + ZnI_2 \\
C_2H_5ZnI + H_2O &= C_2H_6 + Zn(OH)I \\
(C_2H_5)_2Zn + 2H_2O &= 2C_2H_6 + Zn(OH)_2
\end{aligned}
$$

From this it was but a step to prepare a whole range of similar compounds from other metals and metalloids. The result was not merely the establishment of a new branch of chemistry, though that was significant enough. Indeed the very existence of organometallic chemicals testified to the artificiality of separating organic and inorganic chemistry, for if ever hybrids were to be recognized here surely they were. Later on further developments in this field accentuated the anomaly of a divided chemistry. The immediate impact of Frankland's discovery, however, was to open the way towards a theory of valency and structure. As he said "it was evident that the atoms of zinc, tin, arsenic, antimony, etc., had only room, so to speak, for the attachment of a definite and fixed number of atoms of other elements".[15] So from these obscure novelties, themselves an entirely accidental discovery, emerged a doctrine of such power that it was soon found to be applicable to the whole of chemistry.

It is true that Frankland was not the only pioneer to be credited with the theory, for he must share the honours at least with Kekulé. Moreover structure theory was generally accepted only in the 1860s, after the Karlsruhe Conference that paved the way to universally agreed atomic weights. Yet it remains true that at the root of each, with all their implications for chemistry as a whole, lay the organic syntheses of Edward Frankland.

3. Synthesis and Industry

At this point it is appropriate to mention another, quite novel, role that synthesis was to acquire in the second half of the ninteenth century. The fact that it had most obvious economic, social and even political and military implications should not detract from its theoretical importance as well. Once again the story begins with an entirely unexpected discovery. In 1856 the English chemist W. H. Perkin, prompted by Hofmann to attempt a synthesis of quinine, found in fact that chromic acid oxidation of toluidines led to the first synthetic dyestuff, mauve.

Another aniline dye was discovered in 1859 in France, known variously as fuschin, magenta and rosaniline. Meanwhile Peter Griess had discovered the diazotisation reaction, though the first successful azo dye was made by Martius at Manchester in 1863 (Bismarck Brown, a diazotised m-phenylene diamine). But was this synthesis? Only in sense number six, assuming coal tar bases to be "natural". It is also limited in another sense.

After the International Exhibition of 1862 Hofmann wrote:

> England will, beyond question, at no distant day, become herself the greatest colour-producing country in the world; nay, by the strangest of revolutions, she may, ere long send her coal-derived blues to indigo-growing India; her tar-distilled crimson to cochineal-producing Mexico, and her fossil substitutes for quercitron and safflower to China, Japan, and the other countries whence these articles are now derived.[16]

Yet three years later he had himself returned to Germany, and in the 1867 Exhibition Britain did very badly, scooping only ten prizes out of ninety. British syntheses (if such they can be called) were largely pragmatic, almost hit and miss. All eyes were now on Germany where an infrastructure of scientific training was already beginning to serve the new dyestuffs industry. In these advantageous circumstances there could now be synthesis of a more systematic and far-reaching kind. In 1868 Graebe and Liebermann made the first synthetic alizarin (the chief component of madder). Next year, by a variant of their process, it was being manufactured in both Germany and England. Anthracene was treated successively with fuming sulphuric acid and fused alkali, both of which were consumed on an enormous scale. Perkin, visiting the BASF factory observed how convenient it was to have both inorganic and organic industries on the same site—another, if trivial, example of links between the two branches.[17]

By 1881 an aggrieved Henry Roscoe could announce at the Royal Institution: "to Englishmen it is a somewhat mortifying reflection, that whilst the raw materials are produced in our country, the finished and valuable colours are nearly all manufactured in Germany".[18]

Alizarin was followed in due course by indigo, first synthesized by Baeyer in 1880,[19] by the first commercially effective synthesis, developed by Hermann in 1890 from phenyl glycine o-carboxylic acid, obtained via phthalic anhydride from naphthalene (whose oxidation by fuming sulphuric acid was discovered to be catalysed by mercury from a broken thermometer). Variations were invented, partly to overcome patent restrictions and partly to employ fresh outlets for sodamide then being made extensively in Germany.[20] Syntheses were being explored for simple commercial and economic reasons. They also led to the proliferation of dyestuffs and dyestuff intermediates, so that by 1900 no less than 12,000 azo dyes had been made in Germany. And it must be added that, were it not for the dyestuffs production, the pharmaceutical industry of Germany would be more readily credited with spectacular progress.

This immense output of synthetic materials was not without its effect on chemistry generally. Writing in 1918 A.W. Stewart had two particularly relevant things to say:

> Despite the Briarean efforts of the synthetic school, it is safe to say that the latter half of the nineteenth century will be regarded as a time when theoretical speculation played the main part in the development of the subject. Of the hundred thousand organic compounds prepared during that time, the majority were stillborn and their epitaphs are inscribed in Beilstein's Handbook. Compared with the great clarifying process which laid the basis of our modern views, they weigh but little in the balance.[21]

This fairly dismissive reference to organic syntheses has rightly stressed its limited ability to add to theoretical knowledge. It was still possible to do much synthetic work within a very narrow and restricted paradigm. He goes on to add:

> During the last 50 years the flood of synthetic material, principally from the German laboratories, has tended to obscure the genesis of what we still, after respect for tradition, term organic chemistry.[22]

In other words, industrial synthesis had diverted attention from a study of natural products. However it is not clear how this might have been prevented by the non-production of dyes and drugs, nor that the old definition of organic chemistry was necessarily better than that of the twentieth century.

4. Synthesis and Structure

These last remarks are necessary to avoid obscuring an important aspect of organic chemistry in the nineteenth century: the role of synthesis in establishing chemical structures. Examples are legion. To use synthesis for this purpose naturally presupposes a generally accepted set of "rules" and an assumption that rearrangements do not take place (or if they do that they are predictable). By the beginning of the twentieth century it was commonly accepted that structural arguments based on synthesis were at least as compelling as those based on decomposition products. In 1908 Stewart introduced the first edition of his *Recent Advances in Organic Chemistry* with the remark that "when the synthesis of a substance is known, its constitution has been deduced from the method of formation rather than from its decomposition products".[23] By that time chemists had a bewildering array of synthetic methods available.

High on the list of reagents employed would be acetoacetic and malonic esters demonstrated (respectively) by Frankland and Duppa[24] and by Max Conrad.[25] This was the period of the famous condensation reactions named after Claisen,[26] Knoevenagel,[27] Michael[28] and others, to say nothing of heterocyclic syntheses bearing the names of Baeyer, Knorr, Skraup and Hantzsch, or of the use of organometallic reagents by Barbier, Reformatsky and (from 1900) Victor Grignard. Nor will time permit us to tell of Friedel and Crafts, Diels and Alder, Reimer and Tiemann, or even of Hell, Volhard and Zelinsky. It is an interesting commentary on social values of the day that so many synthetic reactions should bear the names of those who discovered them, almost as their personal property. One feels rather sorry for Wurtz whose "aldol reaction" is always thus cited anonymously,—but then he already had one reaction named after him even if it was less important. The record for plurality probably rests with Hofmann, whose three reactions have created understandable confusion amongst generations of undergraduate students.

The use of syntheses to determine structure could be part of a number of different strategies. Most obviously they were employed to elucidate structures of naturally occurring products,—the ultimate task, as was once thought, for the organic chemists. In that connection Ladenburg has pointed out that "the synthetical method constitutes a necessary complement" to the older analytical procedures.[29] He goes on to enumerate syntheses of alanine (Strecker, 1850), mustard oil (Zincke, 1855), glycine (Perkin and Duppa, 1858), racemic acid (the same, 1860), malic acid (Kekulé, 1860), taurine (Kolbe, 1862), guanidine (Hofmann, 1861), choline (Wurtz, 1867), vanillin (Reimer and Tiemann, 1876), allantoin (Grimaux, 1877), tyrosine (Erlenmeyer, 1883), indigo (Baeyer, 1885), coniine (himself, 1886) and much else. Above all, perhaps, one must record the epic work of Emil Fischer on carbohydrate synthesis, following his discovery of phenylhydrazine in 1875 and its exploitation in the elucidation of sugar structures. Nor can one ignore his later synthetic work on peptides and the labours of others leading to syntheses of terpenes, alkaloids, plant pigments, rubber and other natural products. Liebig's prophecy of 1838 that "we shall yet make sugar and quinine and uric acid" [30] was not fulfilled in his lifetime and the words of Fischer in 1907 have a more realistic ring:

> "Laboratory synthetic methods will be indispensable for a long time to come, not only for preparative purposes but also as the means of elucidating the structure of complex substances of natural origin'.[31]

The growing use of syntheses to determine structure was not of course limited to naturally occurring materials. It was also part of the strategy of those who wished to study the nature of materials newly made in the laboratory. The principles are the same, though the motivation may have been slightly different. An example must suffice. Körner worked on the orientation of substituents in the products from aromatic substitution. By allowing (for example) all dichlorobenzenes to be chlorinated further, and comparing the products, it is possible to elucidate not only the structures of those products but also of the starting materials (if unknown). Thus the only dichlorobenzene to give a single trichlorinated derivative must have a 1 : 4 configuration and the product of further chlorination must be 1, 2, 4-trichlorobenzene.[32]

Another strategy well served by synthesis was a testing of general theories. A classic case was the work undertaken by W. H. Perkin Jr. to prove whether or not small carbocyclic rings could exist, in defiance of Baeyer's strain theory. By using the sodium salt of malonic ester and a series of α, ω-dibromo alkanes he was in fact, able to synthesize three, four and five-membered rings, to the discomfiture of those who predicted it impossible. Much later Leopold Ruzicka and Max Stoll were able to prepare very large rings by pyrolysis of α, ω-dicarboxylates or Ziegler condensations of α, ω-dinitriles,—again in contradiction of strain theory but in confirmation of Ruzicka's conclusion that civetone had a seventeen membered carboxyclic ring.[33]

5. Synthesis and Ideology

That syntheses should have had the varied role as described above is hardly surprising to a modern chemist. What must seem remarkable is that they should have been undertaken with the open and explicit purpose of establishing an ideology that, in its essence, was not scientific at all. Yet such was the case of Marcellin Berthelot who in 1860 published his book

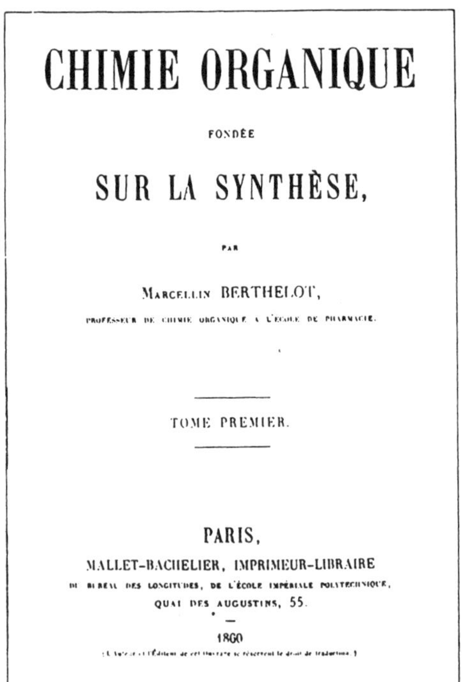
CHIMIE ORGANIQUE

FONDÉE

SUR LA SYNTHÈSE,

PAR

MARCELLIN BERTHELOT,

PROFESSEUR DE CHIMIE ORGANIQUE A L'ÉCOLE DE PHARMACIE.

TOME PREMIER.

PARIS,

MALLET-BACHELIER, IMPRIMEUR-LIBRAIRE

DU BUREAU DES LONGITUDES, DE L'ÉCOLE IMPÉRIALE POLYTECHNIQUE,

QUAI DES AUGUSTINS, 55.

1860

(L'Auteur et l'Éditeur de cet Ouvrage se réservent le droit de traduction.)

Organic Chemistry, founded on Synthesis. Here was no painstaking, detailed and often slow extension of organic chemistry by synthesis in the general manner. Berthelot regarded synthesis as the ultimate weapon with which to destroy a doctrine that he deeply detested: the doctrine of vitalism. Synthesis, and particularly total synthesis, was for him the only way to demonstrate "the identity of the forces which act in mineral chemistry with those which act in organic chemistry, by showing that the former suffice for producing all the effects and all the compounds from which the latter are born"[34]

Berthelot's efforts were concentrated upon total syntheses of hydrocarbons and alcohols from which other organic compounds could be obtained without difficulty. Many of these syntheses still appear in text books. By his "circle of metamorphoses" he was able to reverse the usual order of procedure and was thereby encouraged to believe that one day we might imitate life itself in reversing the classical analysis of chemists and might synthesize instead. One such "circle" is given below:

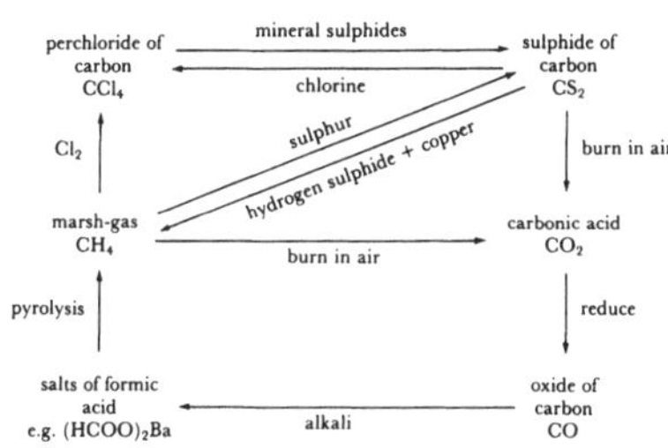

Another cluster of reactions centred around the hydrocarbon acetylene. In 1862 Berthelot showed that this could be produced by passing an electric spark between carbon electrodes in an atmosphere of hydrogen.

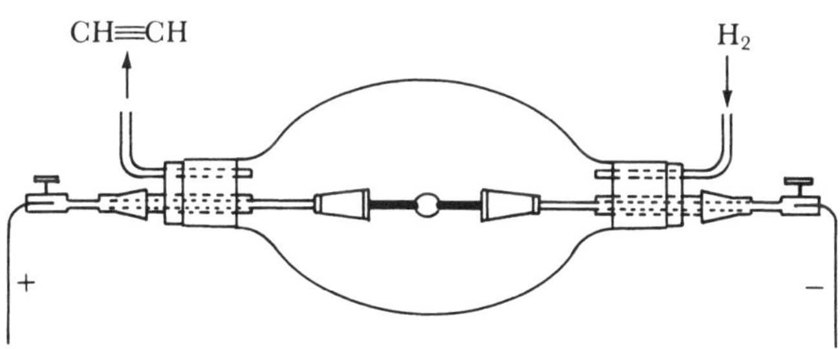

Reduction of acetylene was shown to give both ethylene and ethane, sparking in a mixture of acetylene and nitrogen led to the production of hydrogen cyanide, and pyrolysis of acetylene gave rise to a number of polymeric materials including benzene. Some of those reactions are given below:

$$C_2H_2 + H_2 = C_2H_4 \qquad C_2H_4 + H_2 = C_2H_6$$

$$C_2H_2 + N_2 = 2HCN$$

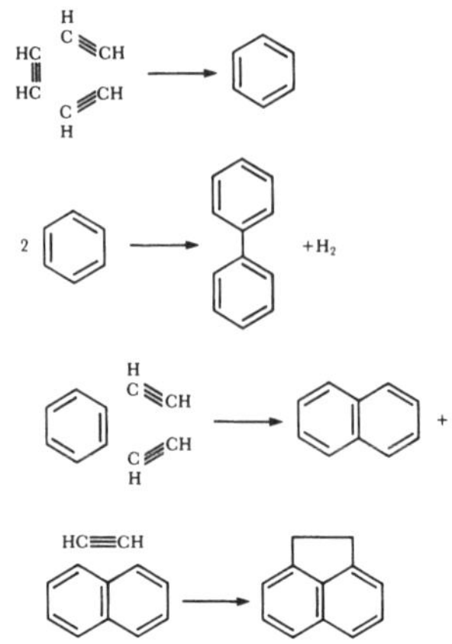

Thus Berthelot was able to obtain synthesis in the first two senses of the terms we have used. His ambition, "to banish life from all explanations relating to organic chemistry", must have seemed strange to those who felt that vitalism was a problem, not for organic substances but for organisms themselves. In fact it was possible to criticise Berthelot's work simply because the conditions under which he performed his syntheses were so very different from those of living organisms. In fact his ideology was almost inverted if one took into account the great differences between processes in laboratories and the living organism. Moreover many of these products were obtained in extremely low yield and were therefore unlikely to prove of practicable (as opposed to ideological) importance. They stand in stark contrast with the "technical" syntheses of industrial Germany where yields and easy reaction were all important. This is doubtless why Partington observed that "modern synthetic organic chemistry stems from Frankland and Kolbe rather than Berthelot".[35]

Whether Berthelot's anxiety to rid chemistry of vitalism sprang from deeper philosophical considerations one cannot be sure. It was certainly consistent with much of the positivist philosophy then permeating Paris, and it does seem likely that there was a connection (though not a necessarily logical one) between his anti-vitalism and a religious agnosticism.

Berthelot received some support from chemists in his own country and elsewhere but, on the whole, his grand ideological programme was unsuccessful. When Wöhler read his book he commented that it might convey the impression that scientific organic chemistry and chemical synthesis originated with Berthelot. He of all people would have been sensitive on this point. But it is interesting that Frankland, one of the very few chemists to cite extravagant claims for Wöhler's urea synthesis as early as 1858, was also a professed sceptic in matters religious. And a similar claim was made by his great friend and disciple H. E. Armstrong in his textbooks of 1880. If this is the case then an ideology affects not only the way one does one's chemistry but how it can be interpreted historically.

Synthesis and ideology have been connected in other ways also. A strong reaction to German synthetic chemistry was voiced, as we have seen, by A. W. Stewart. Writing in September 1918, he sought to counter an objection of ideological prejudice. He said "it is possible that readers may detect a certain bias against the flood of synthetic material which pours chiefly from the German laboratories", although he denies that it "should be attributed to a reconsideration of the value of German science in the light of the war".[36] He says the same views were expressed in earlier editions, but that is hardly relevant since long before the war the chemical community in Britain was becoming aware of the inferiority of British chemistry as compared to German and an understandable antipathy to German chemistry had existed for some years.

In 1958 a distinguished English chemist was testifying before a legal tribunal concerning import duty chargeable on certain key chemicals. The question at issue was whether they could properly be called "synthetics". In the opinion of Professor Christopher Ingold that term went back to the 1840s or a little earlier, since when *it had hardly changed its meaning*.[37] With historical awareness, however, the facts appear to be otherwise. For "synthesis" has had not merely a wide variety of meanings but has had connections with chemical theory at a range of quite different levels. It has been developed in response to many different needs—not all of them obviously chemical—and has itself undergone profound metamorphosis since its first hesitant steps in the early years of the nineteenth century.

References

1. A. Lavoisier, *Traité élémentaire de Chimie*, 2nd edition (Cachet, Paris, 1793), p. 193.
2. M. Berthelot, *Chimie organique, fondée sur la synthèse*, (Mallet-Bacheher, Paris, 1860) pp. xv–xvi.
3. The distinction between direct and indirect synthesis has been stressed by J. H. Brooke, "Organic Chemistry and the unification of chemistry: a reappraisal", *Brit. J. Hist. Sci.*, **5** (1971), 363–92.
4. Graebe, *Geschichte der organischen Chemie*, (Julius Springer, Berlin, 1920) p. 57.
5. D. McKie, "Wöhler's 'synthetic' urea and the rejection of vitalism: a chemical legend", *Nature*, **153** (1944) 608–10.
6. (a) T. O. Lipman "Wöhler's preparation of urea and the fate of vitalism", *J. Chem. Educ.*, **41** (1964), 452–7; (b) J. H. Brooke, "Wöhler's urea, and its vital force? A verdict from the chemists", *Ambix*, **15** (1968), 84–114; (c) J. H. Brooke, *op. cit.* note 3.
7. A. Wurtz, *J. Chem. Soc.*, **15** (1862), 387.
8. By Trautschold, reproduced, with key, in Siegfried Heilenz, *Das Liebig Museum in Giessen* (Giessen, 1986), p. 29.
9. *Sketches from the life of Sir Edward Frankland*, 2nd ed., (privately printed, London, 1902), pp. 69, 170–2, 177, etc.
10. *Annalen*, **38** (1841), 89.
11. *Medical Times & Gazzette*, **6** (1853), 131.
12. *Ibid.*, p. 134.
13. W. H. Perkin, Hofmann Memorial Lecture, *Chemical Society Memorial Lectures 1893–1900* (Gurney & Jackson, London, 1901), p. 602.
14. Frankland (*op. cit.* note 9), p. 173.
15. *Phil. Trans.*, **142** (1852), 440.
16. C. Schorlemmer, *The rise and development of organic chemistry* rev. ed., (Macmillan, London, 1894), p. 249.
17. W. A. Campbell, *The chemical industry* (Longman, London, 1971), p. 94
18. *Ibid.*, pp. 97–8.
19. *Berichte*, **13** (1880), 2254.
20. See e.g., E. Baümler, *A century of chemistry*, (Econ Verlag, Dusseldorf, 1968), pp. 23–30.
21. A. W. Stewart, *Recent advances in organic chemistry*, 3rd ed., (Longmans Green, London, 1918), p. 1.
22. *Ibid*, p. 2.

23. *Ibid*, 1st ed., 1908, p. i.
24. *Pro. Roy. Soc.*, **14** (1865), 198, etc.
25. *Annalen*, **186** (1877), 228, etc.
26. *Berichte*, **14** (1881), 2460, etc.
27. *Annalen*, **281** (1894), 25, etc.
28. *J. prakt. Chem.*, **35** (1887), 349.
29. A. Ladenburg, *Lectures on the history of the development of chemistry since the time of Lavoisier*, (trans. L. Dobbin, Alembic Club, Edinburgh, 1905), p. 289.
30. *Aus Justus Liebig's und Friedrich Wöhler's Briefwechsel in den Jahren 1829–1873*, (Vieweg, Brunswick, 1888), vol. **i**, p. 124.
31. *J. Chem. Soc.*, **91** (1907), 1756–7.
32. *Gazz. Chim. Ital.*, (1874), 305, etc.
33. See K. T. Finley, "The synthesis of carbocyclic compounds: a historical survey", *J. Chem. Educ.*, **42** (1965), 537.
34. Berthelot (*op. cit.* note 2), p. xxv.
35. J. R. Partington, *A history of chemistry* (Macmillan, London, 1964), vol. iv, p. 467.
36. Stewart (*op. cit.* note 21), p. xii.
37. *Chemical Age*, **80** (1958), 902.

XVI

The Origins of Conformational Analysis

1. The Major Landmarks

The concepts of conformational analysis have so deeply penetrated into the fabric of theoretical organic chemistry that it is difficult to realise that, in some respects at least, the subject is not yet a quarter-century old. The terms "conformation analysis" (1) and "conformational analysis" (2) occur in the early 1950s, but it was Barton's classic paper (3) on "The Conformation of the Steroid Nucleus" (1950) that is generally regarded as launching the subject. In 1956 he and Cookson reported:

> "In the last five years the importance of the existence of preferred conformations in organic chemistry has become widely recognised under the title of 'conformational analysis'" (4).

The magnitude of Barton's achievement in this field has been widely - and rightly - recognised. But it must not be allowed to blind us to the importance of events preceding 1950, what we might call the pre-history of conformational analysis. Few, if any, scientific concepts appear, like Minerva, fully-formed in the brain of their creator, and most have their origins traceable in earlier scientific writings. Conformational analysis is no exception, and the development of its rudimentary ideas may have something to tell us about scientific growth in general. Certainly it presents several features of remarkable interest.

At the very heart of present thinking about conformational analysis lie several basic assumptions. Two are of fundamental importance:

(a) The recognition of rotational energy-barriers for carbon-carbon single bonds, with preferred relative arrangements of substituent groupings.

(b) The postulation of so-called "strainless rings".

The next paper deals with an aspect of the first, and my purpose is to draw attention to the strange history of the second. In a recent paper (5) Prof. Ramsay has given a valuable synopsis of the course of events in the late 19th and early 20th centuries, while Prof. Eliel has chronicled the period 1950-1973 (6). It would be presumptuous for me to try to reproduce their work. Instead, I shall give a brief indication of the principal milestones in the journey to 1950, and shall then offer a tentative analysis of the curious story that emerges.

As a point of departure we may take the famous paper by Hermann Sachse "On the geometrical isomers of hexamethylene derivatives" (7). Baeyer had reported two forms each for hexahydromellitic acid (8) and hexahydroterephthalic acid (9). These had been explained in terms of planar rings (10), e.g.

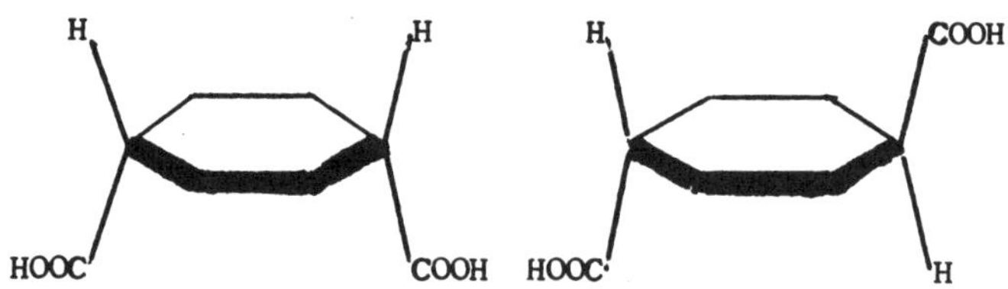

and by their analogy with maleic and fumaric acids

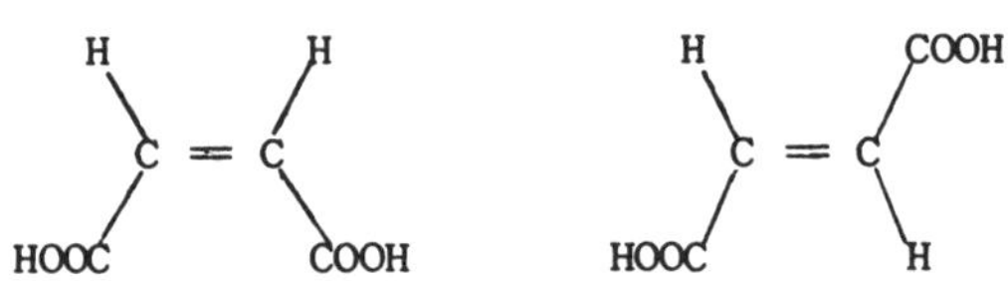

had been designated respectively "maleinoid" and "fumaroid". The ready interconversion of the two forms of each acid caused Sachse to question the rigidity of the 6-membered ring and (by implication) the analogy with the unsaturated acids. He pointed out that Baeyer's insistence on a slightly strained planar ring for cyclohexane was not strictly necessary. Given a regular tetrahedral disposition for all valencies of the ring carbon atoms one could alternatively postulate two arrangements, each quite free from strain. The one he designated "normal" would have its atoms in a zig-zag pattern, and this would be symmetrical and relatively rigid. The other form would be less symmetrical and more flexible.

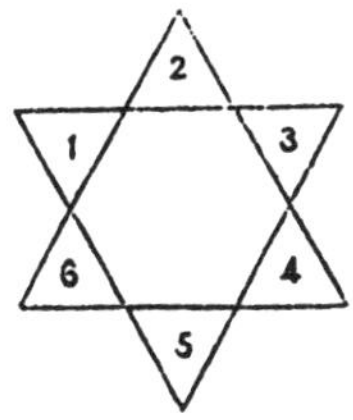

Figure 1. Sachse's representation of "normal" cyclohexane

Apart from an "end-on" view of the "normal" form (Fig. 1), Sachse did not depict these primeval "chairs" or "boats" but did offer instructions for constructing models of them out of cardboard tetrahedra. They were in fact derived from an earlier suggestion he had made for models of the benzene molecule (11). A more detailed mathematical treatment followed shortly afterwards (12).

Sachse himself died in 1893 hardly 31 years old, a relatively unknown assistant at the famous Charlottenburg Institute (13). With these papers Sachse disappears from the pages of chemical history, and so, we are often told, do his avant garde proposals for the cyclohexane ring - at least for many years. How literally true that is we shall see, but certainly it was not until 1915 that the first determined attempt was made to rescue Sachse's theory from its undeserved oblivion. The author of this enterprise was Ernst Wilhelm Max Mohr (1873-1926), Professor of Chemistry at Heidelberg (14).

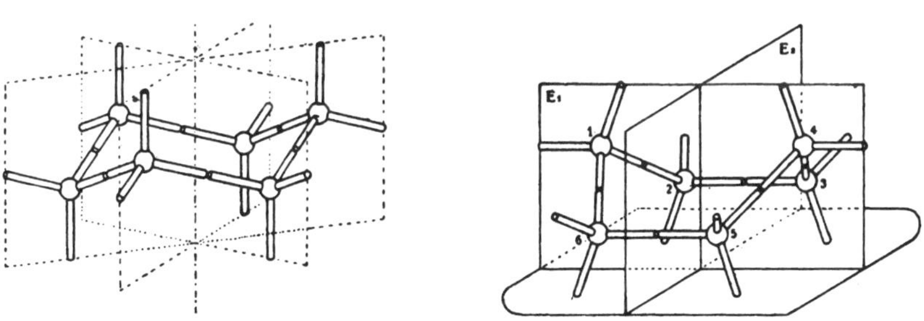

Figure 2. Mohr models for cyclohexane

In several papers over the next few years (15), Mohr elaborated on the Sachse concept of strainless, 6-membered rings, but made his point more clearly with illustrations of ball-and-stick models (rather than solid tetrahedra). But he also went further in another respect. Sachse had recognised the possibility of interconversion between forms, being especially easy in the varieties of "boat" conformations. In the absence of strong heat chair-chair interconversion would be less likely, so two varieties of "normal" monosubstituted cyclo-hexanes should exist:

One of the major objections to his theory had been that unambiguous proof for this was not forthcoming. Mohr, however, asserted that ring-inversion of chair-forms should be expected at ordinary temperatures through the impact of molecular collisions. Had this been all, the theory would have been rendered finally sterile as no empirical evidence could ever be expected to verify it. But it was saved from this ultimate humiliation by Mohr's recognition that ring-flip could not occur in one kind of situation: where the ring was locked into a larger molecule by some kind of ring-fusion. This, he asserted, would be the case with diamond, whose three-dimensional structure had been recently elucidated by Bragg (16).

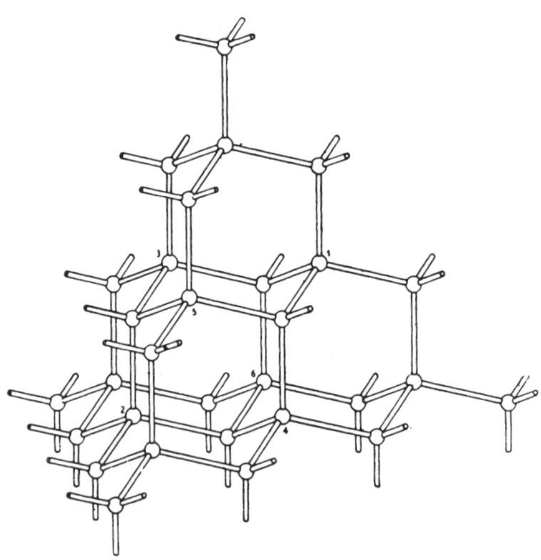

Figure 3. Mohr model for diamond

In fact the chair could be deduced from Bragg's X-ray diagrams. More simple than diamond, decahydronaphthalene should exist in 2 forms and these could not be interconvertible (unless a chemical bond was actually ruptured in the process).

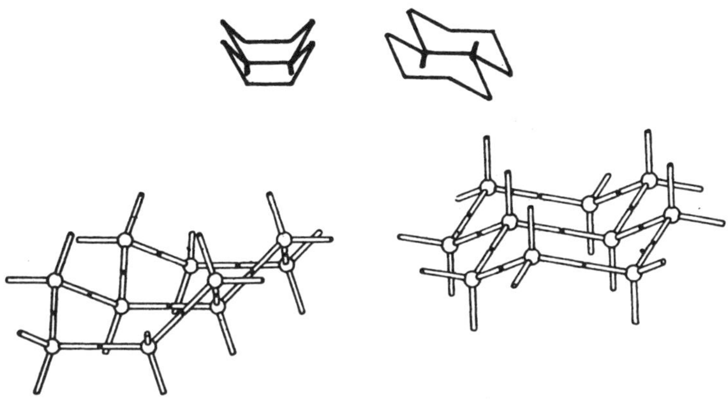

Figure 4. Mohr models for cis- *and* trans-*decalin*

Here was a further challenge to look for experimental evidence for a stereochemical hypothesis. It was accepted by W. Hückel who, in 1925, reported that decalin could be separated by fractional distillation into two isomers (17). By their relation to oxidation products they were termed cis- and trans- and allocated the 2 formulae above.

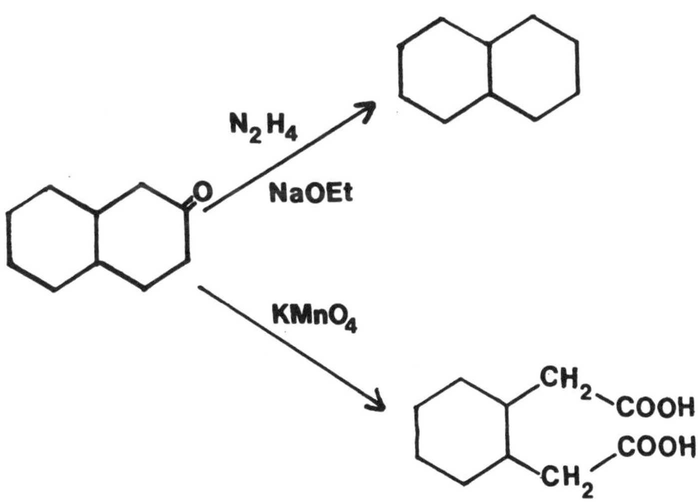

The impact of this work we shall discuss later. It is sufficient to note that Hückel's interpretation of his own work was not, in the end, correct. The first suspicions were voiced in 1943 by Odd Hassel (18). Realising the stereochemically unfavourable H-H interactions inherent in a single boat form (as opposed to a chair), he concluded that a 2-boat structure (for cis-decalin) was even more improbable, having no less than 5 eclipsed pairs and 2 flagpole-flagpole interactions. It was pointed out that a cis ring fusion could also be accomplished by means of two chairs:

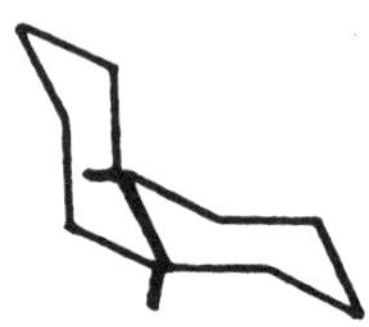

Three years later he was able to establish this by electron diffraction (19) and other evidence confirmed it (20). But, back in 1943, he drew attention - almost for the first time - to an even more fundamental feature of the cyclohexane ring: the distinction between what we should call (21) axial and equatorial bonds but which he designated ϵ and κ. He emphasised the readiness with which ring-inversion converted a κ-substituent into an ϵ one and *vice versa*, and he made the profoundly important point that, generally speaking, monoderivatives have their lowest energy when the substituent occupies an equatorial position.

Seven years later Barton produced his paper in *Experientia* (3).

2. A Forgotten Hypothesis?

In this preliminary account I have simply recorded the principal milestones; or, to change the metaphor, the major peaks, ignoring altogether the fine structure of the early history of non-planar rings. One is often given the impression that Sachse and Mohr were very largely voices crying in a wilderness, prophets before their time, progenitors of a still-born hypothesis. Even Hückel, and to a less extent Hassel, do not significantly contradict that view.

Now if this is so we have a major problem on our hands. How could concepts of such immense value have failed to make their impact? Sixty years from Sachse to Barton, broken roughly halfway with Mohr's contribution, is a long time for hypotheses to lie dormant, even though the history of science has several parallel cases to offer. But the prior question, of course, is whether the impact of Sachse and Mohr *was* as minimal as that. Was it true, as is commonly said, that

Sachse's "theory was completely disregarded until 1918"? We must first establish that this was not the case.

Even a cursory glance through the literature reveals that Sachse's work was far from unknown. He had published in two well-known journals (7,11) and abstracts had appeared in English (22). In 1894 Bischoff's textbook on stereochemistry (23) acknowledged the merit of Sachse's concept of dynamic isomerism as applied to cyclohexane. At about the same time Arnold Eiloart was informing his audience at Cornell (24) of Sachse's suggestion of "another space-formula for hexamethylene" to account for the easy conversion of cis-dicarboxylic acids to trans.

There had been some early adverse reactions on rather general grounds. Hermann, in 1890, opposed the uses of tetrahedra in constructing models of 6-membered rings (25). A few years later Lachman commented on Sachse's theory that, "while geometrically excellent, it is incompatible with our views of valence, and may therefore lead to interesting developments of our valence conceptions" (26). It belonged more to the future than the present - a particularly shrewd comment, one may think.

Specific objections were also raised. An important contribution here was a paper (27) by Werner and Conrad in 1899 - just about the time that Werner was renouncing organic research for good. After referring to, and illustrating, the two strainless arrangements for cyclohexane proposed by Sachse they wrote:

> "These advantages of the steric formulae must be weighed against this disadvantage: they allow one to predict definite isomers for mono- and di-substitution products for which there has been so far no factual confirmation".

The existence of several mono-substituted cyclohexanes was a consequence of buckled ring structure. Taking only the relatively rigid "normal" form of Sachse, one would expect to find two isomeric forms, but these had not been unequivocally shown to exist in any one case, - though it does not seem that many intensive searches were originated. At one stage it had looked as though an example would appear. Sachse, in his 1892 paper, referred to the apparent existence of two acids C_6H_{11}.COOH, one from Caucasian naphthenic acids and the other from hydrogenation of benzoic acid. This had been the work (28) of the Finnish chemist O. Aschan (29) who had worked in 1890-1 under both Wislicenus and Baeyer. However Aschan's work was criticized by Markovnikov (who claimed exclusive rights to research on Baku petroleum acids and warned Aschan off) (30). The result was therefore inconclusive, with severe doubts remaining about the purity (and therefore the identity) of the products. Moreover, though Sachse invoked Aschan's name in support of his view, the Finnish chemist did not return the compliment and claim theoretical justification for his results. All that could be said with much certainty was that, so far, Sachse's hypothesis was an open question.

This certainly is the impression one gains from Hantzsch's Elements of Stereochemistry (1901). We are told that:

> "According to this hypothesis which is due to Sachse, the symmetrical configuration would indicate two mono-substituted stereoisomers, and as a matter of fact, two hexahydroxybenzoic acids are known, one obtained by the reduction of benzoic acid, the other being the hexanaphthalenecarboxylic (sic) acid". (31)

Yet despite this the subsequent treatment of the geometrical isomerism of cyclic compounds is entirely in terms of planar rings!

To return to Werner and Conrad, they proposed an alternative approach. Again making the minimum assumption of "chair" forms, they observed that "hexahydrophthalic acid" should exist as a resolvable trans-form and a non-resolvable cis-form. They came to this conclusion from models which were not described but were presumably like this:

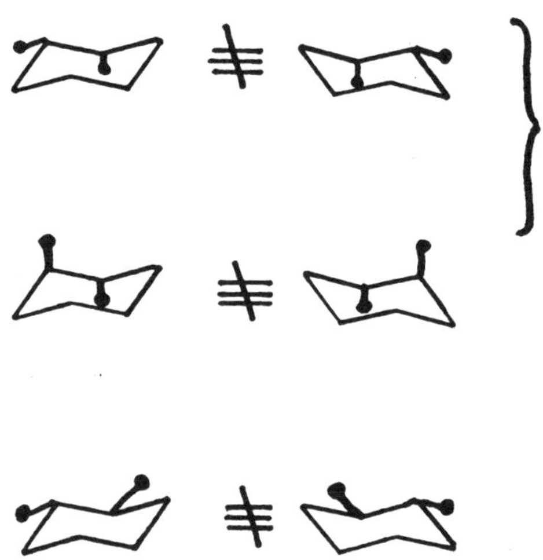

Trans

Cis

Both cis and trans isomers should be resolvable, and this was attempted using fractional crystallisation of quinine salts. They concluded:

> "Despite numerous attempts there has been no success in separating from the cis form an active modification, whereas it is quite easy to resolve trans-hexahydrophthalic acid into its optical antipodes. If the negative result of our attempted resolution of the cis acid can be utilized for one valid deduction with regard to the question under debate, that result (taken in conjunction with the positive result for the trans acid) is that the assumption of a stable, steric arrangement of the carbon atoms in cyclohexane is not very probable" (32).

What is challenged, therefore, is not the shape but the stability of the strainless rings. As time went on the monosubstituted isomers became less credible than ever. In 1905 Aschan himself admitted "the non-existence of two forms of hexahydrobenzoic acid already weakens Sachse's idea" (33).

This was in his great textbook of alicyclic chemistry for which he received the Valbruch Prize - a bienniel award for the best scientific work in German (29). Its influence was considerable and helped materially to shape opinion on the Sachse hypothesis. He went on:

> "The fact that this view of the configuration of the hexamethylenes is not verified is probably to be explained on the grounds that Sachse's formulae represent only different phases of the movement within the molecules of the hexamethylene" (34).

In other words, for practical purposes, Sachse's strainless rings had little relevance. In 1907 Pattison Muir called them "an interesting chemico-geometrical development of the fundamental hypothesis of stereochemistry, especially to benzene and its derivatives" (35). Eight years later Beesley, Ingold and Thorpe wrote significantly that:

> "It is, of course, assumed there is no distortion of the angle of the hexagon caused by the attachment of groups to any one carbon atom of the cyclohexane ring" (36).

There was no question of any other value than 120° for each ring angle, - "of course"! Thorpe himself, who admittedly had his own axe to grind, wrote later that by 1915 it was almost universally held that the cyclohexane ring was planar (37).

So, in summary, it seems that though Sachse was not entirely forgotten his work was largely ignored. Ring-flip might or might not occur; even if it did, the net result was a planar average.

After Mohr's entry on the stage things could never by quite the same again. As we have seen, he exposed the implications of ring-inversion more thoroughly than any before him, proposed a specific test for his theory and related it all to the data and insights of physical chemistry. Yet for all that, conformational analysis had another 30 years of gestation.

Now, however, there appeared to be greater confidence in the reality of those chairs and boats. Occasionally they might be discovered as unwillingly rigid prisoners in a fused-ring molecule. Usually, however, they were transient entities of little concern to the working chemist. In 1919 Stewart could accept Sachse only "under the condition that his two configurations are vibration phases of the hexamethylene molecule" (38). Interestingly enough, he was saying the same thing nearly 30 years later: cyclohexane was "a labile system" (39).

In 1929 Haworth introduced the term "conformation" and applied it to models of glucopyranose and other monosaccharides. A diligent reader of Sachse (40), he had however to conclude that "at present there is insufficient experimental evidence available for any kind of opinion" on strainless rings in polysaccharides (41). He referred to the X-ray work of Spronsler and Dore on cellulose (42), composed of glucopyranose units thus:

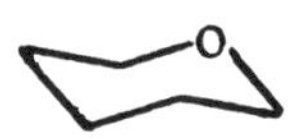

Even this was not decisive enough for some. Haworth himself observed that if the decalin molecules could "be regarded as being built up by the fusion of two uniplanar cyclohexane rings, then, whereas the resulting *cis*-form is free from abnormal strain, the corresponding *trans*-form is seen to be considerably (though perhaps not unbelievably) strained" (43). That was the kind of price to be paid for a planar cyclohexane molecule.

It was also in the 1920s that Hückel separated the two decalins (17). At about the same time the work of Boëseken and the Delft school began on the boric acid complexes formed by condensation between acetone and cyclic 1,2-diols. (44).

Here the researches of Derx were specially noteworthy (45). He showed that cyclohexane-1,2-diol reacted with acetone only in the *cis* form, and illustrated his arguments with models of the type used by Mohr (though in his earlier work he was unaware of Mohr's papers). Yet his valuable insights were largely lost to view in the next 25 years - a further token of the enormous difficulty associated with the concept of three-dimensional strainless rings. Even more astonishing was a paper by Hendricks and Bilicke concluding, on the basis of X-ray data, that "β-benzene hexabromide" had a molecular symmetry consistent with a puckered cyclohexane ring, but asserting

> "Mohr's theory of 'Strainless Rings' as applied to cyclohexane and its derivatives is not compatible with our conclusions in so far as the solid state is concerned. His three-dimensional formulae... have a center of symmetry in one case, but not a plane of symmetry" (46).

This is so at variance with the facts that one can only assume that Mohr's proposal was totally misunderstood. Two years later Bilicke, with R.G. Dickinson (47), gave a clear diagram of the chair-form and specified a tetrahedral ring angle but gave no acknowledgment that Mohr could be right.

In general there is no doubt that most organic chemists would have agreed with these words of Thorpe in 1931 (48).

> "The multiplanar character of the rings beyond those containing six carbon atoms is now fairly generally accepted, as is also the uniplanar character of rings below those containing six carbon atoms. But the six-membered ring has still to rid itself of or adopt the Sachse-Mohr "arm-chair" or "boat" hypothesis.... At present there is no evidence either chemical or physical which supports the Sachse-Mohr hypothesis".

3. Obstacles to Progress

These words bring into sharp focus a major historical problem. Why did conformational analysis take so long to emerge after the original suggestion in 1890? This phenomenon of an apparently stillborn hypothesis is by no means unique in chemistry. Even more famous is the adoption of Avogadro's hypothesis, enunciated in 1811 but not generally used until the 1860's. In fact there is a remarkable parallel between the two cases. The remainder of this paper, inevitably more speculative than the first part, will attempt to identify some of the factors at work in the case of the Sachse-Mohr hypothesis.

3.1 Experimental Ambiguity. It is, of course, extremely obvious that one great difficulty was that of formulating clear and unambiguous experimental tests for Sachse's multiplanar rings. As with Avogadro's hypothesis what were deemed to be simple tests for the theory proved later to be concerned with a whole complex of variables, not just one. This has been termed the Duhemian pitfall (49) after the philosopher of science who drew attention to it (50). Thus the identity or non-identity of all samples of the hexahydrobenzoic acid depends not only upon what we should call axial or equatorial substitution but whether or not interconversion by ring-flip could easily occur. Similarly, the non-resolvability of cis- hexahydrophthalic acid (27) tested not only its ring shape but also the optical stability of such a system. Where the case for rejection of a hypothesis rests upon purely negative evidence, as here, history should warn us to be careful.

This was the point perceived so clearly by Mohr, and after his papers the nature of the historical problem slightly altered. But, for the first 25 or 30 years, the Sachse theory was very much a victim of the tests devised to prove it. Its reception was made even harder by the strength of the older alternative, associated with the name of Baeyer. Assuming planar rings he had predicted that the most stable members would be those with 5 or 6 carbon atoms. The discovery of a stable cyclopentane-1,2-dicarboxylic acid (51) shortly afterwards lent strong support to the theory while, in 1892, heats of combustion for tri-, tetra-, penta- and hexamethylenes again pointed to greatest stability in the last two members (52). Later on, numerous ring-contractions from 6 to 5 members were cited as further evidence, for a cyclopentane should be marginally less strained than a planar cyclohexane (53). The fact that ring-expansion could also occur but did not seriously prejudice the Baeyer theory's popularity emphasises perhaps that one sees what one is looking for.

However, as we have seen, after Mohr's papers the issues were clarified. The concept of a strainless 6-membered saturated ring then became obfuscated not by difficulties in testing but by sheer lack of interest. Its second phase was marked by the problem of relevance.

3.2 Relevance Factors. There seems no inherent reason why substantial progress could not have been made if chemists had been concerned to pursue the matter. On the whole they were not, and the problem of another 30 years' neglect resolves itself into a cluster of inter-related factors which I shall try to identify.

First there was the factor of scientific theory. How did the Sachse-Mohr hypothesis mesh in with the generally accepted theoretical framework of organic chemistry? The answer is a curious one. It seems impossible to separate the fate of cyclohexane from that of benzene. In the closing years of the last century theoretical organic chemistry was dominated by the unresolved problem of the benzene formula. One elegant solution had been the prismane structure of Ladenburg (54) but this had never gained much favour because, it was suggested (55), chemists treated such excursions into three dimensions with "extreme distrust". Its wedge-shaped shadow must have excluded a good deal of light from Sachse's 3-dimensional ring. But the alternative Kekulé structure led to the prediction of two ortho-disubstituted isomers and these, like the hexahydrobenzoic acids, had never been found. One was therefore forced to concede some kind of "oscillation" for benzene; but by so multiplying hypotheses one tended to bring an air of unreality to the whole argument. That is why, in the last three decades of the 19th century, chemical periodicals are littered with benzene formulae that attempted in various unconventional ways to express what was, in classical terms at least, inexpressible. At the turn of the century the dominant influence was not Kekulé but Thiele with his residual affinities (56).

With this background it is scarcely surprising that chemists were reluctant to lose their planar rings or to associate themselves too closely with hypotheses about "oscillation". In conformity with this last point it may be observed that another area of stereochemistry suffered in exactly the same way. This concerns the non-resolvability of amines of the type Nabc. They were regarded as being planar (57) until theoretical considerations were reinforced by other experimental data. It has been claimed that in the 1920s Meisenheimer first proposed that oscillation of the nitrogen atom was responsible.

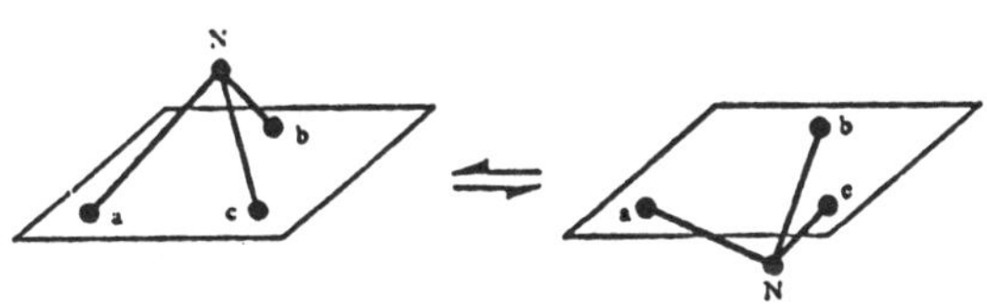

Yet an identical view was proposed by Walter for phosphorus and nitrogen and recorded in Berichte in 1873 (58). This was another unnoticed hypothesis; we do not even know the author's initials!

If the experience of benzene helped to discourage interest in Sachse's hypothesis, the effect would be magnified by a general reluctance to theorise on the fine structure of organic compounds. In Germany the great commitment was to synthesis, and organic chemistry was becoming top-heavy with accumulated facts. During the last half of the century about 100,000 new compounds were synthesised, of which "the majority were still-born and their epitaphs are inscribed in Beilstein's Handbook" (59). Partly this arose from demands from industry - dyestuffs at first, then pharmaceuticals and explosives. It was also a result of the German Ph.D. system where "every Ph.D. in organic chemistry represents so many new compounds" (60).

To chemists preoccupied with these practical issues the speculations of Sachse must have had little apparent relevance. On the other hand those chemists at the turn of the century with a taste for theory were more likely to find satisfaction elsewhere than in the organic branch.

As the 20th century advanced various events within organic chemistry helped to turn the tide of opinion a little more in Sachse's favour. We can only touch on some of these. Around 1900 the worldwide challenge of camphor was at its peak and the recognition of its tricyclic structure:

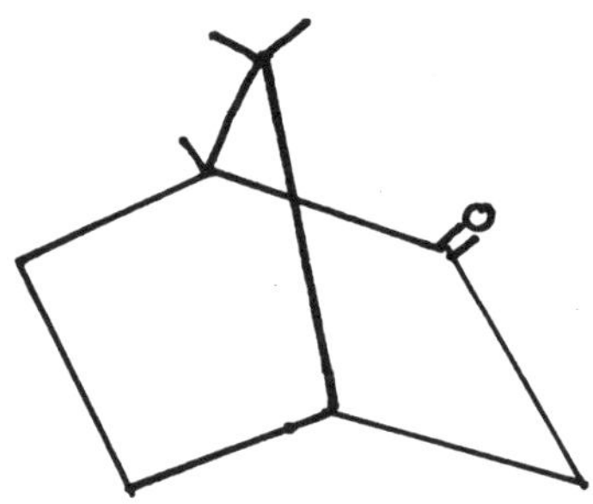

made credible the concept of buckled rings. But in camphor they were kept rigid. The concept of mobile systems was also emerging as one of general importance (61). As Thorpe and Ingold observed in 1922 they "are special only in so far as the phenomenon which they exemplify is related to experimental and human limitations. Magnitudes which cannot be measured are no less important philosophically than those that can" (62). Then, again, the collapse of Baeyer's strain theory with its planar rings was inadvertently engineered by the syntheses of fairly stable large ring compounds by Ruzicka and Stoll. The latter remarked that Ruzicka complained to him:

> "Now we are faced with a terrible difficulty. A long time ago, Adolf von Baeyer demonstrated that rings of carbon atoms in excess of six in number are difficult of formation and that the difficulty increases with an increase in the number of carbons. And here we have a ring with 17 carbon atoms. How shall we ever succeed in synthesizing it?" (63).

This work began in 1926, after a 20 year gap.

Perhaps the most important moves towards general recognition of multiplanar rings came in the new 20th century emphasis on natural products. Interest in camphor and the terpenes and alkaloids branched out into steroids and natural macromolecules (cellulose, proteins etc.). The widespread occurrence of saturated rings in nature led to a gradual awakening of interest in their shapes and reactivities. As Barton has remarked of the obscurity of much of Boëseken's work: "I think myself that a theoretical treatment becomes well accepted only if there is a real need for a large body of chemists to use it" (64). The same consideration must apply to the Sachse-Mohr hypothesis itself. Gradually, over 60 years, the critical number of interested chemists (whatever that was) was being approached; by 1950 it had certainly been reached.

But we must look beyond the confines of organic chemistry itself, and that brings us to a factor of scientific organisation. It is obvious that the organic chemist does not usually employ concepts from astronomy or psychology, though he does draw upon results from physical chemistry. But that is true today, and 50 years ago was far from general. At that time the demarcation between the two areas was fairly strong. The position is well summed up by the characterisation of Thorpe's attitude to physical methods as one of "genial tolerance" and nothing more (65). It is said that W.H. Perkin Jr. held the view that "physical chemistry is all very well but it does not apply to organic substances" (66).

Nor was this approach to physical measurements a transient feature of pre-World War II organic chemistry. It deeply affected stereochemical thinking even after that. Here is Barton's analysis:

> "On the whole organic chemists in the 1940s did not pay much attention to stereochemistry. They knew of course how many stereoisomers there should be when they carried out a reaction, but they had very little idea of why you would get more of one isomer than of the other when you carried out an asymmetric synthesis. Stereochemistry was not treated with the attention it should have received and in particular nobody bothered very much about the shapes of molecules" (67).

Thus right through the first half of the present century the Sachse-Mohr hypothesis suffered from the neglect of physical data. Since it was only through physical data that the theory was to be fully validated, this attitude would clearly be unhelpful. But physical chemistry did not give only validation - it also offered relevance to the new doctrines of stereochemistry. Most of the early physical work was not done to prove the Sachse structures, but it gave the needed evidence as by-products of other work. So the buckled ring entered the consciousness of increasing numbers of chemists and became a live issue at last when someone was able to unite the two fields of knowledge.

In the 1930s diffraction by X-rays and by electron-beams was to be followed in the next decade by infra-red and Raman spectroscopy which, though controversial at first, led to fairly firm conclusions about chair and boat forms of cyclohexane (68). Another factor of importance was the application of thermodynamics to organic chemistry. This owed much to A. Michael who vigorously opposed Baeyer's strain theory as being too mechanical and deficient in the attention paid to energy relationships (69). Since then the thermodynamic approach has had much to offer to the study of molecular conformation.

In conjunction with spectroscopic data it led to a major advance in the 1930s: the recognition of hindered rotation about the carbon-carbon bond in ethane. Perhaps the most interesting case is afforded by the experience of Hassel whose interest in cyclohexane stereochemistry dates back to the 1930s. It was he who introduced into Norway the techniques of dipole moment measurement and electron diffraction. By 1943 he had enough data to lead to a general conclusion (18). Shortly afterwards he was arrested by the Nazis and not able to resume his electron diffraction work until after the war. In 1946 he was able to announce experimental vindication of his thesis that the chair-form was the preferred shape for cyclohexane (19).

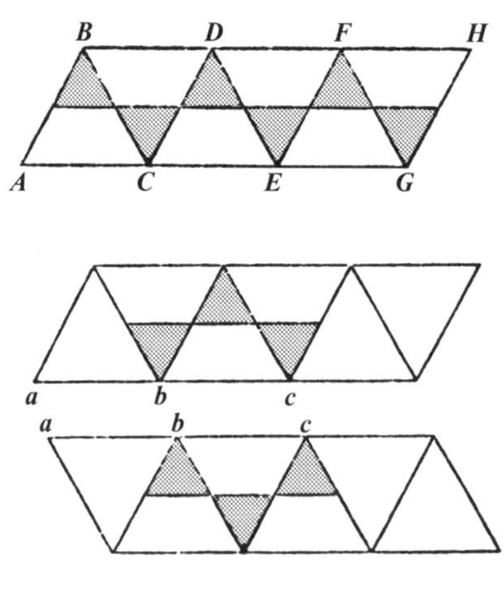

Figure 5. Sachse models of cyclohexane: construction diagrams

Top: Chair Bottom: Boat

Finally we can identify a relevant factor of scientific communication. The more easily ideas are communicated the more readily their relevance is seen. In the case of stereochemistry in general the role of models has always been important (70). My contention is that this was particularly true for the reception of the Sachse-Mohr doctrines. Sachse himself was aware of this and gave instructions for making strainless rings from cardboard tetrahedra (7). For long they dominated the stereochemical scene appearing as the frontispiece to the 6th edition of Stewart's Recent Advances in Organic Chemistry, vol. I (1931). The same author included directions for their construction in both editions of his Stereochemistry, observing "the subject is made much more comprehensible if models be used instead of plane formulae or perspective drawings" (71).

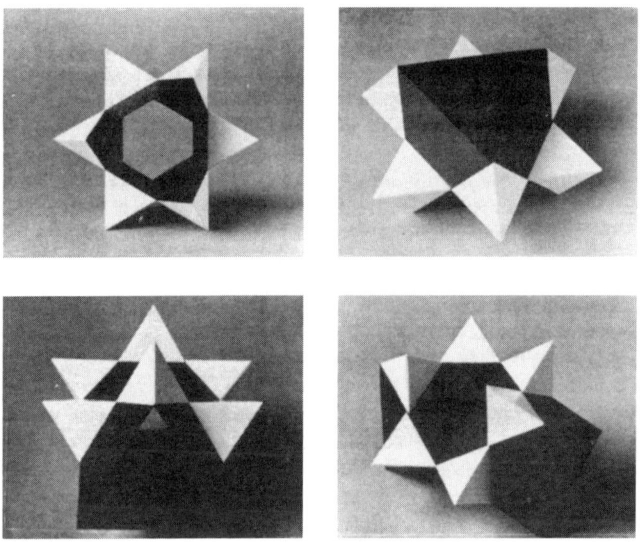

Figure 6. Sachse models of cyclohexane: the completed models

Valuable as these models were they lacked the clarity of tetrahedral ball-and-stick varieties, and,in the long run, they may have helped to obscure the conformational relations that they should have illustrated. Above all they failed to demonstrate rotational effects. It is therefore not without significance that Mohr gave illustrations of a different kind of model, apparently similar to the Dreiding models of today. Shortly afterwards a paper from Leeds by W.A. Wightman (72) reported a similar type of model, constructed for the study of rotations about single bonds. Strangely, he preferred the boat to the chair form for cyclohexane because of its greater response to molecular collisions.

Certainly it would appear no accident that the second phase of rather limited acceptance coincided with the use of greatly improved models. In 1918 Stewart gave his opinion that "if electronic models could be constructed their appearance would stimulate the chemical imagination much more rapidly than any mere written efforts can do" (73). If that was to be true of the electronic theory of organic chemistry it would even be more likely for the concepts of Sachse and Mohr.

As the use of models developed communication was further aided by the adoption of uniformly accepted terminology and by the employment of clearer diagrams in print.

Clear and accurate typography was a problem in the early 1950s as can be seen from the distorted diagrams often encountered. But in time acceptable representations replaced typographical monstrosities with exaggerated shapes and axial bonds emerging at right angles to carbon-carbon bonds. Again, it has been pointed out (6) that Hassel's important work from 1938 until the end of the war was relatively unknown because it has appeared in Norwegian in fairly obscure journals. Like Berzelius over 100 years before another distinguished Scandinavian chemist was the victim of a communications breakdown.

So in these ways the basic ideas advanced as far back as 1890 became slowly seen to be not merely correct but also relevant. As Stewart once wrote:

> "For want of just one connecting link, or even the addition of a few words to a statement which contains the key to a problem, the question may go unanswered for years"(74).

Eventually that connecting link was forged from an alloy of organic chemistry, physical chemistry and chemical physics. The outcome was the paper by D.H.R. Barton in 1950.

Acknowledgment

I am grateful to the Media Photographic Unit and the Media Studio Staff of the Open University for construction and photography of the Sasche models.

Literature Cited

1. Birch, A.J., Ann. Rep. Chem. Soc., (1951), 48, 192.
2. Barton, D.H.R., J. Chem. Soc., (1953), 1040.
3. Idem, Experientia, (1950), 6, 316.

4. Barton, D.H.R., and Cookson, R.C., Quart. Rev., (1956), 10, 47.
5. Ramsay, O.B., Chem. Zeit., (1973), 97, 573.
6. Eliel, E.L., ibid., (1973), 97, 582; see also Science, (1969), 166, 718.
7. Sachse, H., Ber., (1890), 23, 1363.
8. Baeyer A. von., Annalen, (1870), Supp. vol. vii, 1.
9. Idem, ibid., (1888), 247, 103.
10. Idem, Ber., (1885), 18, 2278.
11. Sachse, H., Ber., (1888), 21, 2530.
12. Idem, Zeitsch. phys. Chem., (1892), 10, 203; (1893), 11, 185.
13. Chem. Zeit., (1893), 17, 1929.
14. Stollé, R., Ber., (1926), 59, 39.
15. Mohr, E., (a) Chem. Zeit., (1915), 2, 1065.
(b) J. prakt. Chem., (1918), 98, 315.
(c) Ibid., (1922), 103, 316.
16. Bragg, W.H., Proc. Roy. Soc., (A) (1913), 89, 277.
17. Hückel, W., Annalen, (1925), 441, 1.
18. Hassel, O., Tidsskr. Kjemi, (1943), 3, 32.
19. Bastiansen, O., and Hassel, O., ibid., (1946), 6, 96.
20. E.g., Le Fèvre, C.G. and R.J.W., J. Chem. Soc., (1956), 3549.
21. Barton, D.H.R., Hassel, O., Pitzer, K.S., and Prelog, V., Nature, (1953), 172, 1096.
22. J. Chem. Soc. (Abs.)., (1890), 58, 1386; (1893), 64, ii, 266.
23. Bischoff, C.A., Handbuch der Stereochemie, Frankfurt, (1894), p. 140.
24. Eiloart, A., A Guide to Stereochemistry, New York, (1893), p.13.
25. Hermann, F., Ber., (1890), 23, 2060.
26. Lachman, A., The Spirit of Organic Chemistry, New York, (1899), p. 58.
27. Werner A., and Conrad, H.E., Ber., (1899), 32, 3046.
28. Aschan, O., Ber., (1890), 23, 867; (1891), 24, 1864, etc.
29. Entvist, T., The History of Chemistry in Finland, (1828-1918), Helsinki, 1972.
30. Markovnikov, V.V., Ber., (1892), 25, 370, 3355.
31. Hantzsch, A., The Elements of Stereochemistry, trans. Wolf, C.G.L., Easton, Pa., (1901), p. 128.
32. Ref. 27, p. 3047.
33. Aschan, O., Chemie der alicyclischen Verbindungen, Brunswick, (1905), p. 329.
34. Ibid., p. 331.
35. Muir, Pattison, History of Chemical Laws and Theories, New York & London, (1907), p. 318 (footnote).
36. Beesley, R.M., Ingold, C.K., and Thorpe, J.F., J. Chem. Soc., (1915), 107, 1081.
37. Kon, G.A.R., and Linstead, R.P., British Chemists, ed. A. Findlay & W.H. Mills, London, (1947), p. 397.

38. Stewart, A.W., Stereochemistry, 2nd edition, London, (1919), p. 177.
39. Idem. Recent Advances in Organic Chemistry, 7th edition, London, (1948), vol. III, p. 233.
40. Stacey, M., Chem. Soc. Rev., (1973), 2, 151.
41. Haworth, W.N., The Constitution of Sugars, London, (1929), p. 90.
42. Spronsler, O.L., and Dore, W.M., Colloid Symposium Monograph, (1926), 4, 174.
43. Haworth, W.N., Ann. Rep. Chem. Soc., (1927), 24, 99.
44. See ref. 5 for a much fuller account.
45. Derx, H.G., Rec. Trav. Chim., (1922), 41, 312, etc.
46. Hendricks, S.B., and Bilicke, C., J. Amer. Chem. Soc., (1926), 48, 3007.
47. Dickinson R.G., and Bilicke, C., ibid., (1928), 50, 764.
48. Thorpe, J.F., J. Chem. Soc., (1931), 1020.
49. Causey, R.L., J. Chem. Educ., (1971), 48, 365.
50. Duhem, P., The Aim and Structure of Physical Theory (trans. P.P. Wiener), New York, (1962), p. 187.
51. Perkin, W.H., Ber., (1885), 18, 3246.
52. Stohman, F. and Kleber, C., J. prakt. Chem., (1892), 45, 475.
53. Ref. 33, p.28.
54. Ladenburg, A., Ber., (1869), 2, 140, 272.
55. Ref. 26, p. 45.
56. Stewart, A., Recent Advances in Organic Chemistry, 3rd edition, London, (1918), p. 9.
57. E.g. as late as Sidgwick, N.V., The Electronic Theory of Valency, London (1927), p. 221; it is admitted that "on the theoretical side... the question is open".
58. Walter, Ber., (1873), 6, 1402.
59. Ref. 56, p.1.
60. Stewart, A., Recent Advances in Organic Chemistry, vol. 1, 6th edition, London, (1935), p.15.
61. Baker, J.W., Tautomerism, London, (1934), p.10.
62. Citation from ref. 61, p.10.
63. Stoll, M., Perfumery and Essential Oil Record, (1959), Jubilee edition, p. xviii.
64. Citation from ref. 5, p. 581.
65. Ref. 37, p. 375.
66. Sutton, L.E., Proc. Chem. Soc., (1958), 312.
67. J. Chem. Educ., (1973), 50, 235.
68. Refs. in Hanack, M., Conformation Theory, New York and London, (1965), p.25.
69. Costa, A., J. Chem. Educ., (1971), 48, 243.
70. Ramsay, O.B., Chemistry, (1974), 47, Jan. p.6, Feb. p.6.
71. Ref. 38, p.258.
72. Wightman, W.A., J. Chem. Soc., (1925), 1421.
73. Ref. 56, p.2.
74. Ref. 60, p.11.

XVII

Advances in organic chemistry over the last 100 years

This chapter attempts to cover the last century of organic chemistry, inevitably in an impressionistic way. It highlights the most important developments relating to structure and its determination, stereochemistry, synthesis, new types of molecule that do not occur in nature, a limited selection of natural products, the rise of theoretical organic chemistry and the role of free radicals. The discussion opens with a state-of-the-art picture for 1904 (when Annual Reports made its first appearance) and concludes with some reflections on the position today, particularly in relation to the place of organic chemistry within science as a whole.

1 Introduction

Scope

Any attempt to summarise the chief advances in organic chemistry over the last 100 years must seem like an exercise in frustration. And such it is, given the rate of progress and exponential expansion in many areas. This Report therefore avoids any attempt to offer a summary of even the main discoveries. Instead it attempts to locate the origins of many familiar aspects of the subject today and to give a qualitative indication of how they have subsequently developed. Some kinds of deeper explanation are explored in the conclusion, as the growth of organic chemistry is placed in an even wider context. The Reporter is only too aware that almost every reader will regret that his or her speciality makes such a brief appearance or does not even appear at all. The demands of historical selectivity are usually tough.

Sources

In 1822 a series of 'annual reports' on the whole of physical science, (especially chemistry) was commenced by Berzelius.[1] Even then only a chemist of his gigantic

Annual Reports on the Progress of Chemistry 100 (Section B), 2005, 3–31. Reproduced by permission of the Royal Society of Chemistry.

powers could have undertaken such a Herculean task. When the Chemical Society produced its first *Annual Report* 82 years later there was an editor, a sub-editor and ten authors. Since then chemistry as a whole progressed so fast that, since 1967, the Report on organic chemistry has been published as a separate volume, Section B. To grasp even a little of the century's progress one needs to be aware of source material waiting to be used. First and foremost is the vast accumulation of papers in chemical journals worldwide, next are the published abstracts, and then the compilations like *Annual Reports*. The historian would call these primary sources.

Secondary sources, on the other hand, are lamentably few. There has been very little serious historical analysis of modern organic chemistry. A brief comment may indicate what is available. One book offers a 19th century perspective on organic chemistry,[2] and one other attempts a comprehensive history[3] though it stops early in the 20th century. Some of the advances have been chronicled in Partington's monumental *History of Chemistry*,[4] which includes the first half of the 20th century. The same terminal date applies to a few other volumes, including one valuable source-book.[5] There have been a few general accounts,[6] specialised histories of individual topics, numerous biographies, and personal reflections of participants in the 'Profiles, pathways and dreams' series of the American Chemical Society. Many relevant papers, chapters and books are listed in two general reports.[7]

Organic chemistry in 1904

By 1904 organic chemistry as a science was already a hundred years old. From a mere handful of compounds it grew to encompass over a hundred thousand. With atomic weights settled, a new structure theory emerged, based on the new concept of valency, and by 1870 many simple aliphatic compounds could be given structural formulae that we still accept today. Aromatic and unsaturated compounds presented more problems but Kekulé-type formulae for benzene and its homologues were largely agreed by 1904. Stereochemistry began in 1874 with the work of van't Hoff and le Bel and isomerism no longer presented intractable problems. Synthesis became the chief task of organic chemists from about the 1860s, and the old idea that some "vital force" was necessary to explain organic compounds was finally disposed of by about then. Much has been written about individuals in the 19th century who helped to shape the infant science, with books on Berzelius,[8] Chevreul,[9] Liebig,[10] Kolbe,[11] Wurtz,[12] Hofmann,[13] Frankland,[14] Kekulé,[15] and others.

Unsolved problems of course remained in 1904. Questions of nomenclature and notation continued to cause confusion, there was no real understanding of the rules for aromatic substitution, molecular rearrangements were often passed over in embarrassed silence, and phenomena associated with the Walden Inversion were regarded as "mysterious". There were continuing difficulties with aromatic substitution, as a recent paper suggests.[16] Many of these problems were to be resolved in the century following the first issue of *Annual Reports*, a thin volume containing in total 280 pages and edited by G. T. Morgan. On organic topics there was a chapter each on aliphatic chemistry (H. J. H. Fenton), cyclic chemistry (J. B. Cohen), and stereochemistry (W. J. Pope), 93 pages in all. Reporting among much else the syntheses of dipentene and terpineol by W. H. Perkin, jr. and of nicotine by A. Pictet *et al*,[17] it was a splendid beginning. Some of the major achievements on the next 100 years will be encountered below.

2 Structure

Classical methods

One legacy of the Victorians to organic chemistry was a battery of weapons with which to attack problems of structure. It would be quite wrong to imagine that these were all abandoned in the new century. Indeed some of them are still employed, in conjunction with far more sophisticated instrumental methods. One example may serve best to illustrate a range of these techniques: vitamin C, or L-ascorbic acid. Its structure was investigated in the 1930s, most notably by Haworth and collaborators.[18] Given a molecular formula of $C_6H_8O_6$, the following procedures were applied.

First, there were *quantitative measurements on derivatives*; in this case treatment with diazomethane indicated tetramethylation and therefore four hydroxyl groups. Tests were applied for *characteristic group functions*; thus the suspected presence of carboxyl was disproved by the fact that iodination gave a derivative (dehydroascorbic acid) that was wholly neutral; on the other hand a colouration with iron(III) chloride indicated an enolic substituent. Next, *analogies with known compounds* were sought; the iodination was closely parallel to that of dihydroxymaleic acid, implying an ene-1,2-diol; moreover the dehydroascorbic acid behaved very similarly to lactones, suggesting the parent vitamin might also have a lactone ring. Chemists would also try to *convert to products of known structure*; treatment of the dehydroascorbic acid with concentrated hydrochloric acid led to the immediately recognisable furfuraldehyde, so implying an unbranched chain of at least five carbon atoms. Another common technique was *degradation to simpler products*; here ozonolysis of the tetramethyl derivative yielded only one substance, confirming the original molecule was cyclic, and subsequent tests implied a γ-lactone ring. Finally, in 1933 the whole structure was confirmed by *synthesis* from L-xylose (Scheme 1).[19]

Accounts of this kind could be multiplied indefinitely right through the 20th century. But gradually such classical methods became displaced by instrumental methods. Indeed the Hawarth/Hirst structure has been subsequently confirmed by X-ray crystallography.[20] However it must be said that all modern methods depend on the structural work done before their adoption, as otherwise there would be no logical certainty that they were valid (self-consistency is not enough).

Instrumental methods

It may be surprising to learn that techniques were available in the 19th century for refractometry, polarimetry and absorption spectroscopy. They were looked on as curiosities from physics and had almost no effect in assisting structural determinations in organic chemistry. After 1904 the first progress occurred in refractometry.

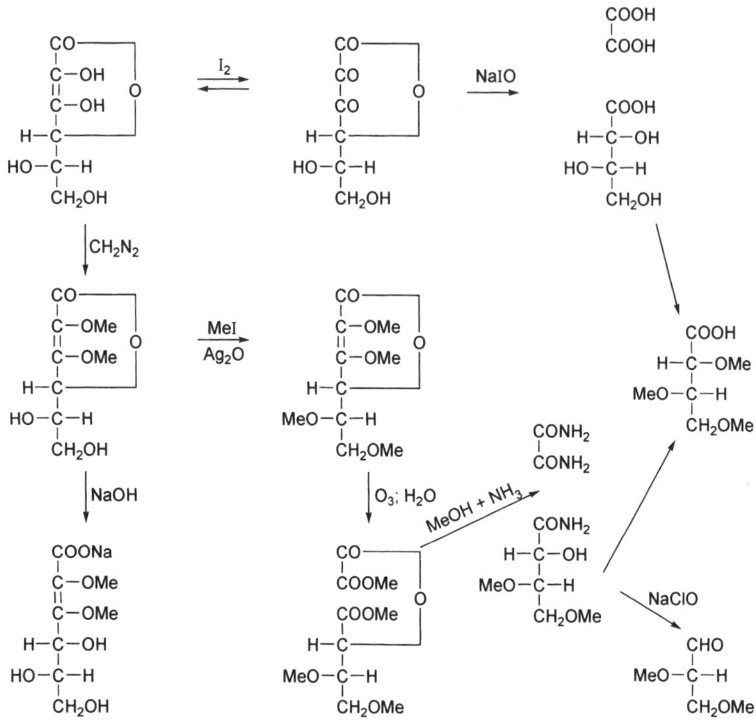

Scheme 1

Refractometry. This technique measured refractivities which were deemed to indicate constitution in a rather complicated way. Each substance had its characteristic refractivity, so the method could distinguish between certain isomers and be of value in assessing the degree of tautomerism present. Little is heard of it after about 1930.

Polarimetry. Polarimetry is discussed below (Section 3).

Absorption spectroscopy in the visible region. This had been known since the time of Bunsen. One of its earliest applications to problems of organic structure had come in 1903 when Tinkler *et al.* noted that the cotarnone alkaloids have quite different visible spectra in water and in ether and proposed different bond arrangements in the heterocyclic ring. In the next few years most studies were concerned with the effect of substituents on a visible spectrum rather than using the technique diagnostically to establish structure.

Ultra-violet spectroscopy. This had been used before World War II in studies of α,β-unsaturated ketones. From a study of some 40 papers on this topic, R. B. Woodward in 1941/2 came out with his 'ketone rules', showing that substituents would produce strong shifts of absorption maxima and suggesting rules for elucidating structure from spectra. A major collation of the UV spectra of substituted benzenes appeared in the 1940s when simple equipment became available.[21] At about the same time other workers were able to deduce steric hindrance from UV spectra and from now on it became a standard participant in structural analysis.

Infra-red spectroscopy. The use of infra-red spectra has been useful to organic chemists in a different way as it can pinpoint certain specific groups in a molecule. Following the recognition of molecular vibration as the cause a few instruments were made for industrial application in Germany in the 1920s and the USA in the 1930s. After the first double-beam spectrometer had been developed Woodward observed that 'infrared spectra have now become an asset of almost incomparable value to all chemists interested in structural problems in organic chemistry'.[22] The advent of commercially available instruments was delayed by the war, but its utility in that conflict (*e.g.* in relation to penicillin and synthetic rubber) brought design improvements and the first low-cost instrument was sold in 1956. In the 1960s the spectral range was extended down into the far infra-red with salt prisms being replaced by those of KBr or CsI.[23] By 1969 there were 16 precision IR spectrophotometers commercially available and 21 simplified instruments.[24] Correlation charts were now freely available and IR spectroscopy had become a routine tool in all major organic laboratories. By 1975 it was reported as showing 'no sign of decline'.[25]

Nuclear magnetic resonance. NMR spectroscopy was developed in the late 1950s following the discovery of nuclear magnetic resonance by Bloch and Purcell in 1946. Through the 1960s ^{13}C NMR became increasingly important and from 1972 was aided by a method known as the Fourier-transform technique. Another variation known as CIDNP (chemically induced dynamic nuclear polarization) appeared in about 1972 and was a valuable probe for free radical reactions of all kinds.[26] Through the 1970s NMR was successfully applied to a vast range of problems in conformational analysis, and in due course extended to biological molecules. The coming of 3D NMR in the late 1980s made these studies much more practical for structural analysis of proteins, nucleic acids and other complex molecules. NMR studies are now so ubiquitous that it is hard to imagine chemistry without them. Their impact on structure determination has been immense.

Mass spectrometry. Although the mass spectrograph dates to 1919 (Aston), mass spectrometry was first applied to organic chemistry in the 1950s by Djerassi, Biemann and others. It was initially used to determine molecular weights, and its application to natural and synthetic products was very successful. From the late 1970s a variety of techniques appeared for involatile or thermally labile products such as chemical ionisation and fast atom bombardment. As for NMR, Fourier-transform methods brought new flexibility.[27] No longer was molecular weight determination the only object; it was now possible to identify the number of atoms like chlorine or nitrogen in a molecule, and further studies of fragmentation patterns gave indications of specific structural features. By 1989 the advances in mass spectrometry were described as 'rapid and staggering',[28] and further dramatic extensions of its use for biological molecules were reported over the next decade.

3 Stereochemistry

The roots of stereochemistry go back to the enunciation of the tetrahedral carbon by van't Hoff and le Bel in 1874. Thirty years later the doctrine was well-established, though the existence of such atoms in compounds like ethylene (the usual assumption at that time) could only be accepted in terms of the strain theory of Baeyer. Some small distortion would also be necessary in the ubiquitous 6-membered rings, but it would be much less than in ethylene. In the field of optical isomerism well-established methods for resolving racemic mixtures led to highly satisfactory results justifying the assignment of optical activity to asymmetric carbon atoms. On the whole stereochemistry seemed to pose few problems for the chemist in 1904. In fact the first challenge came the very next year.

Dissymmetry for structural reasons

Free rotation about a carbon–carbon bond can be inhibited for steric reasons. In 1894 a compound had been resolved that did not possess a formally asymmetric carbon atom, *trans*-cyclopentane-1,2-dicarboxylic acid (**1**). Instead it had an overall *molecular* dissymmetry.[29] Several further examples of optically active disubstituted cycloalkanes were to appear in the next 50 years. But the cycloalkanes (or some of them) were to pose far more fundamental problems of a different kind. It was soon realised that various structural features could have the same effect. The effects of a double bond had been demonstrated as early as 1909 with a resolvable 1-methylcyclohexylidine-4-acetic acid (**2**),[30] other examples soon following. Polysubstituted allenes (**3**) were resolved from the 1930s onwards, the first being a di-α-naphthyl-diphenyl allene.[31] Even earlier were the spiro-compounds (**4**), starting with a resolution in 1921 by Mills and Nodder.[32]

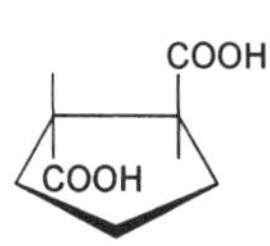

(1)

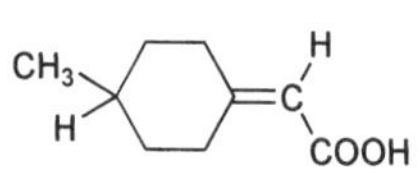

(2)

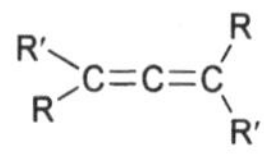

(3)

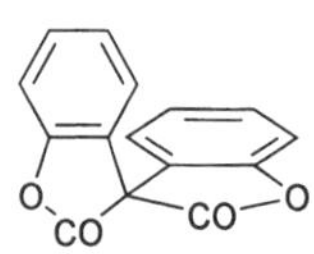

(4)

Optical activity due to elements other than carbon

It had been known before 1904 that other elements than carbon could be 'asymmetric' if bearing suitably different substituents; these included nitrogen and sulfur. A spiro-compound with nitrogen as the atom common to both rings (**5**) was shown to be optically active by Mills and Warren in 1925.[33] Some trisubstituted phosphine oxides had been resolved by Meisenheimer *et al* early in the century.[34] However most progress had to wait until the 1950s when phosphonic acids and their derivatives became more available and interest in phosphorus chemistry had been intensified by industrial uses of phosphorus compounds, as extractants by the atomic energy industry, in systemic insecticides, plasticisers and as phosphate detergents. Within a few years numerous compounds appeared owing their activity to an asymmetric phosphorus atom. Optically active four-bonded arsenic compounds had been known since the 1920s, as methylphenylbenzyl-1-naphthylarsenonium iodide.[34] Campbell prepared the first optically active stibine (**6**) in 1955,[35] but she had slightly earlier resolved a number of compounds with antimony or arsenic atoms unable to invert by being part of a heterocyclic ring. In the 1940s much work was done by Mann on optical activity due to a hetero-atom, including the resolution of analogues of tetralin containing arsenic (**7**), antimony, selenium and sulfur as ring atoms.[36]

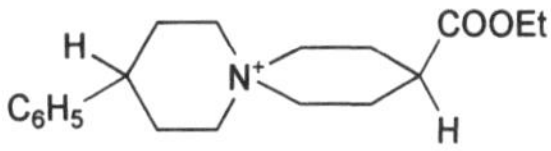

(**5**)

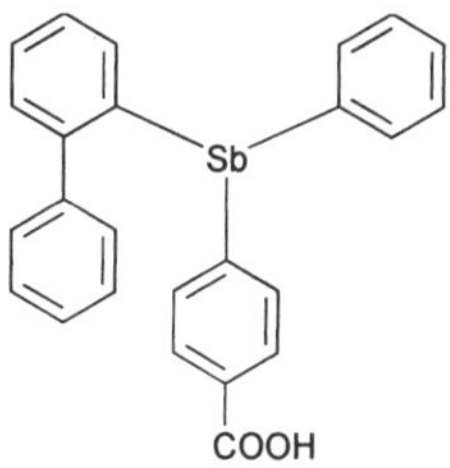

(**6**)

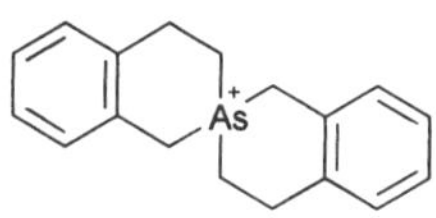

(**7**)

Overcrowded molecules

The phenomenon of 'steric hindrance' had been recognised since the 19th century and for the next 100 years was often invoked to explain anomalous reactivities, acid strengths *etc*. It resurfaced in a different context in the 1920s, when much interest centred on the optical activity shown by certain polysubstituted diphenyls, recently shown to be essentially linear molecules. The explanation was that bulky substituents at positions 2 and 6 in each ring could force the rings away from a coplanar arrangement. Molecular dissymmetry could then exist. By the mid-1930s these cases were well understood and documented.[37] Later work included non-benzenoid diaryls and 4,4′- and 5,5′-diquinolyls.[38] Much later the phrase 'molecular overcrowding' was applied to numerous fused ring compounds which proved to be dissymmetric, and four types were recognised.[39] They also had to be non-planar, therefore. The first such hydrocarbon, with no substituents, was phenanthrophenanthrene (**8**), shown in 1955 to be dissymmetric.[40]

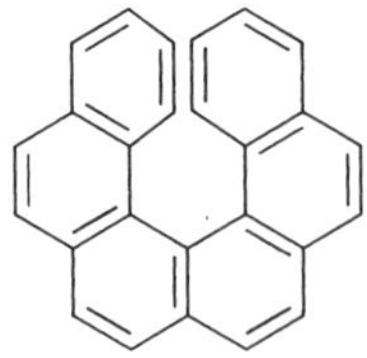

(8)

Conformational analysis

The notion of a non-planar cyclohexane ring had been mooted by Sachse[41] in 1893 but fell on largely deaf ears. It was however revived by Mohr from 1915 onwards,[42] who dealt with the main objection to Sachse: the non-isolation of two monosubstituted cyclohexanes. He argued that ring-inversion would be so easy at room temperature that one should never expect to isolate either form. However, if the ring was fused to another, as in decalin, then ring-inversion should be prevented. The isolation of two distinct decalins by Hückel[43] a few years later established the theory and led to the rise of conformational analysis. Some of the difficulties it continued to face have been discussed.[44] It was a long task. As Barton wrote of the period up to the 1960s, 'On the whole organic chemists did not pay much attention to stereochemistry'.[45] Yet all was not lost. In 1943 Hassel proposed a more plausible structure for *cis*-decalin than the one espoused by Hückel,[46] and it was confirmed by electron diffraction measurements three years later. This difficulty resolved and the wartime shortages over, conformational analysis began to emerge in the form we should recognise. It was effectively launched in 1953 with a paper on the steroid nucleus by Barton[47] and has ever since become a standard part of the way we view molecules. Its application to natural products has been immense.

Determining stereochemical configurations

Determination of the configuration of an asymmetric centre was classically done by chemical methods relating the unknown to a molecule of agreed configuration. The work by E. Fischer on carbohydrates giving a complete relation amongst all the hexose carbohydrates will surely stand for all time as a superb example of this enterprise. It was completed well before 1904, however, although its techniques and methodology have been used well into the modern period and a range of other examples illustrated.[48] All these related configuration to an arbitrary one assigned to glyceraldehyde or serine. Then, in 1951, Bijvoet used X-ray analysis to determine the absolute configuration for (+)-tartaric acid, and showed that the arbitrary configuration assigned to (+)-glyceraldehyde was in fact correct.[49] This not merely obviated the necessity to rewrite all textbooks on organic chemistry but also encouraged the development of other physical techniques and gave great confidence for the exploration of natural products for the next half-century.

Polarimetry has been a powerful tool in problems of stereochemistry. It could determine solution concentrations, could follow a reaction course and could distinguish between substances whose rotatory power was known. However by itself it could not allow a configuration of a given product to be defined, though with the aid of Freudenberg's rules[50] it was of some value in establishing configurations of related substances. Much more important for stereochemistry was the use made of another early discovery (by Biot) that the extent of rotation of the plane of polarisation depends on the wavelength of the incident light. The resultant optical rotatory dispersion, studied by Cotton in 1896, was shown to be very characteristic of substances and of chromophores. This was especially true where the incident radiation was in the far ultraviolet regions. Experimentally this was too difficult without photoelectric polarimeters, one of the first appearing (from O. C. Rudolph) in 1952. It was immediately put to use by Djerassi in his studies of steroids. Shortly afterwards the technique was seized on by Woodward, who devised his octant rule for using ORD measurements to determine the configuration of ketones.[51] Recording spectropolarimeters were available in the late 1950s and the use of ORD was established.

Related to ORD is the phenomenon of circular dichroism, free from certain problems of the former. It is, however, even harder to measure. When suitable instruments became available in the 1960s it became a standard tool in natural product research including steroids, polypeptides, polynucleotides, cobalamins and others.[52] But its superiority was so clear that by 1974 ORD was virtually displaced by CD.[53]

4 Syntheses

Synthesis had been the grand achievement of organic chemistry for over half a century and was performed for a variety of reasons.[54] This was especially true in Germany when thousands of new dyes, drugs and other products were made each year. Partly this reflected the desire of German chemical manufacturers to dominate the market and to remain pre-eminent in the field. It was also a result of competition between each of the many State universities, and synthetic chemistry provided an obvious field for multitudes of young PhD students. However it was not always looked on with sympathy elsewhere. Of the many thousand organic compounds prepared in the 19th century, Stewart observed that 'the majority were stillborn and their epitaphs are inscribed in Beilstein's *Handbuch*'.[55] That was in 1918. In 1931 he still maintained that position, complaining that 'the proportion of interesting substances to mere "gap-filling" compounds is very small indeed'.[56]

Classic methods

The fact behind Stewart's complaint was that classic methods of synthesis were so successful, and possibilities so inexhaustible, that they were still employed until after 1954 (the halfway mark in our account). The student textbooks by Mann and Saunders, Cohen and others were still being employed to train undergraduates well after that time. Laboratory supervisors knew when they were on to a good thing.

Many of their syntheses still used Victorian methods (and Victorian quantities as well!). Probably the most famous examples were based on acetoacetic or malonic esters or on nitration of aromatic species. Multitudes of new organic compounds were still being turned out by classic techniques until the late 20th century.

Some older techniques survived in a changed form. These included ozonolysis with improved apparatus and the famous electrolytic synthesis invented by Kolbe. Even 'Kolbe reaction failures' could be used successfully, as in the provision of new methods for aromatic substitution.[57] The Diels–Alder reaction for [4 + 2] cyclisations was discovered in 1928,[58] but experienced a remarkable resurgence of interest nearly 60 years later.[59]

Organometallic reagents

Most remarkable of all 'resuscitations' was the use of organometallic compounds as synthetic reagents. Dating back to Frankland in the 1850s, alkyls of zinc and other metals had shown enormous potential but their development had been hampered by extreme reactivity and sensitivity to water and air. The perfunctory use of organometallics was transformed almost overnight by the discovery of organomagnesium halides[60] by a young research chemist, V. Grignard.[61] In ethereal solution this reagent was soon to become widely-used for alkylations, arylations and other reactions. By 1905 200 papers had been devoted to the Grignard reagent, and in 1912 they exceeded 700—the year of Grignard's Nobel Prize. When he died in 1935 the number was about 6000. As late as 1965 they were claimed to be 'of more practical importance for the laboratory synthesis of organic compounds than any other single synthetic method'.[62]

After the 1950s other organometallic compounds began to supplement the older Grignard reagents. An early example was the use of organolithium compounds, capable of lithiating even hydrocarbons.[63] One factor favouring the increasing employment of these reagents from the mid-1960s onwards was the availability of polar aprotic solvents such as tetrahydrofuran and dimethylformamide. Shortly afterwards organothallium compounds, prepared *in situ* from thallium(III) salts, were seen to have a wide synthetic utility.[64] Other main group metals less commonly used included aluminium, tin and lead.

It was, however, the transition metals that opened up the field to the greatest extent. At first this generally involved the carbonyls of Group VIII elements, as in the nickel carbonyl induced coupling of allylic bromides,[65] but was soon extended in many other ways. Nickel carbonyl had been discovered in 1890, but others were much more recent, such as those of chromium (1926), ruthenium (1936), iridium (1940) and rhenium (1941). Their use was stimulated by new structural studies[66] and by the discovery of ferrocene and benzenechromium in the 1950s [see below].[67] Compounds of chromium, iron, palladium and other transition metals became widely used in synthesis from the early 1960s. In 1982 an increased interest was reported in transition metal alkyls like $TiMe_4$ and $MeTi(OPr^i)_3$ as agents for alkylation.[68]

Several notable workers in organometallic chemistry have left personal reminiscences that tell much about the history as seen from the inside; they include

Gilman,[69] Stone,[70]and Nesmeyanov.[71] An important series of papers on historical aspects of the subject has begun to appear in the journal *Organometallics*. Topics relating to the period under review include bis(benzene)chromium,[72] (cyclobutadiene)-iron tricarbonyl,[73] and tetraethylead.[74]

Organometallic chemistry has of course become an extremely important branch of chemistry in its own right, not merely as providing agents for synthesis. However by 1983 the opinion was expressed that 'organometallic chemistry seems to be going through a phase that is largely consolidatory',[75] a state that occurs in all fields of synthetic endeavour from time to time. Six years later a more optimistic Report inclined to the view that now 'it continued to flourish'.[76]

Reagents containing non-metal hetero elements

Perhaps the best-known synthetic tool that is not strictly organometallic is that derived from organphosphines. The Wittig olefination reaction was: discovered in 1953,[77] when methylenetriphenylphosphorane converted Ph_2CO to $Ph_2C{=}CH_2$. It was soon much used with aldehydes and ketones and since then with other carbonyl compounds also.[78] Analogous ylides of other elements (N, As, Sb, S) were also used in synthesis from about the late 1950s.[79]

Organoboron compounds were also recruited for purposes of synthesis, particularly after the discovery of hydroboration.[80] By 1982 it was possible to write a long review dealing only with carbon–carbon bond formation by organoboron compounds.[81]

The organic chemistry of silicon owed much to the early work of F. S. Kipping, from 1904 onwards. Summaries of this work[82]and of the ensuing development of silicone synthesis[83] have since appeared, together with a recent history of methylchlorosilanes and their role in synthesis.[84] In the years 1968–1978 organosilicon chemistry underwent an 'explosive growth' accompanied by a growing awareness of how organosilicon compounds could be used in synthesis.[85] It seems it was all triggered in 1968 by several independent discoveries and the publication of a book from an industrial chemist.[86]

Oxidising and reducing agents

The sledgehammer techniques of the 19th century have continued, often in modified form. Permanganate oxidations, sometimes with buffering, may be frequently found throughout the 20th century, and manganese dioxide had a spell of popularity in the 1960s. Greater specificity was sought, and a catalyst could sometimes supply this. The use of aluminium alkoxide can promote both the Oppenauer oxidation of an alcohol and the reduction of a ketone or aldehyde discovered in 1925/6 by Meerwein, Ponndorf and Verley independently (and now bearing all their names). A history of the Baeyer–Villiger oxidation has been given.[87]

The use of nascent hydrogen generated by sodium and alcohol has continued, but its non-specificity told against it. Lithium aluminium hydride, a much more specific reductant for carbonyl groups, was introduced by Finholt *et al* in 1947, while the milder sodium borohydride had been used to generate hydrogen during the war but was available shortly afterwards for non-military uses.[88]

An entirely different angle on organic oxidations and reductions has been provided by the petrochemical industry. The use of a nickel catalyst for hydrogenations has been exploited at both low and high pressures, and has also of course been employed in the manufacture of high-melting vegetable oils. Oxidation has also usually been catalytic, to yield such diverse molecules as ethylene oxide, carboxylic acids and ketones. The most remarkable non-catalytic oxidation that has little parallel in laboratory work has been the low temperature oxidation by oxygen of cumene to its hydroperoxide (which then gives phenol and acetone on acidification). Apart from the Sabatier–Senderens method which predated World War I, most of the relevant industrial progress took place just before or after World War II.

Stereoselective synthesis

A major trend in modern synthesis is towards stereoselectivity. But the origins are much earlier. Production of enantiomers from nonchiral species had, of course, long been effected by enzymes (as in the production of many early drugs). What was so elusive was a means of doing this without agents from nature, a true laboratory-based asymmetric synthesis.

Attempts to synthesise optically active substances from inactive or racemic source material had long been in vain; chiral molecules could not be obtained from non-chiral. Then, in 1935, Maitland and Mills obtained an optically active allene by dehydration of a racemic enol using (+)-camphor-10-sulfonic acid. The product contained a slight excess of the (+) over the (−) isomer.[89] Not only was this the first optically active allene; it was also the first example of an asymmetric synthesis, and since then several other allenes have been similarly obtained. The directing effect of an existing asymmetric centre on introduction of another has been examined by V. Prelog[90] and D. J. Cram.[91]

Following the notable advances of the 1950s asymmetric synthesis (as it is now known) has been greatly extended by using asymmetric metal complexes which by virtue of their solubility can give homogeneous catalysis. The first case was decomposition of ethyl diazoacetate to yield phenylcyclopropane carboxylic acids.[92] Yields were rarely high and efforts were made to introduce new techniques, as the use of chiral lithium amide bases, RR′NLi.[93] In 1994 spectacular improvements were obtained in asymmetric epoxidation of styrenes with chiral manganese catalysts, hailed in one Report as ‘reagent of the year’.[94]

Asymmetric synthesis has now become big business, of which a large-scale synthesis in Japan of (−)-menthol is one of the earliest. In the field of drugs it is important because almost all chiral drugs act as desired only in one enantiomeric form. In America the pharmaceutical manufacturers are moving into asymmetric synthesis because of guidelines from the U.S. Food and Drug Administration on testing and marketing chiral drugs. The chiral catalysts are often derived from rhodium or ruthenium.

Combinatorial synthesis

This is a phenomenon of the last decade of the 20th century. The intention is to make as many closely related products as possible. When a mixture of 6 amines reacted with one of 8 epoxides the number of possible new amines was 48. Given suitable purification techniques each of these could be isolated at the end.[95] Several advantages arose from using a solid support, as in the synthesis of peptides or oligonucleotides. Everything depended on sophisticated and automated methods for separation, but given these this approach promises to be most fruitful.

General

The above account touches only the tip of an immense iceberg, and merely selects a few of the many highlights in this century of organic synthesis. Nor has it been possible even to mention the steady improvement in techniques for handling and examining materials. These would have had to include the development of micro- and semi-micro-scale working, the use of high vacuum methods, chromatography in all its forms (especially GLC), and many sophisticated instrumental aids for following the course of a reaction and separating its products. It may not be too simple to observe that in this period cork stoppers were gradually replaced first by rubber bungs and then by ground glass joints. Except perhaps in schools the latter reigned supreme in Britain from about the late 1960s. Apparatus could then be assembled more quickly, was more leak-proof and greatly facilitated operations under reduced pressures.

The growth of synthesis owed much to all these innovations. In addition it was able to diversify by the availability of many new reagents, and these in turn depended on such factors as bulk provision of fairly 'new' solvents. It had to overcome a certain prejudice against ringing the changes on well-worn procedures but had the advantage of much better communication between chemists, particularly after 1945.

An optimistic conclusion might be that 'The view is becoming more frequently expressed that synthetic methodology has advanced to such a state that almost any stable structure can be made'.[96]

5 New types of non-natural molecule

In 1904 there were very few molecules of a fundamentally new kind that could not be easily accommodated into the existing paradigm (though of course there were many whose structure was then entirely unknown but recognised to be so). That happy state of affairs was not to last for long. The efforts of the synthetic chemists began to throw up strange unheard-of molecular species quite early in the 20th century. One item reluctantly excluded here is the fullerenes ('Bucky-balls') since, despite their intense interest, they are not organic compounds but new forms of carbon. The following were some of the most puzzling compounds in their day.

Large ring compounds

Small rings of 3, 4 and 5 members had been obtained at the end of the 19th century and had posed little problem. Rings of over six atoms were however almost unknown. Then, in 1905, Willstätter prepared cyclooctane by exhaustive methylation of the alkaloid pseudo-pelletierene, obtaining *en route* a highly explosive cyclooctadiene.[97] He also made from the same starting material an even more enigmatic molecule, cyclooctatetrene.[98] Expected to be an analogue of benzene, its shocking reactivity and readiness to add halogen, to be oxidised and even to isomerise into styrene left observers baffled. This was long before physical evidence for its non-planarity, the Hückel Rule and modern concepts of aromaticity.

Larger rings appeared many years later, with the work of Ruzicka *et al.* Dry distillation of thorium salts of dicarboxylic acids having at least eleven carbon atoms led to cyclic ketones with 11–30 atoms in a ring.[99] Rather better yields were later obtained by Ziegler, using imido-nitriles as intermediates.[100] The existence of such substances (like the naturally occurring perfumes civetone and muscone) is inexplicable in terms of planar rings, but extending the hypothesis of Mohr beyond 6-membered rings explained much and further paved the way for conformational analysis.

Since 1960 cyclic compounds have been made with interlocking rings, the catenanes.[101] A trimethylchroman known as Dianin's compound was prepared in 1914 and shown to retain certain organic solvents, and many other substances. Although the structural formula revealed only three 6-membered rings, X-ray analysis in the 1970s disclosed a cage-like structure where parent molecules are held together by hydrogen bonds. This was an early example of a clathrate, where guest molecules of suitable shape and size could be imprisoned at will. Many other analogous compounds have been discovered. The cyclodextrins consist of 6 or 7 glucose units linking together to form cyclic ethers with 30 or more ring atoms. They also form inclusion compounds and have potential applications in pharmacy and many other fields.[102] In 1967 the crown ethers were accidentally discovered.[103] These are cyclic polyethers derived from glycols, and may have 12, 15 or 18 ring atoms. They 'provided the timely fillip for the rapid development of supramolecular chemistry.'[104] That seems to have been the case as many more species have become available, many more uses found for them and a new branch of the science has emerged: host–guest chemistry. By 1994 this was reported as 'an area of increasing activity'[105] and this still seems to be true.

Carbenes

The first suggestion of a simple carbene $:CH_2$ was in papers by Norrish *et al* in 1933 where it might have been a transient product from the photolyses of ketene and diazomethane.[106] The existence of such species as chemical intermediates was not proved until 1954 when Doering and Hoffmann examined the familiar alkaline hydrolysis of chloroform to carbon monoxide and found a $:CCl_2$ intermediate that could be trapped by cyclohexene to add across the double bond, forming a cyclopropane derivative. Many new carbene reactions have since been discovered, and it has been recognised in reactions as familiar as the Reimer–Tiemann synthesis of phenolic aldehydes and the notorious 'carbylamine reaction' of chloroform. By 1964 there were many uses for carbenes in synthesis.[107] Even though its structure as a diradical was debated for many years,[108] here for almost the first time was a transient trappable intermediate of whose existence there could be no serious doubt. Later certain heavily-substituted carbenes were stable enough to be isolated. Organic mechanistic theory in general had been strengthened by the carbenes and was better able to accommodate later proposals like the nitrenes and arynes. The latter class had been postulated as intermediates in certain nucleophilic aromatic substitutions.[109]

Metallocenes

The first metallocene was discovered by accident. In the late 1940s S. Miller, an industrial chemist working at BOC, was heating cyclopentadiene and nitrogen in an attempt to synthesise amines. Working with an iron catalyst he found pipes became blocked by an orange solid that appeared to be an iron derivative of cyclopentadiene. Unaware of the novelty of his product Miller did not publish for three years and then his paper,[110] submitted in 1951, did not appear until 1952. Meanwhile the same product had been obtained by P. Pauson and T. Kealy, again by accident. They were attempting to oxidise cyclopentadienylmagnesium bromide with iron(III) chloride. Their report was published in *Nature*.[111] The structure they assigned was a linear cyclopentadienyl iron, but this was perceived as unlikely by Woodward and G. Wilkinson (independently), both at Harvard. They correctly assumed the now familiar sandwich formula,[112] later confirmed by X-ray analysis. Ferrocene was 'a new aromatic system' where substituents would behave normally, and analogues could be obtained with metals from most Groups of the Periodic Table replacing the central iron atom. Their use in synthesis is paralleled by the great contribution they have made to the detailed study of aromaticity.

Valence-isomers

In the 19th century several alternative structures for benzene had been proposed, including a prismatic structure (Ladenburg) and a hexagon with a 1:4 single bond (Dewar). By 1904 they were firmly displaced by the Kekulé hexagon, though other contenders still had their voice. With the coming of electronic theory the situation was transformed and benzene was until at least mid-century seen as a resonance hybrid between the two Kekulé forms. A reinterpretation in terms of molecular orbitals followed but Dewar and (still more) Ladenburg structures were almost forgotten. However as techniques for handling very labile compounds improved an interesting reaction was discovered in 1964 by Viehe *et al.* They found that fluoro-*tert*-butylacetylene spontaneously polymerises into a remarkable set of compounds, trifluor-tri-*tert*-butyl derivatives of 'ordinary' benzene, Dewar benzene, Ladenburg's benzene (prismane) and others.[113] Later on hexamethyl Dewar benzene led to further transformations. Many years afterwards bicycloprop-2-enyl was synthesised, 'the last remaining valence isomer of benzene'.[114]

6 Natural products

Originally 'organic chemistry' meant simply the chemistry of products of living organisms, *i.e.* natural products. In spite of its much wider remit today, its ultimate rationale is still seen by many as a means to a deeper understanding of the chemistry of life. Natural products have in any case challenged the skills of chemists by their very complexity. In the years from 1904 to 2004 their study has blossomed beyond expectation, and only a briefest summary can be offered here of even the chief achievements in three groups, almost randomly selected.

Carbohydrates

By the end of the 19th century the simple monosaccharides had been established as aldo- or keto-hexoses, and from 1884 the pioneering labours of Emil Fischer, using the phenylhydrazine that he had previously discovered (1879), culminated in the relative stereochemical configurations of all 16 aldohexoses. From 1903 Purdie and Irvine introduced methylation of carbohydrates and subsequent use of this process led both them and Fischer to accept a ring structure for methylated hexoses and, by implication, for the free sugars as well. Further work by Haworth[115] confirmed a 6-membered ring ('pyranose') for aldo-hexoses such as glucose and a 5-membered one ('furanose') for keto-hexoses such as fructose. Planar ring diagrams were re-introduced in 1925/6 by Haworth, having previously been suggested by Boeseken (1913). Representation as non-planar chair forms dates from the rise of conformational analysis in the 1950s.[116] By the 1960s physical methods had been brought heavily to bear on the carbohydrates. For example X-ray diffraction measurements had been reported on numerous monosaccharides, establishing the general picture emerging from other studies.[117]

Methylation techniques were employed by Haworth *et al* in the 1920s to establish the structures of disaccharides such as sucrose and maltose. The synthesis of cellobiose was accomplished by Stacey in 1946[118a] and of sucrose by Lemieux and Huber in 1953.[118b] Polysaccharides as starch and cellulose presented more severe problems, though attempts were made on these by Irvine, Haworth and several other workers. They were seen as 'polymers' of D-(+)-glucose, but by 1931 three main questions were seen as unanswered: the relation between the two amyloses in starch, the nature of the hexose linking, and the number of hexose units involved.[118c] The first problem was resolved by methylation studies, and the second by use of enzymes. Hydrolysis of starch (amylose or amylopectin) could be accomplished by enzymes like β-amylase, maltose being a first product and known to be an α-glucoside. Further studies confirmed that amylose is largely made of glucose units joined by α-1,4-glycosidic links. Amylopectin was shown to be a highly branched molecule with glucose units linked by 1,4 or 1,6 α-linkages. However, hand cellulose was hydrolysed by enzymes specific to β-linkages. Molecular weight determinations on each by osmotic pressure measurements indicated the presence is each of several thousand glucose units. Since then much work has proceeded on carbohydrate derivatives, on carbohydrates on chiral templates and in a multitude of other areas.

Plant pigments

The highly colourful materials that exist in the plant world have early attracted chemists. Three of the most prominent groups are the carotenes, anthocyanins and chlorophylls.

β-Carotene. One of the most widely distributed pigments in nature, carotene was isolated early in the 19th century and its empirical formula was determined by Willstätter in 1907. In 1928 it was shown to have eleven double bonds and shortly work by Karrer *et al* elucidated its structure completely.[119] With the advent of chromatography it became clear that 'carotene' was a mixture of at least three isomers, the most important being the β-form, a precursor of vitamin A. The stereochemistry of an 'all *trans*' arrangement was suggested by the UV spectrum, and later confirmed by X-ray analysis. In 1950 Karrer devised a synthesis[120] and since then many other routes have been found. The α- and γ-isomers were synthesised by Karrer and others in the early 1950s and two other carotenes were identified and synthesised in the 1960s.

Anthocyanins. These were first obtained in a crystalline condition in 1903, and shown to be glucosides in 1906. Two years later the simpler γ -pyrones were shown to form salts with HCl, a discovery that paved the way to an understanding of the anthocyanins' strange ability to do the same. The first structural enquiry was by Willstätter and Everest in 1913. In their now classic paper on cyanin chloride,[121] it was shown to be a glucoside of a pentahydroxy-2-phenylbenzopyrylium salt. Next year Willstätter synthesized it from quercitin,[122a] and in the 1920s R. Robinson explored new synthetic methods, condensing *o*-hydroxybenzaldehydes with ω-hydroxyacetophenones. In 1932 he and Todd achieved a complete synthesis of several anthocyanins, including cyanin chloride.[122b]

Chlorophyll. Here we meet Willstätter again who worked on the topic from 1905 to 1913. He separated the two chlorophylls *a* and *b*; determined the empirical formula, identified the substituents, established chlorophyll's relation to the porphyrins and speculated on the linkage to the magnesium atom. The next major advances came from Hans Fischer: 'Fresh from his dramatic conquest of the blood pigment, Fischer hurled his legions into the attack on chlorophyll'. In 1939, he proposed the structure now accepted, but Linstead, who also established the details of its stereochemistry, did not confirm it until after the war in 1955/6. Yet at precisely this time, Woodward was considering the early structural investigations and questioning their conclusions. As he said, they 'had been carried out almost entirely in during the twilight of the classical period of organic chemistry'.[123] And so an attempt was made to synthesise the complete molecule, Fischer's own attempts having been terminated by his death in 1945. Woodward devised a new method of linking four different pyrroles in the correct sequence and with over a dozen assistants succeeded in achieving the first synthesis[124] in 1960.

Proteins

These are fundamental components of living cells but are also amongst the hardest group to study. Partly this springs from their sheer complexity but it is also because of the difficulty in obtaining pure samples. Much progress was delayed until the advent of purification methods involving chromatography, counter-current distribution and electrophoresis. However as early as 1902 Fischer proposed that they consist of long chains of amino-acid residues, and separation of the amino-acids produced on hydrolysis needed the same techniques. From about 1937 these began to be available. Molecular weights were determined and it has been argued that protein chemists were ahead of polymer chemists in establishing the existence of the macromolecular state.[125]

A further advance came in the late 1940s when, for the first time, proteolytic enzymes came to be available as pure reagents. A key problem was how to deal with free amino-groups at the end of protein or peptide molecules, and thus recognise the amino-acid in which they were situated. One of the first effective reagents to react with terminal amino-groups was 2,4-dinitrofluorobenzene, introduced by Sanger *et al* in 1945. An alternative method using phenyl isocyanate was introduced by Edman in 1950.

An early triumph of this methodology was the peptide gramicidin S, having only 10 amino-acid units; the sequence of these was determined in 1946, and its molecular weight a little later. However greater triumphs were on the horizon. These methods enabled Sanger in 1953 to determine the amino acid sequence of the hormone insulin.[126] The two chains are linked together by disulfide bridges, and another bridge links two units of the same chain. Very quickly Sanger was able to use performic acid to oxidise and break these bridges, locating their points of attachment to the chains and thus determining the complete structure of the protein, the first one to be determined.

This work set the tone for the rest of the century. It established that each protein had its own distinctive amino acid sequence. Many other techniques have since been employed, including the use of ^{32}P labelling methods for numerous enzymes. Through intensive research the sequences of over 20 000 proteins are known today. Further developments have included X-ray analysis of the 3-dimensional structure of proteins, as that of myoglobin (Perutz)[127] and haemoglobin (Kendrew),[128] both in 1960. Finally, protein and peptide synthesis has steadily advanced through the use of ever more effective reagents and especially through solid-phase synthesis such where a substrate with some active sites (such as chloromethyl) holds the reactive species in position for each reaction stage. With this technique Merrifield (who devised it in 1962) was able to synthesise the enzyme ribonuclease[129] in 1969. The 11 931 operations required would have been impossible without the fully automated solid-phase apparatus that had become recently available.

7 Theoretical organic chemistry

Origins

Organic chemistry has always had some theoretical basis but this phrase refers to the great advance in theoretical understanding that came about in the 20th century, though its roots were rather earlier.[130] It is one of the few branches of modern organic chemistry that has had serious historical analysis. Three main areas of chemistry, hitherto largely separate, came together. First was the idea of disposable affinity.

Disposable affinity. Thiele[131a] had proposed a degree of 'disposable' affinity' or 'residual valency' on unsaturated carbon atoms, thus explaining both the relative stability of benzene and the 1,4 addition to conjugated dienes:

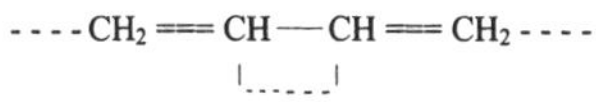

This was extended by Flürscheim[131b] to cover any pair of atoms that had a strong affinity for each other, leaving other atoms less strongly bonded:

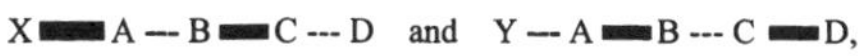

from which he obtained a simple theory of aromatic substitution. In Ingold's opinion 'Flürscheim's theory was about as correct as it was possible for a pre-electronic theory to be'.[132]

Introduction of the electron. Probably the first to introduce the newly discovered electron into organic chemistry was H. S. Fry, who, in 1911, proposed an 'electronic isomerism' in which a molecule might exist in two interconvertible but non-isolable forms

$$\overset{-}{X}—\overset{+}{Y} \quad \text{and} \quad \overset{+}{X}—\overset{-}{Y}$$

Several others, including Vorländer, extended these ideas to benzene substitution but one further development was necessary before the full electronic theory of organic chemistry could be born.

Application of chemical dynamics. The notion that molecules may not be static objects but might be involved in movement had been implicit in Kekulé's oscillation hypothesis for benzene and in the phenomenon of tautomerism[133] recognised by Laar. This was to receive confirmation in 1911 with the separation by Knorr of the keto and enol forms of acetoacetic ester.[134] The question of rates of reactions was rarely raised in organic chemistry. Reaction kinetics had existed since Harcourt's work in the later 19th century,[135a] but (apart from Wilhemy's isolated studies of sucrose inversion half a century earlier) it had not been applied to organic reactions until 1903. It was heralded by Lapworth's classic study of the action of cyanide on camphorquinone, very slow in acid solution, almost instantaneous in alkaline.[135b] Concluding that an initial action must be due to cyanide ions, he virtually invented the 'multi-step' reaction mechanism. After examining cyanide addition to αβ-unsaturated ketones he extended the system of Flürscheim to a notion of alternating polarities induced by some 'key atom' (often oxygen or nitrogen). Full electric charges were not implied but rather an indication of the polarities at the moment of reaction:

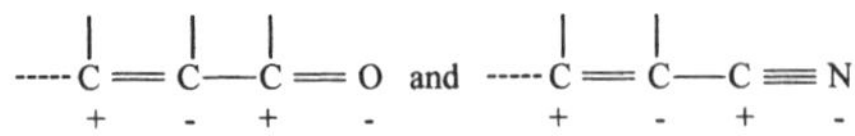

Several workers in addition to Robinson and Ingold made distinctive contributions to the kinetics and mechanisms of organic reactions.[136] It has been argued[137] that many of the features of later mechanistic chemistry were already present by 1910 in Sidgwick's enormously influential *Organic Chemistry of Nitrogen*.[138] Two British physical chemists who also contributed significantly to the new physical organic chemistry were Moelwyn-Hughes and Hinshelwood.[139]

Robinson and electronic theory

A colleague of Lapworth who generously acknowledged indebtedness to him was Robert Robinson,[140] a founder of modern theoretical organic chemistry.[141] He combined the views of Thiele and Flürscheim in understanding the phenomena of conjugation and molecular rearrangements but his first major advance was to apply the new Lewis–Langmuir octet theory and to conclude that polarities on atoms could be induced and then pass down a conjugated chain, in accordance with the octet rule.[142a] In this way he was able to explain addition to conjugated systems and the weakness of aniline bases. He was a pioneer in the use of curly arrows to indicate electron-pair displacements in a molecule. These symbols are still endemic today—despite early criticisms like that of H. E. Armstrong that 'a bent arrow never hit the mark'![142b] Robinson also spoke of the 'directive power' of substituents attached to a benzene ring. He introduced the terms 'cationoid' and 'anionoid'. However he quarrelled with his great rival Ingold and did not accept the application of quantum theory to organic chemistry. By 1927 he had largely left electronic theory for other projects.

Ingold and electronic theory

Ingold[143] arrived at theoretical organic chemistry a little later than Robinson. Ostensibly on the grounds of differences regarding aromatic substitution the two men entered into protracted public controversy,[144] but when that was over Ingold virtually had the field to himself, though questions remain as to their priority.[145] He replaced Robinson's terms with his own 'electrophilic' and 'nucleophilic', and wrote of 'tautomeric' and 'inductive' effects, distinguishing between those that were time-variable and those that were not. Unconvinced that all aliphatic nucleophilic substitutions proceeded in the same way he made the famous distinction between the S_N1 and S_N2 mechanisms. This is still useful today, though in 1972 reports of further progress indicated some alternative mechanisms that had been proposed,[146] and a further review in 1990 discussed subsequent work in nucleophilic aliphatic substitution.[147]

In the aromatic field Ingold focused mainly on electrophilic nitration but he also worked on molecular rearrangements and addition reactions. Some of his early work was jointly with his wife Hilda and much of it with E. D. Hughes.[148] His influence on the subject has been immense, due in no small measure to his textbook *Structure and Mechanism in Organic Chemistry.*[149]

Resonance

During the 1920s and 1930s structural problems continued, but they were insoluble in terms of existing theory. However in 1923 Robinson wrote that 'opinion is crystallising in connexion with the problem of the constitution of benzene'.[150] His own proposal for an 'aromatic sextet' of electrons followed two years later. In facts debates on aromaticity were to continue for over sixty years. In 1986 it was described as 'one of the most nebulous concepts in chemistry' with, perhaps, aromatic properties resulting from symmetric coupling of electron pairs around the ring.[151] Despite such optimism it became increasingly clear that some molecules (such as the nitrate ion or benzene) could be written in several valid forms yet only one substance existed. Its 'real' structure would be somewhere in between the contending structures. To accommodate this difficulty Pauling proposed that the various valence-bond structures ('canonical forms') were in a condition of 'resonance'. This was done in a series of papers in the early 1930s and brought together in his magisterial volume *The Nature of the Chemical Bond.*[152] Through this book and another by Wheland the theory of resonance was disseminated as widely as possible.[153]

Ingold preferred his own term mesomerism but the chief threat to resonance came from a different quarter. In 1949 a book appeared that avoided all double-headed arrows between canonical forms, and eschewed a valence-bond approach almost completely.[154] Its author, M. J. S. Dewar,[155] proposed instead to use a molecular orbital approach with each relevant molecule rewritten in terms of its p- or π-orbitals. He was not the first to do so, but now his unified approach enormously began to exert an influence that lasted for many years. Only in 1961 was C. A. Coulson reported as saying of quantum organic chemistry that in the previous 30 years 'many of the plums have now been picked', the reporter agreeing with respect to π-electron theory.[156]

Dewar's commendation by Robinson helped to polarise the organic community, and the schools associated with Wheland, Ingold and Dewar tended to go their own ways. However battles continued to arise, and a colourful conflict of the 1960s involving the non-classical 2-norbornyl cation has recently been described.[157] It was at just that time that interest in long-lived carbocations was accelerating and the simplest example, and also the simplest aromatic species, was discovered: the cyclopropenyl cation.[158] Today the arguments often seem rather sterile and most organic chemists use whichever model of a molecule is most relevant to their own needs.

8 Free radical chemistry

The first free radicals

The importance of free radicals for organic chemistry was unrecognised for many years, despite some occasional early appearances in the literature. By the mid-1930s

Ingold and Robinson had at least agreed the importance of ion-pairs and of heterolytic bond-cleavage. This became the essential orthodoxy; indeed as late as 1953 Ingold could write 'The general pattern of organic chemistry, as we know it, is based largely on heterolytic reactions'.[159]

Nevertheless the study of homolytic reactions had not been confined to inorganic chemistry. In 1900 Gomberg isolated the first stable organic radical (triphenylmethyl). By the 1930s its stability was being explained in terms of resonance. In 1929 methyl and ethyl made a transient appearance in Paneth's pyrolysis of lead alkyls. Further studies of gas-phase reactions between alkyl chlorides and sodium confirmed the intermediary existence of alkyl radicals.[160] Not all reactions of metals with alkyl/aryl halides involved a free radical mechanism, but some (including the Ulmann reaction) appeared to do so.[161]

Homolytic aromatic substitution

About 1934 some abnormal behaviour in aromatic substitution was observed by D. H. Hey. The anomaly recognised was the unusual ratio of *o*-, *m*- and *p*-isomers in certain cases of aromatic phenylation. Even when a nitro-group was present *o*/*p* substitution predominated. Hey suggested that free phenyl radicals might be involved as the attacking species, in contrast to all the accumulated research on electrophilic and (to a less extent) nucleophilic aromatic replacements. Much evidence was soon available to confirm this view. For many years the most thoroughly investigated class of homolytic aromatic substitution was arylation, a wide variety of sources being available for aryl radicals. These included the decomposition of benzoyl peroxide, already suspected to yield phenyl radicals in other circumstances. Much work continued by Hey *et al.* for the next twenty years, probing further the specificity of homolytic substitution, with the conclusion that there is 'a whole series of free radical reagents of varying electrophilic and nucleophilic character.'[162] Since that time aromatic homolytic substitution has had a continuing role in synthesis.

Alkene addition reactions

Recognition that free radicals may have a role in addition reactions came with the work of M. S. Kharasch (whose important work in homolytic chemistry is commemorated in a memorial volume[163]). When hydrogen bromide was added to allyl bromide the expected Markovnikov addition did not take place and 1,3- rather than 2,3-dibromopropane was formed. This abnormal addition was facilitated by traces of peroxide and the Kharasch 'peroxide effect' was born.[164] Events surrounding the discovery have been reported.[165] Four years later Hey and Waters,[166] and Kharasch and Mayo independently,[167] observed that peroxides were capable of homolytic fission and that the resultant aryl radicals could liberate bromine atoms from HBr and these would then attack position 3 in allyl bromide.

Subsequent work has explored many other radical-catalysed additions to alkenes, together with the direct additions of polyhalogenomethanes, carbonyl compounds and esters. Progress up to 1961 has been described,[168] and much has been reported since.

More recent developments

It has been argued[169] that the Hey and Waters paper was the effective initiator of the whole broad field of free radical chemistry. With the availability of electron-spin spectroscopy and other techniques the field greatly expanded toward the end of the 20th century. Lead tetraacetate has been used since the 1960s for radical oxidations and ring-closure reactions. Other generators of free radicals have been called into service as synthetic reagents, including dimethyldioxirane for oxidation of alkanes and thiohydroxamic esters as sources of alkyl or aryl radicals. It was remarked that in 1985 'synthetic reactions involving free radicals attained a new maturity',[170] moving beyond the cyclisations for which they had hitherto been mainly used.

Oxidations, rearrangements and further additions and substitutions were reported and at least two individuals reflected on their own half-century with free radicals, D. H. R. Barton[171] and C. Walling.[172] Many familiar reactions, hitherto suspected to be entirely heterolytic in character, have been shown to employ transient free radicals, at least on occasion. One of the most remarkable of all such discoveries relates to that monument to electrophilicity, aromatic nitration. Using CIDNP techniques it has been shown that even here radicals may have a role in reactions of the nitronium ion.[173] Free radicals have truly become part of the fabric of organic chemistry.

9 Conclusion

It must be painfully obvious that much work of outstanding value has been omitted. Space will not allow me to write of tautomerism, hyperconjugation and anchimeric assistance, of the wonders of heterocyclic chemistry, of the detailed chemistry of steroids, terpenes, alkaloids and fats, of the syntheses by Woodward of quinine and strychnine, of the total syntheses of cephalosporin C and other complex biologically active molecules, of the pioneering work of Todd on nucleotides, or above all of the triumphant research leading to DNA. Much else is also unmentioned. There are historical studies elsewhere of nomenclature in organic chemistry[174] and of industrial organic chemistry.[175]

It remains to ask: what events or situations have shaped the path of organic chemistry between 1904 and 2004? While it is clearly true that progress in organic chemistry depends on an immense amount of laboratory work conducted within complex theoretical frameworks, there are certainly other factors that can influence its progress.

One of these, of course, is the question of apparatus and equipment. During this period, as has been mentioned, glass laboratory apparatus has changed considerably, with glass and rubber bungs being replaced by ground-glass joints. These have been around since the late 18th century but not commercially available until well after 1904. In many laboratories they did not gain routine use until the 1950s. At the same time came the transition from soda-glass to higher melting silicate glass and eventually silica tubes.

It is, however, instrumentation that has made such a spectacular difference. Chromatography and simple optical instruments had been available well before 1939, but little else. In the USA some spectrophotometers were being made by Beckmann in the 1940s (and were used by Woodward), but in Europe none was commercially available until after the War. Then there began a virtual explosion of instrumentation in all the techniques mentioned earlier. This was a legacy of research in electronics in connection with wartime radar; it was also a consequence of the invention of the transistor in the late 1940s with all the attendant developments in microcircuitry. This is surely one explanation of the rapid growth of organic chemistry in post-war Europe.

A further point, which is not trivial, is that after about 1950 many chemicals were obtainable for the first time (at least at an affordable price). To take but one example out of hundreds, the metal lithium was to find extensive use in organometallic syntheses as well as in reducing reactions of various kinds. From World War I its main use had been in the bearings of German railway vehicles. Other demand (and therefore supply) was very small. It was to find further military uses in World War II, and new electrolytic methods were invented to produce it. In the years 1950–1954 production of lithium minerals in the USA more than quadrupled. From then on the metal, and later some of its organic compounds, became more readily available. A curious side-effect of the surge in marketing was the publication by the Lithium Corporation of America of 'Annotated Bibliographies' of lithium applications,[176] and, in those pre-Internet days, their regular arrival greatly stimulated at least one research group in Britain. In a similar way publication by another chemical manufacturer seems to have triggered rapid development of organosilicon chemistry.[86]

The two world wars have had other effects on organic chemistry. In World War I (sometimes called 'the chemists' war') research was commissioned in Germany, Britain and elsewhere on production of toxic gases and more powerful high explosives. The war from 1939 to 1945 promoted studies on polymers and insecticides as well as more conventional war-gas and explosives production. In both conflicts the numbers of published papers on chemistry as a whole dropped dramatically, as may be seen from the appropriate volumes of *Annual Reports*.

To conclude, there is one other consideration that might prove to be more important than any of the above. That concerns the relation of organic chemistry specifically to science in general and to the rest of chemistry in particular.

It is not hard to show that over the last 200 years or so organic chemistry has had a changing relationship with the rest of the science. Within chemistry one can recognise six phases of alternating convergence and divergence, the last one being our own. At first those who dealt with what we should call inorganic topics had little in common with those who were interested in gums, fats, blood *etc.*, but from the 1820s it became apparent to all that the atomic theory of Dalton should apply to organic substances. The key figure was Berzelius and he believed 'the laws of combination of inorganic nature' are the only guides to organic nature.

This happy state of convergence was not to last for long, however, as it soon became apparent that attempts to interpret organic chemistry in terms of what we should call today polarities or ions simply did not work. At the same time organic chemists were 'doing their own thing' in syntheses and did not need the help of their inorganic colleagues. So, once again, there was a divergence, lasting for perhaps 15 or 20 years till the middle 1850s. When A. W. Hofmann loudly declared 'there is but one chemistry' in the early 1850s he was not describing the state of affairs at that time but rather presenting an idealised situation which would be much more likely to receive government support.

The fourth phase of this mercurial relationship began shortly after this, when both branches were enriched by the discovery of valency and from that the concept of structure (which had never existed before). So now we had a degree of integration which lasted till almost the end of the 19th century. A phrase describing organic chemistry, 'biological modes of thought in a physical science',[177] may be particularly apt here.

By 1900 strains were beginning to show, as the hitherto unifying concept of valency was unable to cope with many new discoveries in all branches, from aromatic to co-ordination chemistry. And there was now a third complication: the physical chemists had appeared and required attention. So, yet again, the organic chemists tended to find solace in each other's company and avoid the distraction of the other branches. In some universities physical chemists were in a separate department, and Perkin's alleged remark says it all: 'physical chemistry is all very well, but it does not apply to organic substances'! There was indeed, for the most part, little that a physical measurement could do for an organic chemist apart from determining molecular weight. This phase was from about the end of the 19th century.

The phase of divergence came to an end in about the 1950s. By now it really was apparent that physical measurements could be of immense importance; the ideas of a few enthusiasts for MO theory were at last seen as plausible, and the old rivals inorganic and organic chemistry had come together in organometallic chemistry which, by almost any criterion, emerged as a discipline in its own right in the 1960s. The rise of modern theoretical organic chemistry is an indicator of the new spirit of convergence. In matters of synthesis the question of 'biological modes of thought' has re-emerged, with the evocative phrase 'bioinspired chemistry'.[178]

If this analysis[179] has anything to commend it, a brief reflection on the history of chemistry will indicate that it was when organic chemistry was most closely related to the other branches that it was most successful. Perhaps this is why several opportunities were missed when ideas 'born before their time' were not taken up for decades. Such was the case of Sachse's non-planar cyclohexane rings[41] and Sidgwick's ideas on reaction mechanisms.[137] Equally, when chemistry has been most in touch with other disciplines everyone has had most to gain. This applies to physics, with the rise of materials science, to medicine, with the explosion in pharmaceutical chemistry, and to the life sciences with molecular biology being of supreme importance.

Therefore, if any lesson is to be learnt from 'this strange, eventful history' it is surely that organic chemistry will most thrive in association with others in science. In the light of a recent discussion at the Royal Society of Chemistry Organic Division it could hardly be more timely. Many other chemical communities will be facing similar questions. In Britain, the issue was 'the need for a separate organic community within the RSC'.[180] Outside the laboratory there is hardly a more important chemical question to be faced.

References

1 J. J. Berzelius, *Jahres-Bericht über die Fortschritte Physischen Wissenschaften, 1822–1848.*
2 C. Schorlemmer, *The Rise and Development of Organic Chemistry*, Macmillan, London, 1879, rev. ed. (ed A. Smithells), 1894.
3 C. Graebe, *Geschichte der organischen Chemie*, Springer, Berlin, 1920.
4 J. R. Partington, *A History of Chemistry*, vol. IV, Macmillan, London, 1964.
5 H. M. Leicester, *Source Book in Chemistry, 1900–1950*, Harvard University Press, Cambridge, MA, 1968.
6 *Essays on the History of Organic Chemistry*, ed. J. G. Traynham, Louisiana State University Press, Baton Rouge, 1987; D. S. Tarbell and T. Tarbell, *Essays on the History of Organic Chemistry in the United States, 1875–1955*, Folio Publishers, Nashville, TN, 1986.
7 *Recent Developments in the History of Chemistry*, vol. I, ed. C. A. Russell, Royal Society of Chemistry, Cambridge, 1985; vol. II, ed. G. K. Roberts and C. A. Russell, Royal Society of Chemistry, Cambridge, 2005, in press.
8 S. Lundgren, *Berzelius och den Kemiska Atomteorin*, Almqvist & Wiksell, Uppsala, 1979.
9 A. B. Costa, *Michel Eugene Chevreul, Pioneer in Organic Chemistry*, University of Wisconsin, Madison, 1962.
10 W. H. Brock, *Justus von Liebig: the Chemical Gatekeeper*, Cambridge University Press, Cambridge, 1997.
11 A. J. Rocke, *The Quiet Revolution: Hermann Kolbe and the Science of Organic Chemistry*, University of California Press, Berkeley, 1993.
12 A. J. Rocke, *Nationalizing Science: Adolphe Wurtz and the Battle for French Chemistry*, MIT Press, Cambridge MA, 2001.
13 *Die Allianz von Wissenschaft und Industrie: August Wilhelm Hofmann (1818–1892)*, ed. C. Meinel and H. Scholz, VCH, Weinheim, 1992.
14 C. A. Russell, *Edward Frankland: Chemistry, Controversy and Conspiracy in Victorian England*, Cambridge University Press, Cambridge, 1996 (pbk, 2003).
15 *The Kekulé Riddle*, ed. J. H. Wotiz, Cache River Press, Vienna, IL, 1993.
16 C. A. Russell, in *The Kekulé Riddle*, ed. J. H. Wotiz, Cache River Press, Vienna, IL, 1993, p. 153.
17 J. B. Cohen, *Annu. Rep. Prog. Chem.*, 1904, **1**, 116–119.
18 E. L. Hirst, *J. Soc. Chem. Ind.*, 1933, **52**, 224; W. N. Haworth, *J. Soc. Chem. Ind., 482.*
19 W. N. Haworth and E. L. Hirst, *J. Soc. Chem. Ind.*, 1933, **52**, 645.
20 J. Hvoslef, *Acta Chem. Scand.*, 1964, **18**, 841.
21 L. Doub and J. M. Vandenbelt, *J. Am. Chem. Soc.*, 1947, **69**, 2714; L. Doub and J. M. Vandenbelt, *J. Am. Chem. Soc.*, 1949, **71**, 2414.
22 R. B. Woodward, *J. Am. Chem. Soc.*, 1950, **72**, 3327.
23 W. J. Lehmann, *Atomic and molecular structure: the development of our concepts*, Wiley, New York, 1972, 397.
24 A. D. Cross and R. A. Jones, *An introduction to practical infra-red spectroscopy*, Butterworths, London, 3rd edn., 1969.
25 G. A. Newman, *Annu. Rep. Prog. Chem., Sect. B, Org. Chem.*, 1975, **72**, 29.
26 S. H. Pine, *J. Chem. Educ.*, 1972, **49**, 1664.
27 M. Jarman, *Annu. Rep. Prog. Chem., Sect. B, Org. Chem.*, 1981, **78**, 3.
28 G. B. Gill and S. A. Whiting, *Annu. Rep. Prog. Chem., Sect. B, Org. Chem.*, 1989, **86**, 1.
29 W. H. Perkin, *J. Chem. Soc.*, 1894, **65**, 572.
30 W. H. Perkin, W. J. Pope and O. Wallach, *J. Chem. Soc.*, 1909, **95**, 1785.
31 P. Maitland and W. H. Mills, *J. Chem. Soc.*, 1935, 987.
32 W. H. Mills and C. R. Nodder, *J. Chem. Soc.*, 1921, **119**, 2094.
33 W. H. Mills and E. H. Warren, *J. Chem. Soc.*, 1925, **127**, 2507.
34 G. J. Burrows and E. E. Turner, *J. Chem. Soc.*, 1921, **119**, 426.
35 I. G. M. Campbell, *J. Chem. Soc.*, 1955, 3116.
36 F. G. Mann and F. G. Holliman, *J. Chem. Soc.*, 1943, 551; F. G. Mann and W. C. Davies, *J. Chem. Soc.*, 1944, 276; F. G. Mann and F. G. Holliman, *J. Chem. Soc.*, 1945, 45.
37 H. B. Watson, *Modern Theories of Organic Chemistry*, Oxford University Press, Oxford, 1949, 215.
38 M. Crawford and I. F. B. Smyth, *J. Chem. Soc.*, 1952, 4133.
39 E. Harnik, F. H. Herbstein, G. M. J. Schmidt and F. L. Hirshfield, *J. Chem. Soc.*, 1954, 3288.
40 M. S. Newman, W. B. Lutz and D. Lednicer, *J. Am. Chem. Soc.*, 1955, **77**, 3420.
41 H. Sachse, *Ber.*, 1888, **21**, 2530; H. Sachse, *Ber.*, 1980, **23**, 1363.
42 E. Mohr, *Chem. Zeit.*, 1915, **2**, 1065; E. Mohr, *J. Prakt. Chem.*, 1918, **98**, 315; E. Mohr, *J. Prakt. Chem.*, 1922, **103**, 316.
43 W. Hückel, *Annalen*, 1925, **441**, 1.
44 C. A. Russell, in *van't Hoff – le Bell Centennial*, ed. O. B. Ramsay, American Chemical Society, Washington DC, 1975, p. 159.
45 D. H. R. Barton, *J. Chem. Educ.*, 1973, **50**, 235.

28

46 O. Hassel, *Tidsskr. Kjemi*, 1943, **3**, 32.
47 D. H. R. Barton, *Experientia*, 1950, **6**, 316.
48 W. Klyne, *Stereochemical correlations*, Royal Institute of Chemistry Lecture Series, 1962, no. 4.
49 J. M. Bijvoet, A. F. Peerdeman and A. J. van Bommel, *Nature*, 1951, **168**, 271.
50 K. Freudenberg, *Ber.*, 1933, **66**, 177.
51 W. Moffitt, R. B. Woodward, A. Moscowitz, W. Klyne and C. Djerassi, *J. Am. Chem. Soc.*, 1961, **83**, 4013.
52 L. Velluz, M. Legrand and M. Grosjean, *Optical circular dichroism*, Weinheim, Verlag Chemie, 1965.
53 P. M. Scopes, *Annu. Rep. Prog. Chem., Sect. B, Org. Chem.*, 1974, **71**, 35.
54 C. A. Russell, *Ambix*, 1987, **34**, 169.
55 A. W. Stewart, *Recent advances in organic chemistry*, 3rd ed., Longmans Green, London, 1918, p. 1.
56 A. W. Stewart, *Recent advances in organic chemistry*, 6th ed., vol. I, Longmans Green, London, 1931, p. 406.
57 J. H. P. Utley, *Annu. Rep. Prog. Chem., Sect. B, Org. Chem.*, 1968, **65**, 231.
58 O. Diels and K. Alder, *Ann.*, 1928, **460**, 98.
59 P. A. Chaloner, *Annu. Rep. Prog. Chem., Sect. B, Org. Chem.*, 1986, **83**, 263.
60 V. Grignard, *Compt. Rend.*, 1900, **130**, 1322.
61 C. V. Courtot, *Bull. Soc. Chim. France*, 1936, **3**, 1433.
62 C. R. Noller, *Chemistry of Organic Compounds*, W. B. Saunders, Philadelphia, 1965, 125.
63 C. G. Screttas and J. F. Eastham, *J. Am. Chem. Soc.*, 1965, **87**, 3276.
64 F. C. Taylor and A. McKillop, *Acc. Chem. Res.*, 1970, **3**, 338.
65 E. J. Corey et al, *J. Am. Chem. Soc.*, 1967, **89**, 2755–2757.
66 E. W. Abel, *Q. Rev. Chem. Soc.*, 1963, **17**, 133.
67 E. W. Abel and F. G. A. Stone, *Q. Rev. Chem. Soc.*, 1969, **23**, 325.
68 M. Bochmann, R. A. Head and M. D. Johnson, *Annu. Rep. Prog. Chem., Sect. B, Org. Chem.*, 1982, **79**, 250.
69 H. Gilman, *Adv. Organomet. Chem.*, 1968, **7**, 4.
70 F. G. A. Stone, *Leaving no stone unturned: pathways in organometallic chemistry*, Profiles, Pathways and Dreams Series, American Chemical Society, Washington DC, 1993.
71 A. N. Nesmeyanov, *Adv. Organomet. Chem.*, 1972, **10**, 1.
72 D. Seyferth, *Organometallics*, 2002, **21**, 1520–2800.
73 D. Seyferth, *Organometallics*, 2003, **22**, 2.
74 D. Seyferth, *Organometallics*, 2003, **22**, 2346–5254.
75 A. G. Davies and P. J. Garratt, *Annu. Rep. Prog. Chem., Sect. B, Org. Chem.*, 1983, **80**, 2.
76 G. B. Gill and D. A. Whiting, *Annu. Rep. Prog. Chem., Sect. B, Org. Chem.*, 1989, **86**, 2.
77 G. Wittig and G. Geissler, *Liebig's Ann. Chem.*, 1953, **580**, 44.
78 P. J. Murphy and J. Brennan, *Chem. Soc. Rev.*, 1988, **17**, 1.
79 S. Trippett, *Q. Rev. Chem. Soc.*, 1963, **17**, 406.
80 H. C. Brown, *Hydroboration*, Benjamin, New York, 1972.
81 A. Pelter, *Chem. Soc. Rev.*, 1982, **11**, 191.
82 H. W. Post, *Silicones and other Organosilicon Compounds*, Reinhold, New York, 1949, p. 13.
83 E. G. Rochow, *An Introduction to the Chemistry of the Silicones*, 2nd ed., Chapman & Hall, London, 1951; G. G. Freeman, *Silicones: an Introduction to their Chemistry and Applications*, Iliffe, London, 1962, p. 15.
84 D. Seyferth, *Organometallics*, 2001, **20**, 4978.
85 E. W. Colvin, *Chem. Soc. Rev.*, 1978, **7**, 15.
86 I. Fleming, *Chem. Soc. Rev.*, 1981, **10**, 83.
87 M. Renz and B. Meunier, *Eur. J. Org. Chem.*, 1999, 737.
88 H. G. Heal, *Recent studies in boron chemistry*, Royal Institute of Chemistry Lectures, Monographs and reports, 1960, no. 1.
89 P. Maitland and W. H. Mills, *Nature*, 1935, **135**, 994; P. Maitland and W. H. Mills, *J. Chem. Soc.*, 1936, 987.
90 V. Prelog, *Helv. Chim. Acta*, 1953, **36**, 308.
91 D. J. Cram and F. A. A. Elhavez, *J. Am. Chem. Soc.*, 1953, **74**, 5828.
92 H. Nozaki, S. Moriuti, H. Takaya and R. Noyori, *Tetahedron Lett.*, 1966, 5239.
93 N. S. Simpkins, *Chem. Soc. Rev.*, 1990, **19**, 335.
94 J. A. Joule and C. I. F. White, *Annu. Rep. Prog. Chem., Sect. B, Org. Chem.*, 1994, **91**, 1.
95 A. J. Shuker, M. G. Siegel, D. O. Matthews and L. O. Weigel, *Tetrahedron Lett.*, 1997, **38**, 6149.
96 G. B. Gill and D. A. Whiting, *Annu. Rep. Prog. Chem., Sect. B, Org. Chem.*, 1990, **87**, 1.
97 R. Willstäter and H. Veraguth, *Ber.*, 1905, **38**, 1975.
98 R. Willstäter, *Ber.*, 1911, **44**, 3428.
99 L. Ruzicka, M. Stohl and H. Schinz, *Helv. Chim. Acta*, 1926, **9**, 249; L. Ruzicka and W. Brugger, *Helv. Chim. Acta*, 1926, **9**, 339 and 389.
100 K. Ziegler, H. Eberle and H. Ohlinger, *Annalen*, 1933, **504**, 94.

101 C. O. Dietrich-Buchecker and J. P. Sauvage, *Chem. Rev.*, 1987, **87**, 795.
102 D. D. MacNicol, J. J. McKendrick and D. R. Wilson, *Chem. Soc. Rev.*, 1978, **7**, 65.
103 C. J. Pedersen, *J. Am. Chem. Soc.*, 1967, **89**, 2495–7017.
104 J. F. Stoddart, *Annu. Rep. Prog. Chem., Sect. B, Org. Chem.*, 1983, **80**, 353.
105 J. A. Joule and C. I. F. Watt, *Annu. Rep. Prog. Chem., Sect. B, Org. Chem.*, 1994, **91**, 2.
106 R. G. W. Norrish and F. W. Kirkbride, *J. Chem. Soc.*, 1933, 119; R. G. W. Norrish, H. G. Crone and O. Saltmarsh, *J. Chem. Soc.*, 1933, 1533.
107 A. Ledwith, *The chemistry of carbenes*, Royal Institute of Chemistry Lecture Series, London, 1964.
108 W. v. E. Doering and A. K. Hoffmann, *J. Am. Chem. Soc.*, 1954, **76**, 6162.
109 E. Jenny and J. D. Roberts, *Helv. Chim. Acta*, 1955, **38**, 1248; J. D. Roberts, D. A. Semenow, H. E. Simmons, jun. and L. A. Carlsmith, *J. Am. Chem. Soc.*, 1956, **78**, 601; J. D. Roberts, C. W. Vaughan, L. A. Carlsmith and D. A. Semenow, *J. Am. Chem. Soc.*, 1956, **78**, 611.
110 S. Miller, J. A. Tebboth and J. F. Tremaine, *J. Chem. Soc.*, 1952, 632.
111 P. Pauson and T. Kealy, *Nature*, 1951, **168**, 1039.
112 G. Wilkinson, M. Rosenblum, M. C. Whiting and R. B. Woodward, *J. Am. Chem. Soc.*, 1952, **74**, 2125.
113 H. G. Viehe, R. Merényi, J. F. M. Oth and P. Valange, *Angew. Chem., Int. Ed. Engl.*, 1964, **3**, 746; H. G. Viehe, R. Merényi, J. F. M. Oth, J. R. Senders and P. Valange, *Angew. Chem., Int. Ed. Engl.*, 1964, **3**, 755.
114 G. B. Gill and D. A. Whiting, *Annu. Rep. Prog. Chem., Sect. B, Org. Chem.*, 1989, **86**, 1.
115 W. N. Haworth, *Nature*, 1925, **116**, 430.
116 R. J. Ferrier and W. G. Overend, *Q. Rev. Chem. Soc.*, 1959, **13**, 265.
117 E. J. Honeyman, *Annu. Rep. Prog. Chem.*, 1961, **58**, 335; E. J. Honeyman, *Annu. Rep. Prog. Chem.*, 1962, **59**, 359.
118 (*a*) V. E. Gilbert, F. Smith and M. Stacey, *J. Chem. Soc.*, 1946, 622; (*b*) R. U. Lemieux and G. Huber, *J. Am. Chem. Soc.*, 1953, **75**, 4118; (*c*) A. W. Stewart, *Recent advances in organic chemistry*, 6th ed., vol. II, Longmans Green, London, 1931, p. 89.
119 P. Karrer and R. Morf, *Helv. Chim. Acta*, 1931, **14**, 1033.
120 C. H. Eugster and P. Karrer, *Helv. Chim. Acta*, 1950, **33**, 1172.
121 R. Willstätter and A. E. Everest, *Annalen*, 1913, **410**, 189.
122 (*a*) R. Willstätter and H. Mallison, *Sitzungsber. K. Akad. Wiss. Berlin*, 1914, 769; (*b*) R. Robinson and A. Todd, *J. Chem. Soc.*, 1932, 2494.
123 R. W. Woodward, *Pure Appl. Chem.*, 1961, **2**, 383.
124 W. A. Ayer, J. M. Beaton, L. D. Celmer, B. L. Shapiro and R. B. Woodward, *J. Am. Chem. Soc.*, 1960, **82**, 3800.
125 C. Tanford and J. Reynolds, *Chem. Intelligencer*, 1999, **5**(3), 24.
126 F. Sanger and E. G. P. Thompson, *Biochem. J.*, 1953, **53**, 353–366.
127 M. F. Perutz, M. G. Rossmann, A. F. Cullis, H. Muirhead, G. Will and A. C. T. North, *Nature*, 1960, **185**, 416.
128 J. C. Kendrew, R. E. Dickerson, B. E. Strandberg, R. G. Hart, D. R. Davis, D. C. Phillips and V. C. Shore, *Nature*, 1960, **185**, 422.
129 B. Gutte and R. B. Merrifield, *J. Am. Chem. Soc.*, 1969, **91**, 501.
130 C. A. Russell, *Nat. Prod. Rep.*, 1987, **4**, 47.
131 (*a*) F. J. K. Thiele, *Annalen*, 1899, **306**, 89; (*b*) J. Shorter, *Ambix*, 2003, **50**, 274.
132 C. K. Ingold, *Nature*, 1955, **176**, 191.
133 J. W. Baker, *Tautomerism*, Routledge, London, 1934, p. 1.
134 L. Knorr, *Ber.*, 1911, **44**, 1138–2767.
135 (*a*) M. C. King, *Time and chemical change*, Open University PhD thesis, 1980; (*b*) J. Lapworth, *J. Chem. Soc.*, 1903, **83**, 995.
136 J. Shorter, *Chem. Soc. Rev.*, 1998, **27**, 355.
137 P. Laszlo, *Ambix*, 2003, **50**, 261.
138 N. V. Sidgwick, *The Organic Chemistry of Nitrogen*, Oxford University Press, Oxford, 1910.
139 J. Shorter, *Chemtech*, 1985, 252.
140 (*a*) R. Robinson, *Memoirs of a minor prophet*, Elsevier, Amsterdam, 1976; T. I. Williams, *Robert Robinson, chemist extraordinary*, Clarendon Press, Oxford, 1990.
141 M. D. Saltzmann, *Nat. Prod. Rep.*, 1987, **4**, 53.
142 (*a*) W. O. Kermack and R. Robinson, *J. Chem. Soc.*, 1922, **121**, 427; (*b*) R. Robinson in J. V. Eyre, *Henry Edward Armstrong, 1848–1937*, Butterworth, London.
143 K. T. Leffek, *Sir Christopher Ingold: a Major Prophet of Organic Chemistry*, Nova Lion Press, Victoria, 1996.
144 J. Shorter, *Nat. Prod. Rep.*, 1987, **4**, 61.
145 I. Eilks and J. Friedrich, *Prax. Naturwiss. Chem.*, 1999, **48**, 29.
146 T. W. Bentley, *Annu. Rep. Prog. Chem., Sect. B, Org. Chem.*, 1974, **71**, 111.
147 A. R. Katritzky and B. E. Brucki, *Chem. Soc. Rev.*, 1990, **19**, 83.
148 F. M. Akeroyd, *Found. Chem.*, 2000, **2**, 99.

149 C. K. Ingold, *Structure and Mechanism in Organic Chemistry*, G. Bell, London, 1953.
150 R. Robinson, *Annu. Rep. Prog. Chem.*, 1923, **20**, 112.
151 M. Godfrey, *Annu. Rep. Prog. Chem., Sect. B, Org. Chem.*, 1986, **83**, 20.
152 L. Pauling, *The Nature of the Chemical Bond*, Cornell University Press, Ithaca, NY, 1940.
153 B. S. Park, *Br. J. Hist. Sci.*, 1999, **32**, 21.
154 M. J. S. Dewar, *Electronic Theory of Organic Chemistry*, Clarendon Press, Oxford, 1949.
155 M. J. S. Dewar, *A Semiempirical Life*, Profiles, Pathways and Dreams Series, American Chemical Society, Washington, DC., 1992.
156 R. McWeeny, *Annu. Rep. Prog. Chem.*, 1961, **58**, 138.
157 S. J. Weininger, *Bull. Hist. Chem.*, 2000, **25**, 123.
158 R. Breslow, J. T. Groves and G. Ryan, *J. Am. Chem. Soc.*, 1967, **89**, 5048; D. G. Farnum, G. Mehta and R. G. Silberman, *J. Am. Chem. Soc., 5048.*
159 C. K. Ingold, *Structure and Mechanism in Organic Chemistry*, G. Bell, London, 1953, p. 206.
160 M. Polanyi, E. Horn and D. W. G. Style, *Trans. Faraday Soc.*, 1934, **30**, 179.
161 W. A. Waters, *The Chemistry of Free Radicals*, Clarendon Press, Oxford, 1946.
162 J. I. G. Cadogan, D. H. Hey and G. H. Williams, *J. Chem. Soc.*, 1955, 1425.
163 *Vistas in Free-radical Chemistry*, ed. W. A. Waters, Pergamon Press, London, 1959.
164 M. S. Kharasch and F. R. Mayo, *J. Am. Chem. Soc.*, 1933, **55**, 2468–2531.
165 F. R. Mayo, in *Vistas in Free-radical Chemistry*, ed. W. A. Waters, Pergamon Press, London, 1959, p. 139.
166 D. H. Hey and W. A. Waters, *Chem. Rev.*, 1937, **21**, 202.
167 M. S. Kharasch, F. Engelmann and F. R. Mayo, *J. Org. Chem.*, 1937, **2**, 288.
168 J. I. G. Cadogan, *Recent Developments in the Chemistry of Free-radical Addition Reactions*, Royal Institute of Chemistry Lecture Series, 1961, no. 6.
169 R. S. Cooke and G. S. Hammond, in *Essays on Free-radical Chemistry*, ed. R. O. C. Norman, Chemical Society Special Publication no. 24, London, 1970, p. 1.
170 D. Griller, *Annu. Rep. Prog. Chem., Sect. B, Org. Chem.*, 1985, **82**, 75.
171 D. H. R. Barton, (with S. I. Parekh), *Half a Century of Free Radical Chemistry*, Cambridge University Press, Cambridge, 1993.
172 C. Walling, *Fifty Years of Free Radicals*, Profiles, Pathways and Dreams Series, American Chemical Society, Washington, DC., 1995.
173 J. H. Ridd, *Chem. Soc. Rev.*, 1991, **20**, 149.
174 P. E. Verkade, *A History of the Nomenclature of Organic Chemistry*, Reidel, Dordrecht, 1985.
175 *Chemistry, Society and Environment: a new History of the British Chemical Industry*, ed. C. A. Russell, Royal Society of Chemistry, Cambridge, 2000 (chapters 8 and 9).
176 Lithium Corporation of America, 'Annotated Bibliographies', from 1949.
177 M. J. Nye, *Before Big Science: the Pursuit of Modern Chemistry and Physics, 1840–1940*, Twayne, New York, 1996, p. 120.
178 B. A. Armitage, *Annu. Rep. Prog. Chem., Sect. B, Org. Chem.*, 2000, **96**, 187.
179 C. A. Russell, *The Structure of Chemistry*, Units 1–3 of Course S304, Open University, Milton Keynes, 1976.
180 R. Jones, *Organic News, Summer*, 2004, 1.

XVIII

Chemical Bonds 1841—1991: 150 Years of the British Chemical Community*

1 Introduction

The Chemical Society existed for 20 years before one of its members (Edward Frankland) introduced the novel idea of a 'chemical bond'.[1] Although this concept eventually became central to the science of chemistry it was at first received with a mixture of scepticism (by people like Kolbe) and amusement (by leading Chemical Society fellows in the 1870s).[2] The subsequent development of this suggestion led to its adoption as one of the central themes of chemistry. However, it is not with such bonds that I am concerned today. Instead I would like to draw your attention to a different kind of bonding, to those equally intangible, but nevertheless real, bonds of social companionship and common cause that unite the community of chemists, and have united it for 150 years in this country.

Today we celebrate the sesquicentenary of the Chemical Society,[3] and you have given me the great privilege and responsibility of delivering this first plenary lecture. Fifty years ago our mono-centenary passed almost unnoticed during the dark wartime days of 1941. But 100 years ago the Jubilee *was* celebrated, with typically Victorian abandon. A Jubilee Book[4] was compiled to record the great and the good of the first 50 years. Festivities commenced with a eulogy by the President of the Society, W. J. Russell, and concluded with another Victorian custom: a dinner of gargantuan proportions (Figure 1). One feels it is not only the 18th century that can be justly called 'the age of indigestion'.

So 100 years after the last major celebration nothing would be easier for me than to launch into a fulsome, congratulatory account of the Society's remarkable achievements (and they have been truly remarkable) over the last 100 or even 150 years, and leave it at that. Yet such a eulogy would be devoid of analysis, and

* This is a slightly expanded version of the Plenary Lecture delivered on 8th April 1991 at the Opening Session of the 150th Anniversary Annual Chemical Congress of the Royal Society of Chemistry at Imperial College London.

1 E. Frankland, *J. Chem. Soc.*, 1866, **19**, 377.

2 C. A. Russell, 'The History of Valency', Leicester University Press, 1971.

3 General accounts of the history of the Royal Society of Chemistry or its predecessors include: (*a*) T. S. Moore and J. C. Philip, 'The Chemical Society 1841—1941', Chemical Society, London, 1947. (*b*) R. Bud, *Chem. Brit.*, 1991, **27**, 230. (*c*) C. A. Russell, N. G. Coley, and G. K. Roberts, 'Chemists by Profession', Open University Press/Royal Institute of Chemistry, Milton Keynes, 1977. (*d*) D. W. Whiffen, 'The Royal Society of Chemistry: The First 150 Years', Royal Society of Chemistry, London, 1991.

4 'The Jubilee of the Chemical Society of London, 1891', Chemical Society, London 1896.

Chemical Society Reviews 20, 1991, 425–40. Reproduced by permission of the Royal Society of Chemistry.

THE DINNER.

MENU.

Huîtres au Citron.

Chablis.

POTAGES.
Consommé à la d'Orléans. Crême d'Orge à l'Allemande.

Dry Sherry.

POISSONS.
Saumon, Sauce Hollandaise. Eperlans Frits, Sauce Tartare.

Niersteiner.

ENTRÉES.
Médaillons de Volaille à la Princesse. Ris de Veau à la Provençale.

G. H. Mumm & Co. Extra Quality, Extra Dry.

RELEVÉS.
Selle de Mouton de Galles. Jambon Braisé au Champagne.

Deutz and Geldermann's Gold Lack.

LÉGUMES.
Haricots Verts Sautés. Pommes de Terre Dauphine.

RÔTI.
Cailles Rôties au Cresson.
Salade.

ENTREMETS.
Pouding à la Rossini. Crême Rubanée.

Fine Champagne Liqueur Brandy.

Canapés Norvégiens.

Bombe à la Vanille. Gâteaux Condé.

DESSERT.

Cantenac. Cockburn's Old Bottled Port.

Café Noir.

Johannis Natural Mineral Water.

Figure 1 *Menu for the Jubilee Dinner, 1891*

therefore unworthy of the occasion, the Society, and this distinguished audience. In the belief that the past has lessons for the future, I want to look first at what actually happened in 1841, to draw some general conclusions, and to see what relevance these may have for the later and possibly future adventures of the Society.

2 Founders of The Chemical Society, 1841

So, what were these Victorian chemists up to 100 years ago? In a word, *they were celebrating the formation of the world's first national chemical society.* True, in Britain there had been a handful of student societies, a few short-lived chemical clubs and one abortive attempt in 1824 to establish a 'London Chemical Society'.[5] But until 1841 the only permanent home for chemists had been beneath the comprehensive umbrella of the Royal Society. There had, it is true, been a gathering of chemists in Section B of the British Association for the Advancement of Science. Much later (1860) this led to the formation of the 'B-Club', an informal, convivial, and light-hearted gathering, often at picnics in the Surrey countryside (chemists have always been sociable creatures). It is possible that the scientific meetings of Section B 'suggested the formation of a formal and central society ten years later'.[6] However that may be, on 23 February 1841, at the Society of Arts in the Adelphi, London, 'a meeting was held of gentlemen desirous of uniting themselves for the purpose of forming a Chemical Society'. But who were these men, and what brought them together?

They were a mixed bunch, and they can be classified in all kinds of ways. *Socially* they were a mixture of fairly well-to-do manufacturers or entrepreneurs, a few poorer chemists struggling to make a living (like the artisans who tried unsuccessfully to unite in 1824), and some of a new and rising class who have been called 'professionals' because they were paid (often quite generously) to act as consultants, expert witnesses, *etc.*[7] Or one can divide them *geographically*, in which case most were in London, though Cumming was Professor at Cambridge and Graham had recently come from Scotland. Moreover, of the first 77 members Scotland had the largest share (14) after London (40). Nevertheless one potential recruit declined to join on the grounds of his distance from London; he lived in Islington!

A better division is *functionally*, according to their occupations. Some were industrialists, like John Mercer, a calico printer, who wrote to the promoters, saying that such a society was 'much wanted'. But most were academics like C. G. B. Daubeny of Oxford or W. H. Miller from Cambridge (both later Presidents). All except a handful are forgotten now. Let us look at the 14 founding committee members, when it will be apparent that many of them did more than one thing in any case (Table 1).

Of these 14 men, eight taught in some capacity. Furthest from the metropolis was the Rev. James Cumming, a relatively little-known professor at Cambridge. Associated with the new University of London were Thomas Graham (since 1837 at University College and to become the Society's first President), and John Daniell (since 1831 at King's College London). Another London teacher was William Brande who worked for 40 years at the Royal Institution; of him it was said that his emphasis on teaching rather than research 'reflected the labours

[5] W. H. Brock, *Ambix*, 1967, **14**, 133; also reference 3(*c*), p. 56.
[6] A. Scott, *J. Chem. Soc.*, 1916, **109**, 342.
[7] R. Bud and G. K. Roberts, 'Science versus Practice: Chemistry in Victorian Britain', Manchester University Press, Manchester, 1984.

Table 1 *Members of the first Committee of The Chemical Society, 1841*

A. Aikin (1773—1854)	T. Graham (1805—1869)
W. T. Brande (1788—1866)	W. R. Grove (1811—1896)
H. J. Brooke (1771—1857)	H. Hennell (?—1842)
J. T. Cooper (1790—1854)	G. Lowe (1788—1868)
J. Cumming (1777—1861)	R. Phillips (1778—1851)
J. F. Daniell (1790—1845)	R. Porrett (1783—1868)
T. Everitt (1803—1846)	R. Warington (1807—1867) (*Hon. Sec.*)

necessary to keep a wife and family in style'. William Grove taught physics at the London Institution, a kind of poor relation to the Royal Institution. Thomas Everitt contrived to be Professor of Chemistry simultaneously at Putney College of Engineering and Middlesex Hospital. Others who lectured from time to time were Cooper and Phillips.

Four of these chemists were committed to research. Daniell and Grove were especially active in electrochemistry (though the latter is alleged to have said that science ceases to be interesting as soon as it becomes applied). Thomas Graham will always be remembered for his researches in gaseous diffusion,[8] while Robert Porrett as chief clerk in the Ordnance Department at the Tower of London worked on explosives, electroendosmosis, and complex iron cyanides.

Not all members of this committee shared Grove's dismissive view of applied chemistry. At least five of them were engaged in manufactures (usually as employees). Explosives manufacture occupied Porrett, and also (at the more surprising location of Apothecaries' Hall) Henry Hennell, who eventually blew himself up with 3 kilos of mercury fulminate that he was making for the Afghan War. George Lowe worked for the Gas Light & Coke Company in the days when engineers like himself were required to engage in all manner of chemical operations connected with the production and purification of coal gas. Henry Brooke was a wealthy manufacturer, and the variegated career of John Cooper had included a spell of chemical production.

Finally it must be stressed that probably half of the Committee also enjoyed from time to time the new rôle of consultant. In addition to Brande and other academics, their numbers included Arthur Aikin of Guy's Hospital, who was also Secretary to the Society of Arts and became Treasurer of the Chemical Society and its second President. Cooper and Phillips frequently combined lucrative consultancy work with their other activities, Cooper also selling chemicals and apparatus, as well as giving popular lectures, while Phillips was chemist at the Museum of Economic Geology.

The four sets of overlapping activities marked the founding members of the Chemical Society and go some way to offering a first explanation for its formation.

[8] R. J. H. Clark, *Chem. Soc. Rev.*, 1991, **20**, 405.

3 Strengthening the Bonds

The Chemical Society was thus a coming together of people with widely differing interests. But why did they meet in the first place? A clue is provided by Robert Warington, first Secretary of the Society:

> 'Such a Society was much needed, not only to break down the party spirit and petty jealousies which existed, but to bring Science and practice into closer communication, and to bring the experience of many to bear in discussing the same subject.'[9]

What does that really say? Beneath the rhetoric is the unspoken assumption that unity is a good thing in itself, that bonds between chemists *need* to be forged. But is that not obviously and always true? If so it is necessary to enquire what was special about 1841, and about the London of that time. To examine his words more closely is to peer below the surface of events and to begin to discern something of the underlying forces at work. Warington is in fact suggesting three separate reasons:

—*to break down the party spirit and petty jealousies*: that refers to fragmentation and internal disorder.

—*to bring Science and practice into closer communication*: this implies new external pressures involving industry ('practice') as well as chemistry ('Science').

—*to bring the experience of many to bear in discussing the same subject*: suggesting a new climate of 'togetherness', of sharing experiences, of institutional unity.

In the formation of the Chemical Society, and in all its subsequent history, these considerations seem to be of great importance. Amongst chemists the processes of institutional bond-making (and bond-breaking) are affected by all three elements of their situation:

(1) Internal disorder,
(2) External constraints,
(3) Social and cultural environment.

Somewhat tongue-in-cheek may I suggest an analogy to effects of entropy, temperature, and solvent? There is one other effect which Warington does not mention (and for a very good reason) which we may term catalytic. So we have additionally:

(4) An appropriate catalyst.

We can now see how they contributed to the events of 1841.

A. Internal Disorder.—I refer to the state of chemistry itself. For an understanding of modern chemistry an idea even more fundamental than the concept of a chemical bond is that of an atom. In 1841 Dalton was still alive, yet the atomism of the Cumbrian chemist was still being widely rejected forty years after its birth, for no one had yet seen an atom, and, despite Avogadro, no one knew how to measure its mass. One man, Berzelius, believed in atoms, and clothed them with electric polarities, positive for hydrogen and the metals and negative for all other

[9] Reference 3(*a*), p. 117.

elements. For this he was in deep trouble with the organic substitutions recently discovered in France, for how could a positive hydrogen atom be replaced by chlorine which must be negative? Moreover, if molecules were held together by combination between electrical opposites a diatomic molecule like H_2 was an impossibility.[10] At this very time a propaganda war was raging between those who (like Berzelius) believed chemical forces were fundamentally electrical in nature, and the majority of chemists who were not sure about that but were certain that chlorine could replace hydrogen. Molecular agnosticism was the order of the day, and remained so for many years to come. And with it went a degree of confusion that seems incomprehensible today. A classic illustration is the inclusion in his 'Lehrbuch der organischen Chemie' by Kekulé of no less than 19 alternative formulae for the simple substance acetic acid.[11] That was as late as 1859, just before the value of Avogadro's hypothesis was to be demonstrated by Cannizzaro[12] and a way out of the maze became clear.

In 1841 confusion reigned and, not surprisingly, the subject failed to grow. Thus Fox Talbot declined to join the new Society on the grounds that chemistry wasn't big enough on its own; why not include electricity as well? As if to prove the point, in 1841 not a single chemical paper graced the pages of the *Philosophical Transactions* of the Royal Society.

This malaise in chemistry, this state of internal disorder, might have been expected to fragment the chemical community still further, into sectarian parties and divisions. To some extent this was already happening. But similar causes can sometimes lead to opposite effects; in those circumstances everything depends on other factors which thus become critical. What is quite clear is that while deep ideological divisions remained between (for example) the followers of Berzelius and those of Gerhardt, in Britain the very possibility of deepening divisions caused the chemical community to come together in a quite new way.

Those who seek to 'explain' scientific growth or its lack solely in terms of external, often sociological, factors would do well to take a long, hard look at the internal state of chemistry itself. At the very heart of chemical theory lay one cogent reason for chemists to unite.

B. External Constraints.—Tipping the balance between fragmentation and association were several external influences which led the founders of the Chemical Society to see that a remedy for this internal disorder might lie in a new association between chemists.

First, there was the much-publicised visit to this country in 1837 of the German chemist Justus Liebig. Founder of the *Annalen* and Professor at the university at Giessen, he already had a European reputation. For, despite the smallness of his university, Liebig had created a great school of organic research. Moreover his visit coincided with a dawning awareness that English science was

[10] C. A. Russell, *Ann. Sci.*, 1963, **19**, 117, 127.
[11] F. A. Kekulé, 'Lehrbuch der organischen Chemie', Erlangen, 1861 (circulated in parts from 1859).
[12] J. Bradley, 'Cannizzaros Methode: der Schlüssel zur modernen Chemie', Franzbecker, Bad Salzdetfurth, 1990.

a very amateurish affair compared with that in Germany and promoted a wave of general discontent. His theories of agricultural chemistry were of great interest to the more enlightened landowners, and thus to the chemists who would serve them; the example of his research school at Giessen was to bring A. W. Hofmann to direct the new Royal College of Chemistry in London (1845), forerunner of Imperial College, London. Meanwhile Liebig's visit undoubtedly stimulated formation of the Chemical Society. Not surprisingly the first Foreign Member of the Society was Justus Liebig, whose bust to this day stands like a sentinel outside the Council Room at Burlington House.

Even more important than Liebig's coming was a new British phenomenon. This was the emergence of two groups who were already represented on that first Committee. For these men chemical knowledge was not a luxury but a necessity: the people Warington implied by his reference to 'practice'. First were those oddly assorted individuals who earned their living as consultants to a variety of enterprises, like Richard Phillips, who undertook soil analysis for landowners, W. T. Brande who fought against pollution of air and water, and J. T. Cooper who advised gas companies on the disposal of noxious by-products. Even at this time chemists were being called upon as expert witnesses in litigation, Michael Faraday being one of the most successful at this form of consultancy, earning several hundred pounds a year in this way.

Secondly there were manufacturers of pharmaceuticals, metals, and heavy chemicals. The early membership included several well-known names in this field. Among these was Hugh Lee Pattinson, the lead manufacturer who had in that very year discovered his method for extracting magnesia from dolomite and was about to found the famous Washington Chemical Company. Other manufacturers who were founder members are included in Table 2. All these men needed a wide

Table 2 *Some founder members engaged in chemical manufacture*

William J. Cock: London, platinum manufacture
Walter Crum: Glasgow, calico-printing, dyeing, and bleaching
John Joseph Griffin: Glasgow, chemical apparatus
John Mercer: Oakenshaw, calico-printing
Hugh Lee Pattinson; Gateshead, lead, alkali, and acid manufacture
Robert Porrett: London, explosives manufacture
Thomas Richardson: Newcastle-upon-Tyne, chemical manure manufacture
James Thomson: Clitheroe, calico-printing

acquaintance with chemistry, but how were they to get it? University training was not appropriate to men in mid-career, and even if it had been there was little undergraduate teaching in chemistry available in England, despite the recent advance of King's and University Colleges in London. In Scotland the situation was relieved by the presence in Glasgow of that renowned 'chemist breeder' Thomas Thomson.[13] Otherwise a would-be chemist had to go abroad, probably to Germany, and the visit of Liebig to Britain would certainly have given valuable ammunition to advocates of British chemical training. The need was urgent, the

[13] J. B. Morrell, *Ambix*, 1972, **19**, 1.

heat was on. Bond-making was becoming an economic as well as a cultural necessity. The same driving-forces led to the creation four years later of the Royal College of Chemistry in London. But in 1841 the Chemical Society was to become the embodiment of these hopes and of an academic tradition that was distinctively Scottish.

C. Social Context.—No organic chemist can underestimate the role of solvent in bond-breaking and bond-making processes. In a very similar way the bonding processes that led to the formation of the Chemical Society were sensitive to the social environment.

Imagine yourself in the London of the early 1840s. The Napoleonic Wars were over, the post-war depressions all too familiar and the whole country was in the throes of the Industrial Revolution (which, incidentally, would have been impossible without chemistry). And now a new voice was being heard, the voice of organized labour. Trade Unionism, once suppressed under the Combination Acts, was now legitimate and workers all over the country were *getting together*; if manual workers, why not chemists? One may profitably scan the list of other scientific groups founded near this time (Table 3). All these were in London and

Table 3 *Some scientific societies formed in London before* 1841

Spitalfields Mathematical Society	1788
Linnean Society	1788
Mineralogical Society	1799
Chalcographic Society	1803
Geological Society	1807
Astronomical Society	1820
Meteorological Society	1823
Zoological Society	1826
Entomological Society	1833
Botanical Society	1836
Microscopical Society	1839

must have constituted a powerful incentive for chemists to do the same. Nothing, however, concentrated the mind so much as the regrouping of those other people called chemists into what became on 15 April, 1841, the Pharmaceutical Society of Great Britain. Their action was a response to a newly perceived threat to their profession from some impending legislation. This was a reasonable move to protect the public by insisting on a proper registration of all who practised medicine and pharmacy. The problem lay in the proposed certificating body, where pharmaceutical chemists would be always outnumbered by the doctors. So a Pharmaceutical Society was formed 'for the purpose of protecting the permanent interests and increasing the respectability of Chemists and Druggists'. For the other chemists this was to prove a stimulating example. Only in the distant future was it to constitute a mild threat, and that arose from a clause in its Charter permitting only members of the Pharmaceutical Society to be called 'chemists'.

For now, however, its formation was typical of the era. Gone were the days when apologists for the Royal Society could complain of 'separatists who, under the plea of cultivating with more intensity and attention animal chemistry, tore themselves off from the bosom of their mother society and yet adhered still to its mangled remains'.[14] This was the hour for newly mature specialisms to have their own organizations.

Even that does not exhaust what might be said about the solvent effect, or social environment. Not only was *combination* the order of the day; so also was *communication.* London was poised on the brink of the railway revolution. 1841 saw the opening of both the Great Western and the London & Brighton Railways, while Euston's primitive shed was the point of departure for a journey to Glasgow that from 1841 could be reached within 24 hours (though admittedly a steamship voyage was necessary to get you from Lancashire to Scotland). And only in 1840 had the Penny Post been launched, making scientific and other communication immeasurably easier than before.

It is not hard to see how all these circumstances facilitated the establishment of bonds between chemists. But, as we all know, bond-formation can often be promoted in yet another way, by the right catalyst.

D. The Right Catalyst.—The catalyst for the Chemical Society was Robert Warington, which doubtless explains why he omitted to mention the fact. Robert Warington (1807—1867) had received an apprenticeship in the chemical industry and from 1831 to 1839 had served as chemist to a brewery. Shortly afterwards he became chemist to the Apothecaries' Society of London. However in the early 1840s he confessed that he had time on his hands, being in between jobs, and he became available at just the right moment to become honorary secretary to the committee setting up the new society, and then to its successors for twenty more years. No great chemist himself, he had, as all who knew him acknowledged, an extraordinary capacity for hard, detailed work, and a more than usual gift of patience. As W. J. Russell observed, 'Obviously some special force was required to build up this complicated molecule; that special force was embodied in and exercised by Robert Warington.'[15] We are all in his debt.

4 Crises to Come

And so, for all these reasons, the world's first national Chemical Society was born. It immediately instituted regular meetings, chiefly for reading and discussing papers, held at a variety of venues. It began collecting books and journals for a library, and in 1842 bought a bookcase. Its one short-lived enterprise was a chemical museum, with many curious exhibits, such as 'a specimen of butter, supposed to be 300 years old, discovered in a bog near Downpatrick'. Most importantly it created a new journal: *Memoirs & Proceedings* from 1841 to 1848, then 14 years of a *Quarterly Journal*, and from 1862 the

[14] Reference 3(*a*), p. 12.
[15] W. J. Russell, reference 4, p. 6.

Journal of The Chemical Society. The Faraday Lectureship was instituted in 1869, with its splendid Medal. For the Society's first 30 years membership increased apace and by 1870 was a healthy 551.

In the 150 years since its birth there have been four major crises (landmarks if you prefer) in which bond-making and, sadly, bond-breaking were prominent. In every case the same four principles apply, though in different combinations. The first major challenge arrived in the 1870s.

A. The Challenge of Professionalism.—Despite growth, from the late 1860s internal strains began to appear. This time they were not within chemical theory itself, for (despite lingering doubts about atomism[16]) the science was at last acquiring a new unity through the concepts of valency and structure. Instead the tensions were within the Society itself.

The troubles started as a challenge to the Council which had hitherto regarded as absolute its prerogative to nominate its successors. Now some impudent upstarts in the membership were prepared to exert their constitutional rights and propose alternatives. Worse still, the election of Fellows was being hindered by the process of 'black-balling', whereby a small minority at a meeting could deprive nominees of the necessary 75% majority to secure their election.

To us it all seems a storm in a teacup, a rather squalid power struggle best consigned to the dustbin of history. In fact the rise of the Institute was also a response to genuinely external constraints. It sprang from a growing recognition that membership of the Society was no guarantee of chemical competence, and the letters FCS much used by the Victorians were no indicator of professional status. Jumping on the bandwagon of chemical analysis were all manner of unqualified persons, quacks in fact, whose activities put in peril the jobs of competent analysts. Horror stories abounded. Much analysis of food, water, and air was also being undertaken by medical men whose training, though possibly excellent for its purpose, did not include chemical analysis. Chemical indignation knew no bounds when testing of water supplies and manufacture of gas were entrusted not to chemists but to engineers. Many chemists were now experiencing an uncomfortable rise in temperature and were endeavouring to bring it down; le Chatelier would have been pleased! The result was not a coming together in unity, but a fragmentation and the emergence of a new institution out of and alongside the Chemical Society, the Institute of Chemistry (1877), whose designatory letters would imply genuine chemical achievement. This was another 'first' for British chemists: the first professional *scientific* institute in the world.

There is another reason why, this time, internal strains led to fission rather than fusion. Partly this was because the leadership of the Chemical Society was unable or unwilling to tackle the problem. The days when manufacturing interests were seen to be served by the Society were long over, for now its emphasis was largely academic. It must be added that people with such sympathies are not always likely to be most aware of the pressures on grass-roots

[16] 'The Atomic Debates', ed. W. H. Brock, Leicester University Press, Leicester, 1967.

chemists who work in a strongly competitive environment. A further difficulty lay in the very metropolitan character of the Society, with meetings always in London, and out of touch with the great centres of manufacturing chemistry on the banks of the Mersey and Tyne. An unheard-of precedent occurred in 1868 with the formation of the Society's first provincial rival: the Newcastle Chemical Society.[17]

It held its own meetings, published its own *Journal*, and 14 years later became a founding constituent of the Society of Chemical Industry. It was a provincial manifestation of the same professionalizing tendencies that led to the Institute of Chemistry; it was much in sympathy with the latter and sought to serve and encourage the professional chemists concerned as employers or employees in the great chemical works springing up on both banks of the Tyne. One of the Newcastle leaders, A. F. Marreco, complained of 'that excessive centralisation with which we meet at every turn, and which would aim at concentrating the whole of art, science and literature in one corner of London, for that is what we are fast drifting into'.[18] The whole situation demanded imaginative and flexible leadership from London that was unfortunately not forthcoming.

The tensions within the Chemical Society were also a manifestation of a new phenomenon that was now evident in middle-class British society. As one historian put it, 'the very Victorians who condemned trade unions as vicious, restrictive, futile and unwarrantable interferences with individual liberty, flocked to join professional combinations'.[19] Everyone was doing it, so why not chemists? Here is a classic case of response by chemists to the cultural climate, the 'solvent effect'. And to hand was another and powerful catalyst, this time in the person of Edward Frankland, first President of the Institute. A restless man, peculiarly sensitive to recognition,[20] recently retired from active research he proved to be a master at social engineering. No doubt he had learned the techniques from his fellow agitators in the semi-secret pressure group calling itself the X-Club.[21]

The social reaction induced by these tensions was a concomitant process of bond breaking *and* bond making. True, some forsook the Chemical Society for the new Institute which first appeared in 1877. The interesting thing is that, for all its failure to adjust, the Chemical Society seems to have survived rather well. Of the 54 founding members of the Institute all but two were fellows of the Chemical Society, and many had held or were to hold high office in the parent body. And if Chemical Society membership figures are examined, it seems that formation of the Institute, far from inhibiting further growth may actually have stimulated it (Figure 2). From which I conclude that solidarity amongst chemists may transcend institutional barriers.

[17] W. A. Campbell, *Chem. Ind.*, 1968, 1463.
[18] A. F. Marreco, *Trans. Newcastle Chem. Soc.*, 1874 7, **3**, 233.
[19] F. M. L. Thompson, 'The Chartered Surveyors: The Growth of a Profession', Routledge, London, 1968, p. 149.
[20] C. A. Russell, 'Lancastrian Chemist: The Early Years of Sir Edward Frankland', Open University Press, Milton Keynes, 1986.
[21] R. Barton, *Brit. J. Hist. Sci.*, 1990, **23**, 53.

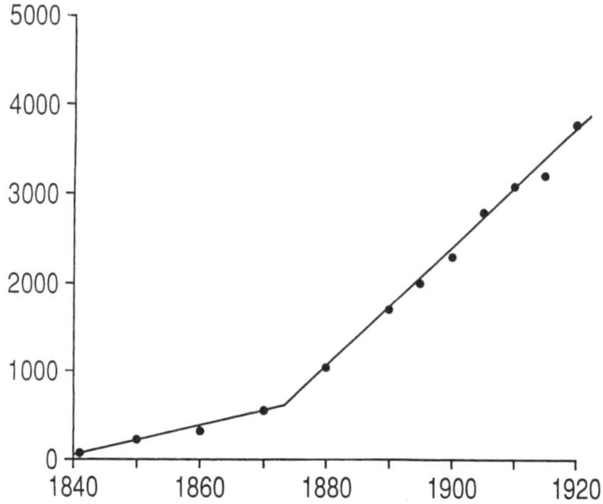

Figure 2 *Chemical Society membership*, 1841—1920

B. The Challenge of Inter-war Decline.—For another half century the Society prospered, its *Journal* increased its international reputation, and the Longstaff Medal joined the Faraday in 1881. From 1857 to 1873 the Society had been lodged (under rather cramped conditions) in the Old Burlington House. In 1874 the House was enlarged and the Society moved into the new East wing, with better accommodation for meeting-room (which some may recall) and the growing library. It successfully weathered the difficulties of World War I, its war memorial being prominently displayed at the foot of the stairs. Some of its members played distinguished advisory and research rôles in the conduct of 'the chemists' war'. Then, suddenly in the mid-twenties, disaster struck. Membership statistics, for the first time in its history, began to fall. Between 1926 and 1932 there was a decline of 12%. Bonds were being broken and (from 1930) not remade (Figure 3). What had happened?

It is clear that Victorian values had long dominated the Society's leadership which was out of touch with modern needs. Thus it was increasingly being criticized for adhering to the old established form of specialist scientific meeting. Only in 1920 did it begin to admit women to its Fellowship.[22] Above all it was still fiercely metropolitan; not till 1926 did it hold its first AGM outside London, and Local Sections were only formed as a response to the crisis. Internal strains were beginning to show strongly. The bond-breaking processes owed much to these.

We need also to glance at its sister establishment, the Institute of Chemistry. The truth seems to be that World War I marked a watershed for chemistry as it did for much else in Britain. But it took some years for the effects to be fully felt.

[22] J. Mason, *Chem. Brit.*, 1991, **27**, 233.

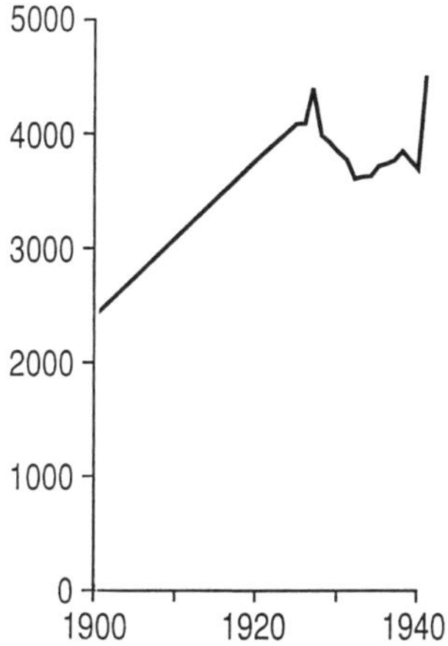

Figure 3 *Chemical Society membership*, 1900—1940

One of these had been predicted as far back as 1919 by William Pope[23]—the severe competition from the German chemical industry. Many German chemical plants, escaping the war unscathed, were poised to flood the market with dyes, fertilizers, and much else, while British wartime over-production of acids and explosives left factories with plant that was redundant as well as obsolescent. The UK industry needed more than the rhetoric of encouragement. In the hope of encouraging the Institute of Chemistry to cater for the needs of industrial members a British Association of Chemists was set up in 1918. It called for wider qualifications for entry, establishment of Local Sections on the lines of those of the SCI, creation of a Benevolent Fund, and more open government with better communication with the membership. The Institute responded positively, with the result that its membership dramatically increased[24] (Figure 4). One reason was the more favourable attitude of its academic leadership who now saw its value for a wider range of chemist than the consultants and analysts who had hitherto formed the backbone of its membership. This was evidently one lesson the war had taught them.

While it does not seem that the Institute's prosperity was at the expense of the Society's, there was a curious connection. For at the very moment of the Institute's reforms other chemical bodies, including the Chemical Society, were beginning to lose their isolationism. In April 1918 the Society's President, W. J. Pope, called for a closer alignment between societies representing pure and applied chemistry. When in the following year the Federal Council for Pure and Applied Chemistry was established, Pope became its first Chairman. It was a

[23] W. J. Pope, *Chem. News*, 1919, **119**, 179.
[24] Reference 3(*c*), p. 330.

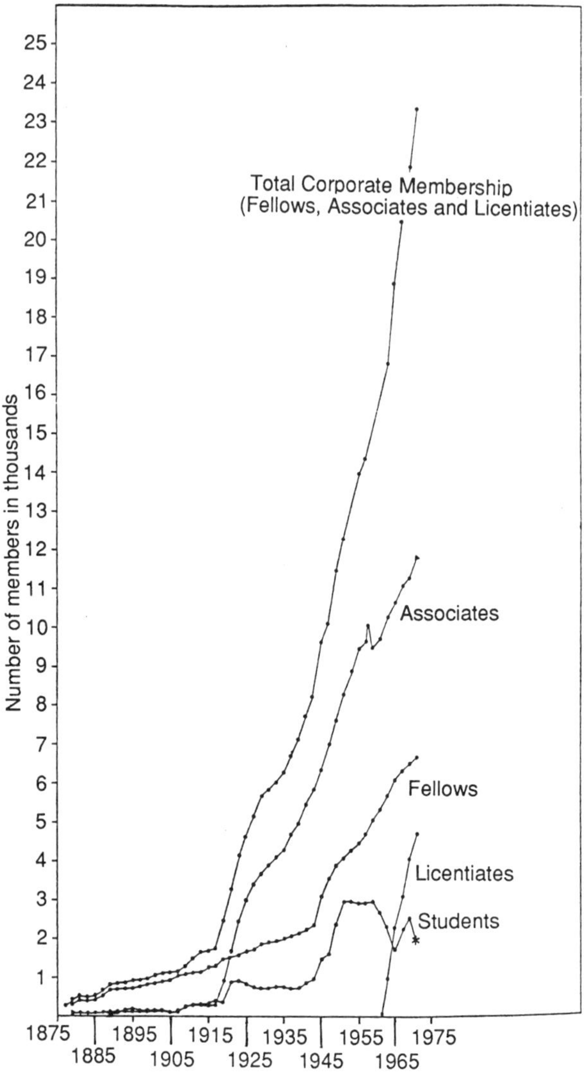

Figure 4 *Institute of Chemistry membership*, 1878—1971

federation of the Chemical Society, the Society of Chemical Industry, and the Association of British Chemical Manufacturers (three seats each), with several smaller bodies each represented by one seat. This was an attempt at creating new bonds between chemists on a wholly new scale. The Institute of Chemistry declined to join, fearing its position as a professional body might be compromised by any loss of independence.

Ironically these very open gestures to the rest of the chemical community led to some of its troubles. The Chemical Society had granted access to its Library to members of other chemical institutions, so the consumer benefits of individual membership were thereby diminished. It was thus the victim of its own generosity. As its *Journal* was now being taken by increasing numbers of company libraries, it was also a victim of its own success, for again individuals saw less reason to subscribe.

By a number of shrewd administrative measures the Society weathered the storm. Having previously tried in vain to set up Local Sections in conjunction with the Institute and SCI, the Society succeeded in doing so independently in 1938, and created a territorial base for its own government. Moreover it accepted an initiative from the Institute for an alternative federation, the Chemical Council, with three representatives each for the Society, the Institute, SCI, and ABCM. This was formed in 1935 and raised considerable sums from industry, one effect of which was to finance the Library, hitherto the sole responsibility of the Chemical Society. The Chemical Council continued until 1966.

Such, in very brief detail, were the complex processes of bond-formation and rearrangement within the British chemical community between the wars.

C. The Challenge of Isolationism.—The Chemical Society survived both the inter-war crisis and the Second World War. But its third major challenge was round the corner. Although the Chemical Council had worked well, Sir Harry Melville, President in 1967, pointed to a series of danger signals that suggested a much closer unity within the chemical community should be sought and achieved. As in 1841 there was some internal disorder within the subject itself. This came about from the fragmentation of chemistry into narrow specialist sub-disciplines, often with their own language and specialist journals. The traditional divisions of chemistry (organic, inorganic, physical) no longer served to meet the needs of people whose interests could be described as physical organic or organometallic, to mention only two of the largest groups. A real danger existed of the chemical community disintegrating.

There were external constraints as well. Since the late 1960s the 'brain drain' of chemists to the USA had caused mounting concern, as did the decline in numbers of school children electing to study chemistry. With the rise of instrumental techniques there came a huge increase in costs and an even greater need to seek support from industry and government. Just as Hofmann 120 years earlier had proclaimed the unity of chemistry, organic and inorganic,[25] in his efforts to gain support for the infant Royal College of Chemistry, so now it was imperative for the world to see a further institutional expression of that kind of creed. So Melville proposed some kind of merger between the three chartered bodies (Chemical Society, SCI, and RIC).[26] Although the SCI eventually declined to join, as we all know this led eventually to amalgamation of the

[25] A. W. Hofmann, *Medical Times & Gazette*, 1853, **6**, 131.
[26] H. Melville, *Chem. Brit.*, 1967, **3**, 212.

Chemical Society with the RIC in 1980, and the emergence of the Royal Society of Chemistry.[27] An entirely new type of bonding came into being, a result of a painstakingly careful response to yet another combination of internal and external pressures. As for the catalysts for this process it would be invidious to mention names for some of the most effective are happily still with us today.

D. The Challenge of Public Approval.—We have come a long way since the simple seal of the 'Chemical Society' was affixed to its documents. Now, at the end of the 20th century, the chemical community throughout the world is beset by further crises, not least among them being a fourth which I would define as the diminishing public image of science and of chemistry in particular. The challenge today is not merely that of pollution and the 'green' response to it. It is a ground-shift in the whole world-view that includes science as an appropriate human activity and nature as an appropriate object of disinterested enquiry. So serious is the threat that the Royal Society set up a standing committee to deal with 'the public understanding of science'.[28] For the anti-science movement generally, chemistry has always been prominently in the front line because of its potential for environmental destruction.

Public approval is invariably important for science, as our Victorian predecessors knew all too well. But now there is public disenchantment with science and the scientific method on a scale unheard of since the Scientific Revolution. External constraints, such as government attitudes, are not favourable; and for the first time the 'solvent effect', the influence of cultural climate ('post-modernism'), is really against us. This reflects changing world-views in an increasingly pagan society, and much of it is beyond our control. But not all.

The need for chemists to present a united front is more obvious than ever. This time it is to engage in another kind of bond-making process: to establish strong links with the world outside. This is not new, of course, but it needs to be addressed with renewed vigour. And the requirement is not a half-hearted apologetic bleat that chemistry is not quite as bad as painted. It is rather, a well-documented denial of historical half-truths about industry and the environment, a scrupulously fair assertion of the positive rôle of chemistry in the past, and a full-throated affirmation of the benefits that a responsible chemistry can provide in the future. An alliance must be forged with the wider scientific community, for this battle is not only for the chemist.

It is surely here that the Royal Society of Chemistry comes into its own. In the tradition of the old RIC, but on a greater scale, it has already initiated bonding processes of the utmost importance with sectors of the world outside. To mention but two: the Sesquicentenary initiative for chemical education and the Parliamentary Links Scheme. Perhaps today offers the greatest challenge yet faced in its 150 years by the RSC. May it learn from the past and face the future with confidence.

[27] Reference 3(*d*), p. 56.
[28] Report 'The Public Understanding of Science', Royal Society, London, 1985.

INDEX

These entries are, in general, those that are deemed most important and helpful. Generally, tabulated names and those in the Notes are excluded, as are names of secondary sources, except for those in the first chapter.

INDEX

INDEX

INDEX

INDEX